## General engineering conversion factors and constants* (*Continued*)



### Power

| | |
|---|---|
| 1 kilowatt | 737.56 foot-p |
| 1 kilowatt | 56.87 Btu per |
| 1 kilowatt | 1.341 horsepo |
| 1 horsepower | 550 foot-pour |
| 1 horsepower | 0.707 Btu per second |
| 1 horsepower | 745.7 watts |

### Heat, energy, and work equivalents

| | cal | Btu | ft · lb | kWh |
|---|---|---|---|---|
| cal | 1 | $3.97 \times 10^{-3}$ | 3.086 | $1.162 \times 10^{-6}$ |
| Btu | 252 | 1 | 778.16 | $2.930 \times 10^{-4}$ |
| ft · lb | 0.3241 | $1.285 \times 10^{-3}$ | 1 | $3.766 \times 10^{-7}$ |
| kWh | 860,565 | 3412.8 | $2.655 \times 10^{6}$ | 1 |
| hp-h | 641,615 | 2545.0 | $1.980 \times 10^{6}$ | 0.7455 |
| joules | 0.239 | $9.478 \times 10^{-4}$ | 0.7376 | $2.773 \times 10^{-7}$ |
| liter-atm | 24.218 | $9.604 \times 10^{-2}$ | 74.73 | $2.815 \times 10^{-5}$ |

| | hp-h | joules | liter-atm |
|---|---|---|---|
| cal | $1.558 \times 10^{-6}$ | 4.1840 | $4.129 \times 10^{-2}$ |
| Btu | $3.930 \times 10^{-4}$ | 1055 | 10.41 |
| ft · lb | $5.0505 \times 10^{-7}$ | 1.356 | $1.338 \times 10^{-2}$ |
| kWh | 1.341 | $3.60 \times 10^{6}$ | 35,534.3 |
| hp-h | 1 | $2.685 \times 10^{6}$ | 26,494 |
| joules | $3.725 \times 10^{-7}$ | 1 | $9.869 \times 10^{-3}$ |
| liter-atm | $3.774 \times 10^{-5}$ | 101.33 | 1 |

### Constants

$e$ 2.7183

$\pi$ 3.1416

Gas-law constants:

$R$ 1.987 (cal)/(g mol) (K) or (Btu)/(lb mol) (°R)

$R$ 82.06 ($cm^3$) (atm)/(g mol) (K)

$R$ 10.73 ($lb/in^2$) ($ft^3$)/(lb mol) (°R)

$R$ 0.730 (atm) ($ft^3$)/(lb mol) (°R)

$R$ 1545.0 ($lb/ft^2$) ($ft^3$)/(lb mol) (°R)

$R$ 8.314 (kPa) ($m^3$)/(kg mol) (K) or (J)/(g mol) (K)

$R$ 21.9 (in Hg) ($ft^3$)/(lb mol) (°R)

$g_c$ 32.17 (ft) (lbm)/(s) (s) (lbf)

### Analysis of air

By weight: oxygen, 23.2%; nitrogen, 76.8%

By volume: oxygen, 21.0%; nitrogen, 79.0%

Average molecular weight of air on above basis = 28.84 (usually rounded off to 29)

True molecular weight of dry air (including argon) = 28.96

### Viscosity

| | |
|---|---|
| 1 centipoise | 0.001 kg/(m) (s) |
| 1 centipoise | 0.000672 lb/(s) (ft) |
| 1 centipoise | 2.42 lb/(h) (ft) |

* See also Tables A6 and A7 in App. A for SI conversion factors and more exact conversion factors.

# ELEMENTARY CHEMICAL ENGINEERING

## McGraw-Hill Chemical Engineering Series



BUILDING THE LITERATURE OF A PROFESSION

Fifteen prominent chemical engineers first met in New York more than 50 years ago to plan a continuing literature for their rapidly growing profession. From industry came such pioneer practitioners as Leo H. Baekeland, Arthur D. Little, Charles L. Reese, John V. N. Dorr, M. C. Whitaker, and R. S. McBride. From the universities came such eminent educators as William H. Walker, Alfred H. White, D. D. Jackson, J. H. James, Warren K. Lewis, and Harry A. Curtis. H. C. Parmelee, then editor of *Chemical and Metallurgical Engineering*, served as chairman and was joined subsequently by S. D. Kirkpatrick as consulting editor.

After several meetings, this committee submitted its report to the McGraw-Hill Book Company in September 1925. In the report were detailed specifications for a correlated series of more than a dozen texts and reference books which have since become the McGraw-Hill Series in Chemical Engineering and which became the cornerstone of the chemical engineering curriculum.

From this beginning there has evolved a series of texts surpassing by far the scope and longevity envisioned by the founding Editorial Board. The McGraw-Hill Series in Chemical Engineering stands as a unique historical record of the development of chemical engineering education and practice. In the series one finds the milestones of the subject's evolution: industrial chemistry, stoichiometry, unit operations and processes, thermodynamics, kinetics, and transfer operations.

Chemical engineering is a dynamic profession, and its literature continues to evolve. McGraw-Hill and its consulting editors remain committed to a publishing policy that will serve, and indeed lead, the needs of the chemical engineering profession during the years to come.

THE SERIES

**Bailey and Ollis:** *Biochemical Engineering Fundamentals*
**Bennett and Myers:** *Momentum, Heat, and Mass Transfer*
**Beveridge and Schechter:** *Optimization: Theory and Practice*
**Carberry:** *Chemical and Catalytic Reaction Engineering*
**Churchill:** *The Interpretation and Use of Rate Data—The Rate Concept*
**Clarke and Davidson:** *Manual for Process Engineering Calculations*
**Coughanowr and Koppel:** *Process Systems Analysis and Control*
**Fahien:** *Fundamentals of Transport Phenomena*
**Finlayson:** *Nonlinear Analysis in Chemical Engineering*
**Gates, Katzer, and Schuit:** *Chemistry of Catalytic Processes*
**Holland:** *Fundamentals of Multicomponent Distillation*
**Holland and Liapis:** *Computer Methods for Solving Dynamic Separation Problems*
**Johnson:** *Automatic Process Control*
**Johnstone and Thring:** *Pilot Plants, Models, and Scale-Up Methods in Chemical Engineering*
**Katz, Cornell, Kobayashi, Poettmann, Vary, Elenbaas, and Weinaug:** *Handbook of Natural Gas Engineering*
**King:** *Separation Processes*
**Klinzing:** *Gas-Solid Transport*
**Knudsen and Katz:** *Fluid Dynamics and Heat Transfer*
**Luyben:** *Process Modeling, Simulation, and Control for Chemical Engineers*
**McCabe and Smith, J. C.:** *Unit Operations of Chemical Engineering*
**Mickley, Sherwood, and Reed:** *Applied Mathematics in Chemical Engineering*
**Nelson:** *Petroleum Refinery Engineering*
**Perry and Chilton (Editors):** *Chemical Engineers' Handbook*
**Peters:** *Elementary Chemical Engineering*
**Peters and Timmerhaus:** *Plant Design and Economics for Chemical Engineers*
**Probstein and Hicks:** *Synthetic Fuels*
**Ray:** *Advanced Process Control*
**Reid, Prausnitz, and Sherwood:** *The Properties of Gases and Liquids*
**Resnick:** *Process Analysis and Design for Chemical Engineers*
**Satterfield:** *Heterogeneous Catalysis in Practice*
**Sherwood, Pigford, and Wilke:** *Mass Transfer*
**Smith, B. D.:** *Design of Equilibrium Stage Processes*
**Smith, J. M.:** *Chemical Engineering Kinetics*
**Smith, J. M., and Van Ness:** *Introduction to Chemical Engineering Thermodynamics*
**Thompson and Ceckler:** *Introduction to Chemical Engineering*
**Treybal:** *Mass Transfer Operations*
**Valle-Riestra:** *Project Evolution in the Chemical Process Industries*
**Van Ness and Abbott:** *Classical Thermodynamics of Nonelectrolyte Solutions: With Applications to Phase Equilibria*
**Van Winkle:** *Distillation*
**Volk:** *Applied Statistics for Engineers*
**Walas:** *Reaction Kinetics for Chemical Engineers*
**Wei, Russell, and Swartzlander:** *The Structure of the Chemical Processing Industries*
**Whitwell and Toner:** *Conservation of Mass and Energy*

*(C. F. Braun and Co.)*

The chemical engineer understands the complete plant including the physical processes and the chemical changes.

| \$ | + | $Cl_2$ | + | $CH_2{=}CH_2$ | $\longrightarrow$ | Stoichiometry<br>Equipment<br>Unit Operations<br>Chemical Technology<br>Design and Economics<br>Production |
|---|---|---|---|---|---|---|
| Dollars | | Chlorine | | Ethylene | | |

$\longrightarrow$ $CH_2ClCH_2Cl$ + \$

Ethylene dichloride   Dollars

# ELEMENTARY CHEMICAL ENGINEERING

Second Edition

**Max S. Peters, Ph.D.**

*Registered Professional Engineer*
*Professor of Chemical Engineering*
*University of Colorado*

**McGraw-Hill Book Company**

New York St. Louis San Francisco Auckland Bogotá Hamburg
Johannesburg London Madrid Mexico Montreal New Delhi
Panama Paris São Paulo Singapore Sydney Tokyo Toronto

This book was set in Times Roman.
The editors were Kiran Verma and J. W. Maisel;
the production supervisor was Leroy A. Young.
R. R. Donnelley & Sons Company was printer and binder.

**ELEMENTARY CHEMICAL ENGINEERING**



1234567890 DOCDOC 89876543

ISBN 0-07-049586-6

**Library of Congress Cataloging in Publication Data**

Peters, Max Stone, date
Elementary chemical engineering.

(McGraw-Hill chemical engineering series)
Includes bibliographical references and index.
1. Chemical engineering. I. Title. II. Series.
TP155.P43 1984 660.2 83-12058
ISBN 0-07-049586-6

*Dedicated to*
LAURNELL STEPHENS PETERS

# CONTENTS

# PREFACE

Modern industrial methods require the services of large numbers of technically trained specialists such as chemists, physicists, electrical engineers, mechanical engineers, engineering technologists, economists, or technical salespersons. These specialists cannot do an effective job unless they have a general knowledge and understanding of many subjects outside their own field. This is particularly true with reference to chemical engineering, a basic knowledge of which is necessary in almost all types of industrial work.

In this book, the principles of chemical engineering are presented in a form that will be easily understood by readers with no previous training in the field. The essential purpose has been to present a unified picture of chemical engineering, with emphasis upon the development, applications, and interrelationships of the basic principles.

The subjects of economics, distillation, flow of fluids, and the rest are not unrelated: they are part of the overall process and, as such, constitute the practical science of chemical engineering. Hence, a composite picture of chemical engineering is presented in the introductory chapters of this book, and the concept of the complete chemical plant and the various principles involved are discussed. This gives the reader an opportunity to become acquainted with the entire profession, and the need for understanding the various principles becomes immediately apparent.

Later chapters deal with stoichiometry, typical equipment, unit operations, chemical technology, economics, and plant design. Each chapter gives clearly and briefly the essential, basic information on the particular subject.

The book is organized to permit its use as a text in a general chemical engineering course for nonmajors, as well as an introductory text for persons majoring in the field. The entire text can be covered in a two-semester three-hour course, and it should be possible to treat most of it more briefly but adequately in a one-semester course. The book can be used as a reference in a short course on chemical engineering for non-chemical engineers, and it should also be of value to practicing chemists, industrial supervisors or executives, technical salespersons, engineers, and engineering technologists.

The author has used the first edition of this book successfully as the text for the introductory first-semester freshman-level course for chemical engineers. For this two-hour course, emphasis was placed on all of Chaps. 2 (Technical Introduction), 3 (Chemical Engineering Stoichiometry), 4 (Industrial Chemical Engineering Equipment), 5 (Fluid Flow), 13 (Chemical Technology), and 14 (Chemical Engineering Economics and Plant Design), with partial coverage of Chaps. 6 (Heat Transfer), 7 (Evaporation), and 8 (Distillation). This coverage was sufficient to show the general field of chemical engineering to the students and allow a presentation that was relatively quantitative. Senior chemical engineering students volunteered to serve as teaching assistant–counselors for the course, grading the homework problems and serving as friend and counselor for the individual freshmen on a person-to-person basis. The seniors enjoyed this activity and always volunteered in sufficient numbers so that each senior would grade and work with only about six to eight freshmen.

A large amount of previous technical training is not required for an understanding of the material as presented. A good mathematical background with some calculus will permit the reader to obtain a grasp of the material rapidly; however, the method of presentation will enable even the reader with no advanced mathematical training to understand much of the subject matter. In some sections, where the mathematics involved would be above the level of many readers, simplified derivations of the mathematical relationships of physical variables have been presented. The fundamental derivations of these relationships using more advanced mathematics have been presented in App. B. References to the chemical engineering literature have been included to show where additional information may be obtained.

Many illustrative examples have been used throughout the text. The author has found this to be one of the best means for conveying to the reader the applicability of theoretical reasoning to actual cases and for showing the importance of clear designation and use of units. Problems have been given at the end of individual chapters to illustrate the information presented in the chapter and to give the reader a chance to test the understanding of the material. Some of the problems are designed to require computer solution for those who wish to try out their skills in computer programming. Answers have been given in App. D, permitting the reader to determine immediately if he or she has attacked the problems correctly.

The study of chemical engineering can be a fascinating experience. However, there is no quick and easy method for becoming proficient in this field. The reader who expects to obtain the maximum benefit from this book must be prepared to make a sincere and concentrated effort to understand the text material and the problems. If the elementary principles are clearly understood, the applications and advanced treatment will be easy to grasp.

The same style and form of presentation used in the first edition have been retained for the second edition, with changes relating primarily to the introduction of SI units and to the updating of discussions of process data. The author repeats his thanks from the first edition to numerous individuals for their critical perusal of the material and for their many helpful suggestions. The author here extends his thanks to his students who have made suggestions for improvements after using the book and to his many colleagues and peers, such as Klaus D. Timmerhaus and Ronald E. West of the University of Colorado, James R. Fair of the University of Texas, and Joseph E. Nowrey and J. E. Ekhaml of Trident Technical College, who have provided useful input for preparing the second edition.

*Max S. Peters*

# ELEMENTARY CHEMICAL ENGINEERING

CHAPTER
# ONE

# CHEMICAL ENGINEERS AND THEIR PROFESSION

Chemical engineering is a science and an art dealing with the relationships of physical and chemical changes. This profession came into existence toward the end of the nineteenth century when the need for technical workers with a combined training in physics, chemistry, and mechanical engineering became apparent. Since that time, the profession of chemical engineering has increased in stature, until it now stands as an integral and essential part of our modern industrial society.

The chemical engineer must have a wide variety of talents. One must understand how and why a given process works, be able to design, set up, and operate equipment to carry out the process, and have the ability to determine the profits obtainable by carrying out the particular process. One must have the theoretical understanding of a physicist combined with the practical attitude of a mechanic or a loan agency. In addition, the chemical engineer is in constant contact with workers of widely varying levels of intelligence and must have the ability to maintain friendly and effective relations with all these persons.

The fundamental working tools of the chemical engineer are the basic laws of chemistry and physics, a logical mind, and an aptitude for the practical application of mathematics, including computer applications. The use of these tools permits the various principles to come to life as finished chemical plants or such simple final products as a cake of soap, a can of paint, or the plastic handle of a toothbrush.

An example of the application of the principles can be found in the final step involved in the production of nitric acid. Nitric acid is formed when nitrogen dioxide reacts with water. This reaction gives off heat, which must be removed if the process is to continue efficiently. One method for removing the heat is to pass cold water through the inside of a tube immersed in the hot nitric acid. The heat passes through the wall of the tube into the cold water and is thereby removed from the system. By applying basic principles, the chemical engineer can determine what length of tube must be immersed in the nitric acid to ensure adequate removal of the heat, the best diameter of the tube, the flow of water necessary, and even the best material for making the tube. These same principles can be used to find how fast a hot ingot of iron will cool or how much heat should be supplied to a room to maintain a constant temperature.

The chemical engineer is trained to take chemical and physical processes from small-scale operation to large-scale operation. The chemist in the research laboratory may find a new method for producing a certain chemical. In the laboratory, the chemist probably carried out the process in glass beakers and transferred the products from one container to another by hand. When the chemical engineer takes over this process to develop it for large-scale operation, many things must be considered which were of no importance to the chemist. A glass beaker may have been suitable for the laboratory work since it was only the chemical reaction that was of importance. However, the chemical engineer must attempt to use for the container a material which has good heat-transfer and corrosion-resistance qualities while being cheap, easily fabricated, and not easily broken. The rate of the reaction must also be considered as well as effects of changes in temperature or pressure. Practical methods for transfer of the material from one container to another must be found for the large-scale operation, since it is ordinarily too expensive to transfer materials by hand. In the practice of chemical engineering, then, it is necessary to understand the basic principles of chemistry as well as the economics and principles involved in physical processes.

There are many different types of work for which chemical engineers are fitted. Some of these are plant and equipment design, plant operation and supervision, technical sales, consulting work, basic and applied research, market development, and teaching.

## THE TECHNIQUES OF CHEMICAL ENGINEERING

The chemical engineer must use fundamental working tools to transform the theoretical principles into practical results. A number of different techniques have been developed for accomplishing this transformation. One of the basic techniques is the use of generalized mathematical expressions for indicating the relationships between physical variables. These expressions are merely shorthand indications showing how the different variables are interrelated. This last statement may appear to be obvious; however, many prospective chemical engineers have fallen by the wayside because they could never attach the true physical significance to a mathematical expression.

Consider the simple example of the pythagorean theorem, which is familiar to all high school mathematics students. This theorem states that the square of the length of the hypotenuse of a right triangle equals the sum of the squares of the length of the two triangle legs. Mathematically, this may be expressed as

$$h^2 = a^2 + b^2$$

where $h$ is the length of the hypotenuse and $a$ and $b$ are the respective lengths of the two legs of the triangle. There is nothing mysterious about this mathematical expression. It is simply a shorthand method for representing the relationship among the three variables of hypotenuse length and two leg lengths. However, even this basic equation has its limitations. It applies only to a complete triangle in which one of the angles is 90°. It also applies only when $h$, $a$, and $b$ are expressed in the same physical units. No high school sophomore would attempt to apply this equation by substituting values for $a$ in inches and $b$ in meters. Perhaps the preceding statements are ridiculously simple. Their correctness is certainly self-evident. Yet many beginning chemical engineers have difficulty because they cannot apply these same reasoning methods to more advanced cases.

A typical example of a shorthand mathematical expression in chemical engineering is found in the study of fluid dynamics. When a fluid flows through a confining medium, the individual particles of the fluid may flow in smooth, straight lines, or they may flow along irregular paths forming vortices and eddies in the flowing fluid. When the particles move along smooth, straight lines, the flow is said to be *viscous* or *streamline*. When the particles move along irregular paths, the flow is said to be *turbulent*. The chemical engineer needs to be able to predict under what conditions the flow will be streamline and under what conditions the flow will be turbulent. For a fluid flowing in a long pipe of circular cross section, the type of flow depends on the inside diameter of the pipe, the average velocity of the fluid, the density of the fluid, and the viscosity of the fluid. It has been found experimentally that the type of flow may be predicted from the magnitude of a dimensionless grouping of these variables called the *Reynolds number*.

$$\text{Reynolds number} = N_{Re} = \frac{DV\rho}{\mu}$$

where $D$ = diameter of pipe, m or ft
$V$ = average velocity of flowing fluid, m/s or ft/s
$\rho$ = density of flowing fluid, $kg/m^3$ or $lb/ft^3$
$\mu$ = viscosity of the flowing fluid, kg/(m)(s) or lb/(ft)(s)

Here, then, is a shorthand mathematical expression which can be used by the chemical engineer to determine whether the flow is turbulent or streamline. One must realize that the Reynolds number has a true physical significance. It is not just a number. It is a physical representation of the relationship of the crucial variables encountered in the flow of a fluid. Since the Reynolds number is designated as dimensionless, the physical units (or dimensions) of the variables must be such that they can cancel one another to yield a dimensionless result. Therefore, any form of consistent units can be used.

The reasoning involved in the preceding example is exactly analogous to that for the simple pythagorean theorem example. The chemical engineer must recognize the physical significance of mathematical expressions. Remember, *mathematical equations are merely shorthand methods for showing how different physical variables are interrelated.*

**Assumptions.** An engineer must be able to work not only rapidly but also effectively and honestly. In order to work rapidly, it is often necessary to make simplifying assumptions. If the work is to be effective, these assumptions must be essentially correct. If the work is to be honest, the assumptions should be clearly indicated. In many cases, it would be impossible to solve a problem without certain basic assumptions. In other cases, if no simplifications were made, the final solution would be extremely complex and require a large amount of extra work with little gain in accuracy. For example, the perfect-gas law applies strictly only to simple gases at low pressures. However, at atmospheric pressure, the common gases show only slight deviations from the perfect-gas law. Thus, the chemical engineer is justified in assuming the perfect-gas law as applicable under these conditions.

One should apply this technique with extreme care. This applies in particular to the beginner who has little experience to indicate what assumptions should or should not be made. Whenever a simplification of this type is made, it should be clearly stated and justified. *The chemical engineer may make simplifying assumptions from time to time, but they are never made unless necessary, essentially correct, and clearly indicated.*

## THE COMPLETE PLANT

The overall operation of a chemical plant may be broken down into a number of small operations such as transfer of materials, filtration, extraction, distillation, and transfer of heat. The chemical engineer must understand each of the individual operations; however, it is important not to lose sight of the unified picture of the complete plant. For example, in the brewing industry the essential products are obtained from a distillation operation. Yet the distillation cannot be carried out without heat being transferred to the feed liquid to make it boil. The correct feed liquid cannot be obtained unless a controlled fermentation reaction occurs. The desired fermentation process cannot occur until the purified raw materials are mixed together in the correct proportions under controlled conditions of temperature and pressure. The purified raw materials are obtained by operations involving mixing, grinding, and filtration. The initial raw materials must be transported to the plant and intermediate materials must be transferred from one piece of equipment to another before the individual operations can occur.

The effective chemical engineer cannot limit consideration to one small portion of the plant. The idea of a complete plant must constantly be kept in mind. Thus, the profession of chemical engineering might be said to be built around technology (or the complete process) rather than around the individual operations. The individual steps are important, they must be understood, and their principles must be mastered; but the chemical engineer must never lose sight of the final applications where all the smaller operations are combined into one smoothly operating plant.

The development of the "complete plant" concept can be compared to the development of a good pole vaulter. A pole vaulter must learn how to carry the pole correctly. A sprinter's speed must be developed on the approach, and the steps must be spaced to permit the correct foot to hit the desired takeoff spot. It is necessary to develop a strong, raising kickoff and a body snap at the top of the lift. All these can be worked on and perfected individually, but the final winning vault does not come until all the individual parts are put together to function as a smooth, complete unit.

Just as the pole vaulter needs to work on the individual parts of that skill, chemical engineers need to study the separate operations involved in their field. The chemical engineer must study and understand the basic laws of chemistry and physics, must know the various types of equipment and the economics involved in the overall plant process, and must understand the operations of heat transfer, flow of fluids, filtration, distillation, absorption, extraction, and the rest. However, one must never forget that these individual parts are all basically interrelated. Alone, they are of little value; combined, they result in a complete plant.

## THE PRACTICAL APPROACH

A chemical engineer must have a practical mind. Would the theoretical brilliance of a special heater design be appreciated if the designed unit required steam at a pressure of 120 $lb/in^2$ for adequate operation when the highest pressure of steam available in the plant was 90 $lb/in^2$? Would a reactor 3 m in diameter and 3 m high be ordered in an assembled form if the reactor were to be used in a brick building in which the largest doorway or opening was 2 m wide? The answer to both of these questions is obviously no. On a theoretical basis, a heater with 120-lb steam is undoubtedly superior to one using 90-lb steam. From an economic viewpoint, it may be cheaper to order a reactor in an assembled form. However, from a *practical* viewpoint, it would certainly be better to design the heater for 90-lb steam rather than put in the extra equipment necessary for increasing the pressure of the present steam. It would be better to order the reactor in sections and assemble it inside the building than to tear down the brick walls; or perhaps the reactor could be redesigned to permit it to pass through the building openings in an assembled form.

A full realization of the importance of the practical approach can come only with experience. However, the practical aspects of chemical engineering should be recognized and kept constantly in mind by the beginning student.

## THE PHYSICAL PROCESS

Any person who drives an automobile knows what the physical and legal speed limits are for the modern cars. The driver knows what a steering wheel looks like and its purpose and probably knows what essential parts are included in the motor and how these parts operate. In a similar manner, chemical engineers know the physical appearance of the equipment they use. They know what the equipment looks like on the inside as well as on the outside, and they know the physical limitations and the best type of equipment for certain purposes.

Some examples of physical processes and types of equipment related to the transfer of materials are given here. Before any chemical plant can operate, the basic raw materials must be transported to the plant area. The physical transportation may be accomplished by means of railroad freight cars or tank cars, trucks, ocean-going freighters or tankers, pipelines, air transports, and even belt-conveyor systems. Liquids may be shipped in drums, barrels, tanks, carboys, and smaller containers. Solids are transported in loose carloads or in smaller containers such as burlap or paper bags, barrels, and cans. Liquids and gases may be moved from one place to another by passing them through pipes. These pipes come in standard sizes of diameter, wall thickness, and length; diameters range from $\frac{1}{8}$ in up to 5 ft or more. The smaller-diameter pipes usually come in standard lengths of 20, 30, or 40 ft. Various types of pumps, compressors, or blowers are used to furnish the driving force necessary for the movement of the fluids.

The preceding examples are just a few of the many physical processes and types of equipment with which the chemical engineer should be familiar.

## THE ECONOMIC VIEWPOINT

No industrial plant can operate for long unless it can realize a profit from the operation. The amount of profit depends on the overall cost of the plant, including operating costs, investment costs, and general-overhead costs. The chemical engineer must understand the economic principles involved in industrial operations and be able to estimate costs and determine the approximate profit which can be obtained by making an investment in a given process.

## THE THEORETICAL CHEMICAL ENGINEER

Theoretical principles are the building stones for the chemical engineering profession. The old-time industrial operators may sneer at the so-called "theoretical gang"; yet these same operators are undoubtedly working with equipment which was developed on the basis of theoretical considerations.

The practical person may know *what* and *how*, but the good chemical engineer goes one step further and knows what, how, and *why*. This additional step is what makes the true technical person. It permits the chemical engineer to apply real knowledge in the development of new processes or new ideas. Theory indicates why a process works—why a physical or chemical phenomenon occurs. Herein lies the foundation of chemical engineering.

## THE PROFESSIONAL CHEMICAL ENGINEER

In addition to understanding theory and practice, the chemical engineer also has a professional responsibility as a person serving society. He or she should be aware of the major organization in the United States for chemical engineering activities, the American Institute of Chemical Engineers (AIChE). The AIChE has been in existence since 1908 and has a membership of about 60,000 persons. Among its many activities, the AIChE has a major program of continuing education and publishes journals to help chemical engineers keep up-to-date on the latest developments in their field. It also has a code of ethics which spells out the basic principles and rules by which an engineer should practice.

One of the fundamental canons in the AIChE code of ethics is that "Engineers shall hold paramount the safety, health, and welfare of the public in the performance of their professional duties." This concern for the public and the environment in general should be of great importance to chemical engineers and should always be taken seriously.

Another of the fundamental canons in the AIChE code of ethics is that "Engineers shall issue public statements only in an objective and truthful manner." This statement underlines the importance to all engineers of learning appropriate communications skills. This applies to learning to organize and present technical information in well-written reports as well as to developing skills in verbal presentations. The importance of developing such skills cannot be overemphasized.

Among other professional activities that are desirable for the overall development of a chemical engineer's career is individual registration as an engineer. All state legislatures now require that engineers who offer their service directly to the public be registered (or licensed) engineers. In addition, many regulatory agencies and other groups require registration as an indication of engineering competence. The title Registered Professional Engineer or Licensed Professional Engineer provides assurance to the public that those with that title have met certain minimum requirements of engineering capability at some point in their careers. While each state has its own rules for registration, basically the procedure is to become an Engineer in Training by taking the EIT examination as soon as is convenient after graduating with a bachelor's degree from an engineering program accredited by the Accreditation Board for Engineering and Technology (ABET). After a certain period of years as an EIT, the engineer can take the final professional engineers' examination for full licensing.

The road to success in the chemical engineering profession is not easy. It is strewn with practical, theoretical, professional, and economic hazards. Diligence and hard work are required for traveling the road. Yet the experience can be exciting. Who can deny the pleasure that comes from seeing a useful product emerge from some basic ideas and theories? The chemical engineer is a member of an alive and growing profession with a solid foundation of practical, theoretical, and professional knowledge and a future of unlimited horizons.

CHAPTER
# TWO

# TECHNICAL INTRODUCTION

The study of chemical engineering is facilitated by dividing the field into four general sections. These are as follows:

1. Stoichiometry
2. Unit operations
3. Chemical technology (or unit processes)
4. Economics and plant design

The four general subjects are all interrelated; however, it is best for the student to consider one at a time in the order given here. When chemical engineering is studied in this manner, one topic leads directly into the next, and a smooth and integrated passage through the entire field can be accomplished.

The ultimate application of a chemical engineer's training lies in the design and use of equipment for carrying out the various operations and processes. A complete understanding of the theoretical principles is of little value if the student does not have a concept of the practical equipment used in applying the theory. A general discussion of chemical engineering equipment is given in this book just prior to the treatment of unit operations. The reader should attempt to retain a visualization of practical equipment throughout the study or application of chemical engineering.

## STOICHIOMETRY

Chemical engineering stoichiometry is the study of material balances, energy balances, and the chemical laws of combining weights as applied to industrial processes. It is very important to have a clear concept of the stoichiometric principles, since they may be considered as the basis of all chemical engineering.

A material balance is based on the *law of conservation of matter*, which states that the total mass of materials entering a system in a fixed period of time must equal the mass of all materials leaving, plus the mass of any accumulation that has taken place. Thus, if 100 lb of coal and air is charged to an empty furnace, the total mass of the gaseous products plus the mass of the ash and residual material left in the furnace must equal 100 lb.

It is possible for mass to be transformed into energy by the emission of radiant energy or by the transmutation of elements. However, the loss of mass in this manner is so small and the occurrence so rare in industrial processes that the law of conservation of matter may be assumed as valid for all practical purposes.

The *law of conservation of energy* is the basis for energy balances. According to this law, energy can be transformed from one form to another but can never be destroyed. If 100 lb of coal is put into a furnace, this coal contains a certain amount of energy determined by its temperature and intrinsic heat-of-combustion content. If there is no heat loss to the surroundings, this same amount of energy must appear in the gaseous and solid products of the combustion. This energy may come out partly as sensible heat due to the temperature of the products and partly as intrinsic energy due to incomplete combustion.

The chemical *laws of combining weights* express the weight ratios in which substances combine when undergoing chemical changes. When the carbon in coal is burned to carbon dioxide, the chemical reaction may be written as

$$C + O_2 = CO_2$$

According to this equation, one atomic weight of carbon can combine with one molecular weight of oxygen to form one molecular weight of carbon dioxide. This may be expressed in the actual units of weight or mass by using the atomic and molecular weights of the materials involved. It can then be said that 12 kg of carbon unites with 32 kg of oxygen to form 44 kg of carbon dioxide. From a consideration of the law of combining weights, it is evident that 44 kg of carbon dioxide must be obtained for every 12 kg of carbon that is burned to carbon dioxide.

## UNIT OPERATIONS

The various physical processes of importance in chemical engineering can be broken down into a series of groups which are called *unit operations*. While chemical reactions or changes may be occurring in the process, the unit operations limit themselves mainly to the physical changes that take place. For example, in the combustion of coal the chemical reaction of oxidation occurs; however, the main unit operations for this process are the transfer of heat and the physical flow of the coal, entering air, and product gases. The unit operation of distillation applies to the physical processes of heat and mass transfer wherein an enriched vapor is obtained. The unit operation of crushing applies to the physical process of breaking a solid material into smaller particles.

The concept of unit operations may be used to simplify the study of chemical engineering. Instead of having to study each chemical industry separately, the chemical engineer can merely learn the basic principles of the different unit operations. These principles can then be applied where the particular unit operation occurs, no matter what the chemical field or industry may be.

Examples of the more important unit operations are fluid flow, heat transfer, humidification and dehumidification, evaporation, distillation, absorption, extraction, filtration, crushing and grinding, and drying.

The various unit operations are not separate and distinct from each other. Many of the same general principles are applicable to different unit operations. For example, the concept of mass transfer is involved in humidification, evaporation, distillation, absorption, extraction, and drying.

In some cases, the general concepts of flow of materials, heat transfer, mass transfer, and mechanical separations have been used in place of unit operations. This generalized breakdown is useful for advanced chemical engineers; however, for an elementary treatment, the presentation of individual unit operations is much more satisfactory since it permits the reader to visualize the immediate applications.

## CHEMICAL TECHNOLOGY

While the chemical engineer learns about physical changes in the study of the unit operations, it is also important to learn something about the chemical changes which occur along with the physical changes. *Chemical technology* deals with the chemical reactions and the physical processes and controls in industrial operations. This is often called unit processes, but in this book it will be called chemical technology to differentiate it clearly from the unit operations.

The production of nitric acid by the oxidation of ammonia can be chosen as a typical example of chemical technology. In this process, chemical engineers are interested in knowing the best catalyst for the oxidation as well as the temperature and pressure for the optimum oxidation yields. They want to know what the oxidation reaction is and what additional reactions are necessary to give the final nitric acid. They are interested in knowing special materials of construction for the equipment as well as the control methods for obtaining the best grade of nitric acid with the greatest ease and least expense. In other words, the chemical engineer wants to know all the details of the actual process involved in the production of nitric acid.

In dealing with the chemical technology of specific processes, the engineer does not regard the information as applicable to only one particular product. It is clear that much of the chemical technology of one operation will be readily applicable to many other processes. The emphasis should be on the broad aspects rather than on the small details. An understanding of chemical technology will round off the chemical engineer's background and point his or her talents in any one of many directions. Through application of chemical technology, the chemical engineer extends the chemist's interest and work in laboratory-scale chemical reactions to a complete operating plant.

## ECONOMICS AND PLANT DESIGN

One of the important applications of chemical engineering is in the design of special equipment and chemical production plants. In this type of work, it is essential to consider the economics of the process. The cost of fabricating the equipment, setting up the plant, operating the plant, supplying raw materials, etc., must be balanced against the income in such a manner as to yield a definite profit on the investment. This economic analysis must be kept constantly in mind during design work; therefore, it is best to consider the combination of plant design and economics as a single field in chemical engineering.

## METHODS FOR SOLVING PROBLEMS

Experience is the best teacher of methods for solving problems in chemical engineering, but there are some general rules which can prove invaluable for obtaining rapid and accurate solutions. Above all, it is necessary to use logic and to think through each step clearly. Before starting with any paperwork, the engineer should read the problem carefully and clearly understand it. If possible, a rough, box-type diagram, showing the flow of materials and the chemical reactions taking place should be drawn, and all important sections should be labeled.

A basis for the calculations should be chosen and adhered to throughout the entire solution. This basis is usually determined by the result being sought in the problem, and there are many different bases which can be chosen. The important thing is to choose one basis and to stick to it. Typical bases would be a unit weight of one of the entering materials, a unit weight of one of the product materials, a given length of time, or a certain area of reacting surface.

Throughout the entire solution of any problem, it is essential to give a considerable amount of attention to the physical units. Each calculated quantity should have its units indicated as well as its numerical value. A complete designation of the units should be shown, e.g., 1 kg of water produced per 100 kg of dry air or 130 lb of carbon dioxide produced per 100 lb of initial coal.

Following is an outline of the general rules to be followed in solving chemical engineering problems:

1. Examine the problem carefully and understand its meaning thoroughly.
2. Use logic in the method of solution.
3. Draw a diagram of the process and label all sections.
4. Write any chemical reactions involved.
5. Decide what must be done to obtain the final result.
6. Pick a basis.
7. Do the work of the problem in a neat form.
8. Label all parts and indicate units.
9. Check the results.

## EMPIRICAL CONSIDERATIONS

Empirical methods are based on repeated experimental observations rather than on theoretical reasoning. A large amount of experimental work has been done on chemical engineering unit operations. This work has resulted in empirical methods for determining certain important variables. The field of heat transfer offers many outstanding examples of empirical methods. Most of the common equations for determining liquid- and gas-film heat-transfer coefficients involve dimensionless groups raised to empirical powers. The field of heat transfer has been carefully investigated experimentally, and most of the empirical heat-transfer equations can be accepted as accurate representations of the true result.

The chemical engineer must be prepared to employ the empirical methods but must be certain the constants used are generally accepted. The combination of theoretical reasoning with experimental data has laid the foundation for many of the principles of chemical engineering.

## UNITS

A considerable amount of confusion exists among students of engineering because of the presence of several different systems of units; these systems are divided into the two general categories of *English* and *metric base.* The English system can be divided into the two categories of American engineering and British engineering, with both of these systems using the units of foot (ft) for length; second (s) for time; pound (lb) for force; British thermal unit (Btu) or foot-pound force (ft · lbf) for energy; atmosphere (atm), pound-force per square inch (lbf/in$^2$), and inches (in Hg) or millimeters (mmHg) of mercury for pressure; and degrees Rankine (°R) or Fahrenheit (°F) for temperature. The primary difference between these two systems is that the unit for mass in the American system is pound-mass (lbm), while the unit for mass in the British system is slug (1 slug is equivalent to 32.17 lbm). The British system also makes use of a force unit of poundal and an energy unit of foot-poundal in the so-called "English absolute" system.

The metric base system is divided into two categories, one being called the *cgs* (centimeter, gram, second) *metric system.* The units for the cgs system are centimeter (cm) for length; second (s) for time; gram (g) for mass; dyne for force; erg, joule (J), or calorie (cal) for energy; atmosphere (atm) or millimeters of mercury (mmHg) for pressure; and kelvin (K) or degrees Celsius (°C) for temperature. This is the system which was commonly used in Europe and by scientists through the nineteenth century and is still the basic system of units used in the teaching of chemistry and physics in the United States.

The second metric base system is the International System (Système International) of Units, or the so-called "SI units." The SI system has been established on the basis of numerous international conferences and is based on the original metric system, with the primary difference being new names for derived terms and regulations that simplify international usage. The units for the SI system are meter

(m) for length; second (s) for time; kilogram (kg) for mass; newton (N) for force; joule (J) for energy; pascal (1 Pa = 1 N/m$^2$) or bar (10$^5$ Pa) for pressure; and kelvin (K, with no degree symbol) or degrees Celsius (°C) for temperature.

Appendix A of this book has a detailed description of SI units, including information on the source of the system and rules for usage. The reader is strongly urged to read this appendix to become familiar with the regulations and procedures called for when pure SI units are used. In this book, the practice will be to use a mixture of American engineering, cgs, and SI units, because this is what the practicing engineer in the United States will encounter, and it is essential that the reader become familiar with all these systems.

The United States is making some progress in adopting SI units, but it will be many years before they will be accepted for common use here. Accordingly, for the next 50 years at least, the practicing engineer in the United States will use and need to be familiar with all four of the systems of units described above.

Table A-6 in App. A gives an exhaustive list of factors for converting to SI units from other systems. The front cover pages of this book and App. C give convenient tables for conversions in the American system and the cgs system, while the back cover page of the book gives a list of SI conversions for commonly used American engineering units. Little difficulty should be encountered in changing from one system to another when these tables are used.

When making a conversion between systems of units, the possibility of error is much reduced if the multiplication form for the conversion is set up showing both the numbers involved and the units involved for each separate number. The units can then be treated as numbers, and the appropriate cancellations can be made. This can be illustrated by determining the pounds per cubic foot equivalent to 3.0 g/cm$^3$. From the table of equivalents in App. C, it can be found that 1 lb equals 453.6 g and 1 in equals 2.54 cm. The following forms of numbers and units can then be set up: 453.6 g/lb, 2.54 cm/in, 12 in/ft, and a starting point of 3.0 g/cm$^3$. The multiplication form for the number is

$$\frac{(3.0) \;\Big|\; \;\Big|\; (2.54)^3 \;\Big|\; (12)^3}{\;\Big|\; (453.6) \;\Big|\; \;\Big|\;} = 187$$

The corresponding multiplication form for the units is

$$\frac{(\text{g}) \;\Big|\; (\text{lb}) \;\Big|\; (\text{cm})^3 \;\Big|\; (\text{in})^3}{(\text{cm})^3 \;\Big|\; (\text{g}) \;\Big|\; (\text{in})^3 \;\Big|\; (\text{ft})^3} = \text{lb/ft}^3 \text{ by cancellation}$$

Therefore, 187 lb/ft$^3$ is equivalent to 3.0 g/cm$^3$. An equivalent and perhaps easier form with numbers and units included on the same line is

$$\frac{3.0\text{ g}}{1\text{ cm}^3}\;\Big|\;\frac{1\text{ lb}}{453.6\text{ g}}\;\Big|\;\left(\frac{2.54\text{ cm}}{1\text{ in}}\right)^3\;\Big|\;\left(\frac{12\text{ in}}{1\text{ ft}}\right)^3 = 187\text{ lb/ft}^3$$

## Molal Units

Another useful consideration in dealing with units is the concept of the molal unit. In chemical reactions, atoms or molecules combine to form other atoms or molecules. One atomic weight of carbon reacts with one molecular weight of oxygen to form one molecular weight of carbon dioxide. Using molal units, this can be expressed as follows: 1 lb atom of carbon reacts with 1 lb mol of oxygen to give 1 lb mol of carbon dioxide. One pound atom of carbon is equivalent to the atomic weight of carbon expressed as pounds; i.e., it is equivalent to 12 lb of carbon. Gram atoms or gram moles may be used in place of pound atoms or pound moles if this is more convenient. Thus, 1 g mol of carbon dioxide is equivalent to 44 g of carbon dioxide, since the molecular weight of carbon dioxide is 44. In the SI system, 1 g mol (or simply 1 mol by SI terminology) has about $6.023 \times 10^{23}$ molecules. This means, of course, that 1 lb mol in American engineering terminology would have about $454 \times 6.023 \times 10^{23}$ molecules, because there are about 454 g in 1 lb. Thus, the mass of 1 lb mol would be about 454 times greater than the mass of 1 g mol.

## Units of Force

Chemical engineers need to have a clear concept of the units of force. A force exerted on a body that is free to move tends always to give that body an acceleration. According to Newton's fundamental law of motion,

$$F = \alpha ma \tag{2-1}$$

where $F$ = force exerted to give a free mass an acceleration
$m$ = mass of body (i.e., quantity of matter contained by body)[1]
$a$ = resulting acceleration
$\alpha$ = a proportionality constant dependent on the units of other terms

In the following definitions of force units, the masses may be considered as on a level, frictionless surface, and the accelerating force is in a horizontal direction.

The unit of force used in the metric system is the *dyne*. A force of 1 dyne will give a 1-g mass an acceleration of 1 $cm/s^2$; i.e., the velocity changes at the rate of 1 cm per second each second. Therefore, when the force is expressed as dynes, mass as grams, and acceleration as centimeters per second squared, the value of the proportionality constant $\alpha$ in Eq. (2-1) must be unity.

The unit of force used with SI is a *newton* (N), which is defined as that force which will give a 1-kg mass an acceleration of 1 $m/s^2$. Therefore, just as in the case of force expressed in dynes, when the force is expressed as newtons, mass as kilograms, and acceleration as meters per second squared, the value of the proportionality constant $\alpha$ in Eq. (2-1) must be unity.

Two different units of force are used in the English system. A *poundal* is defined as that force which will give a 1-lb mass an acceleration of 1 $ft/s^2$. A *pound-force* is defined as that force which will give a 32.17-lb mass (or 1 slug) an accelera-

[1] Mass may be defined as that property of matter by virtue of which it resists any change in velocity. Weight may be defined as the downward force exerted on a body by gravitational attraction. At sea level and 45° latitude, a weight of 1 lb is equivalent to a mass of 1 lb.

tion of 1 $ft/s^2$. When the force is expressed as poundals and the mass and acceleration are expressed as pounds and feet per second squared, respectively, the value of $\alpha$ in Eq. (2-1) must be unity.

The universal constant $g_c$ is the basic factor in determining the definition of a pound-force. This constant is the standard value of the earth's gravitational attraction at sea level and 45° latitude. Thus, at sea level and 45° latitude, the earth's attraction will give a free-falling mass an acceleration of 32.17 $ft/s^2$, assuming no air friction.

The value and units of $g_c$ are 32.17 $(ft)(lbm)/(s^2)(lbf)$. From the definition of a pound-force, it can be seen that the proportionality constant $\alpha$ in Eq. (2-1) must be $1/g_c$ when the force is expressed as pounds-force, mass as pounds, and acceleration as feet per second squared.

The symbol $g$ is used to designate the local gravitational acceleration and equals 32.17 $ft/s^2$ at sea level (9.80 $m/s^2$ in SI units). Pounds-force due to gravitational attraction (or weight pounds) times the ratio $g_c/g$ gives mass pounds. Since $g_c$ and $g$ are practically equal everywhere on the earth's surface, 1-lb weight is often taken as equivalent to 1-lb mass. However, the difference between pounds-weight and pounds-mass should be recognized. As an example, if a mass of 100 lb were weighed on a spring scale at sea level, the weight (or pounds-force due to gravitational attraction) would be 100 lb. If this same mass were weighed on a spring scale located on the moon, the scale reading would be only about 16-lb force or 16-lb weight. However, the mass pounds would still be 100, since the pounds-force must be multiplied by $g_c/g$ to give the correct indication of mass pounds.

By consideration of the units involved, it can be seen that pounds-force times $g_c$ gives poundals; therefore, 1 lbf is equivalent to 32.17 poundals. The pound-force is widely used in ordinary engineering work, and the basic definition of this unit should be thoroughly understood.

Pressure is often expressed as force per unit of area. Thus, in the American engineering system, a common pressure unit is pounds-force per square inch, absolute ($lbf/in^2$, absolute), which represents the total pressure above absolute zero pressure, while pounds-force per square inch, gage ($lbf/in^2$, gage) represents the pressure greater than atmospheric pressure. Pressure is also expressed in other units, such as atmospheres (1 atm = 14.7 $lb/in^2$, absolute) or height of a column of mercury (760 mmHg or 29.92 inHg = 1 atm) or of water (33.93 $ftH_2O$ = 1 atm) that the pressure would support in a vertical tube sealed at the top so that the pressure at the top of the column is absolute zero (a perfect vacuum). Typical SI units of pressure are newtons per square meter ($N/m^2$) or pascals (Pa).

**Example 2-1: Conversion of pressure units from English system to SI** For a pressure of 29.4 $lbf/in^2$, absolute, determine the following:

(*a*) The force in newtons exerted on an area of 10 $m^2$
(*b*) The pressure expressed in pascals
(*c*) The pressure expressed in atmospheres
(*d*) The pressure expressed in bars
(*e*) The pressure expressed in inches of mercury

SOLUTION

(*a*) According to Table A-6 in App. A, 1 lbf = 4.448222 N and 1 $\text{in}^2$ = $6.4516 \times 10^{-4}\ \text{m}^2$. Therefore, 29.4 $\text{lbf/in}^2$ over an area of 10 $\text{m}^2$ is equivalent to a pressure of

$$\begin{array}{c|c|c|c} (29.4) & (4.448222) & & (10) \\ \hline & & (6.4516 \times 10^{-4}) & \end{array} = 2.027 \times 10^6\ \text{N}/10\ \text{m}^2$$

$$\begin{array}{c|c|c|c} (\text{lbf}) & (\text{N}) & (\text{in})^2 & (\text{m})^2 \\ \hline (\text{in})^2 & (\text{lbf}) & (\text{m})^2 & (10\ \text{m}^2) \end{array} = \text{N}/10\ \text{m}^2$$

An alternate form for writing the preceding is

$$\begin{array}{c|c|c|c} 29.4\ \text{lbf} & 4.448222\ \text{N} & \text{in}^2 & 10\ \text{m}^2 \\ \hline \text{in}^2 & \text{lbf} & 6.4516 \times 10^{-4}\ \text{m}^2 & 10\ \text{m}^2 \end{array} = 2.027 \times 10^6\ \text{N}/10\ \text{m}^2$$

(*b*) According to App. A, a pascal is a newton per square meter. Therefore, pressure in pascals equivalent to 29.4 $\text{lbf/in}^2$ is

$$\frac{2.027 \times 10^6\ \text{N}}{10\ \text{m}^2} = 2.027 \times 10^5\ \text{N/m}^2 = 2.027 \times 10^5\ \text{Pa or } 202.7\ \text{kPa}$$

(*c*) According to App. A, 1 atm is $1.013 \times 10^5$ Pa. Therefore, pressure in atmospheres equivalent to 29.4 $\text{lbf/in}^2$ is

$$\begin{array}{c|c} 2.027 \times 10^5\ \text{Pa} & \text{atm} \\ \hline & 1.013 \times 10^5\ \text{Pa} \end{array} = 2.0\ \text{atm}$$

(*d*) According to App. A, 1 bar is $1 \times 10^5$ Pa. Therefore, pressure in bars equivalent to 29.4 $\text{lbf/in}^2$ is

$$\begin{array}{c|c} 2.027 \times 10^5\ \text{Pa} & \text{bar} \\ \hline & 1 \times 10^5\ \text{Pa} \end{array} = 2.027\ \text{bar}$$

(*e*) According to the table of conversion factors on the front cover pages, pressure of 14.7 $\text{lbf/in}^2$ = pressure of 1 atm = pressure of 29.92 inHg. Therefore, a pressure in inches of mercury equivalent to 29.4 $\text{lbf/in}^2$ is

$$\begin{array}{c|c} 29.4\ \text{lbf} & 29.92\ \text{inHg} \\ \hline \text{in}^2 & 14.7\ \text{lbf/in}^2 \end{array} = 59.84\ \text{inHg}$$

**Example 2-2: Buying gold on the moon using earth force pounds** You are buying gold using an earth-calibrated spring balance (which measures force pounds based on gravitational attraction) to get the weight of the gold, and you are making the purchase on the moon where $g$, local acceleration due to gravity, is 5 ft/s$^2$.

(*a*) Using the reading on this balance and regular ounce-mass prices on earth ($350/oz), would you be cheating the seller (who plans to return to earth to sell it if you don't buy it) by buying gold from him on the moon?

(*b*) If you bought 10 oz of gold at $350/oz on this basis on the moon, how much could you sell it for on earth at $350/oz?

SOLUTION

(*a*) Yes, based on earth prices, you would be cheating the seller because the gold would appear to weigh much less on the moon than on earth because of the lower gravitational attraction on the moon.

(*b*) *Basis*: 10 ozf on the moon

$$g_c = 32.17\ \text{(ft)(lbm)/(s}^2\text{)(lbf)} = 32.17\ \text{(ft)(ozm)/(s}^2\text{)(ozf)}$$

$$g = 5\ \text{ft/s}^2 \text{ on the moon}$$

$$\text{Mass} = (\text{force})\frac{g_c}{g} = \frac{10\ \text{ozf} \,\big|\, 32.17\ \text{(ft)(ozm)} \,\big|\, \text{s}^2}{\phantom{10\ \text{ozf}} \,\big|\, \text{(s}^2\text{)(ozf)} \,\big|\, 5\ \text{ft}} = 64.34\ \text{ozm}$$

On earth, $g = 32.17$ ft/s$^2$

So the gold will weigh 64.34 ozf on earth, and the selling price would be

$$\frac{64.34\ \text{ozf or ozm} \,\big|\, \$350}{\phantom{64.34} \,\big|\, \text{oz}} = \$22{,}519$$

## FLOW DIAGRAMS

A flow diagram is used by chemical engineers to show the order in which equipment is employed in an overall process and to allow easy visualization of the full process. Flow diagrams are also used to indicate the amounts of materials involved and general process conditions. Such diagrams can be divided into three general types: (1) qualitative, (2) quantitative, and (3) combined-detail.

A *qualitative flow diagram* indicates the flow of materials, sequence of equipment necessary for the process, and special information on operating conditions. A *quantiative flow diagram* shows the amounts of materials required for various stages of the process operation, such as raw materials fed to the process, materials present in various stages of the processing, and products. Figure 2-1 is an example of a qualitative flow diagram for the production of nitric acid, while Fig. 2-2 is a quantitative flow diagram for the same process.

At the start of the development of a process design, preliminary qualitative and quantitative flow diagrams are prepared. As more details are generated and the

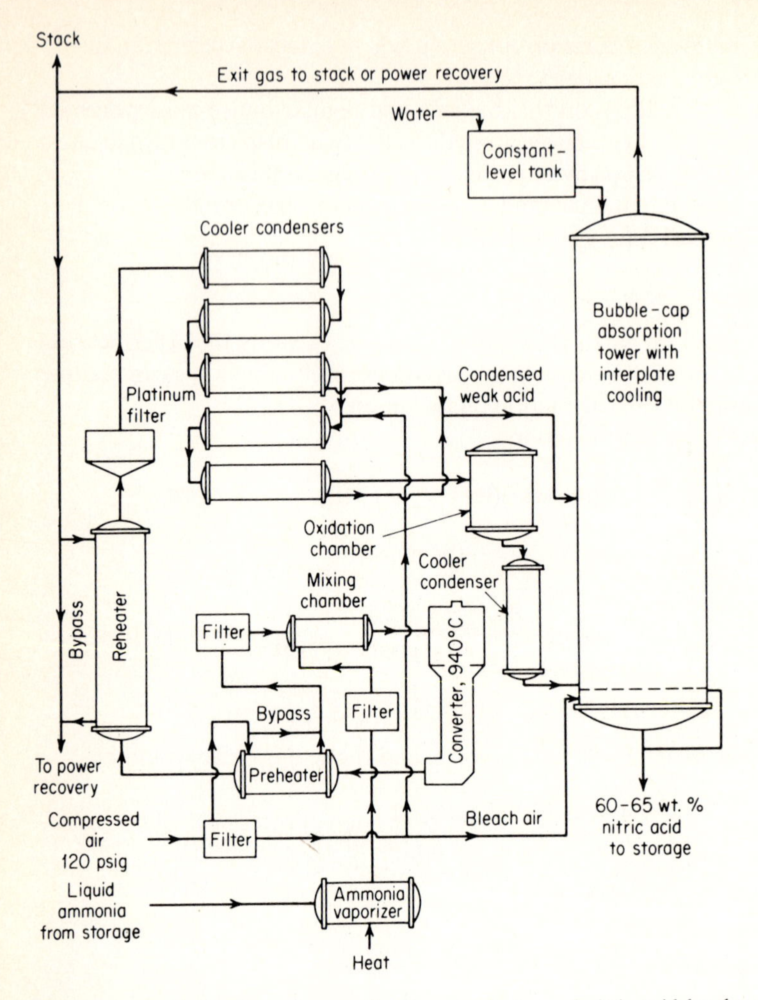


**Figure 2-1** Qualitative flow diagram for the manufacture of nitric acid by the ammonia-oxidation process.

design proceeds toward completion, detailed information on equipment specifications and operating conditions become available, and *combined-detail flow diagrams* can be prepared. This type of diagram is based on the qualitative flow diagram as a starting point, but it gives additional quantitative data and serves as a base reference for presenting equipment specifications, details of quantitative data, and process details. Tables presenting pertinent data on the process and the equipment are cross-referenced to the figure so that qualitative information and quantitative data are combined on one flow diagram. The diagram does not lose its effectiveness by presenting too much information; yet the necessary data are

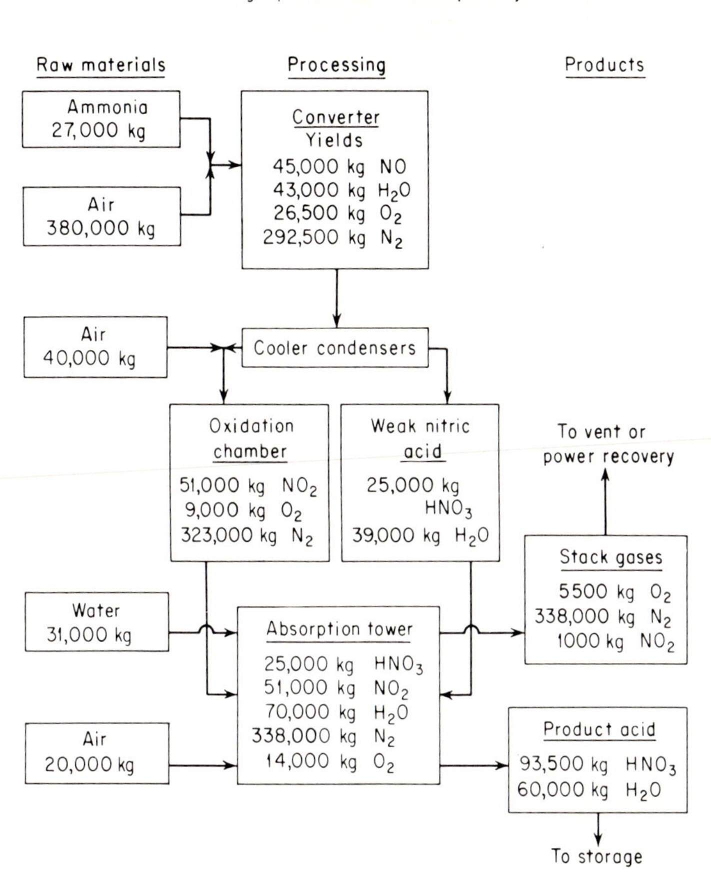


**Figure 2-2** Quantitative flow diagram for the manufacture of nitric acid by the ammonia-oxidation process.

readily available by direct reference to the accompanying tables so that the chemical engineer can get an overall picture of the full process from one coordinated diagram.

In general, a combined-detail flow diagram shows the location of temperature and pressure regulators, as well as the location of critical control valves and special control instruments. Each piece of equipment is shown and is designated by a defined code number. For each piece of equipment, accompanying tables give essential information, such as specifications for purchasing, specifications for construction, type of fabrication, and quantities and types of chemicals involved. A typical example of a combined-detail flow diagram is shown in Fig. 2-3.

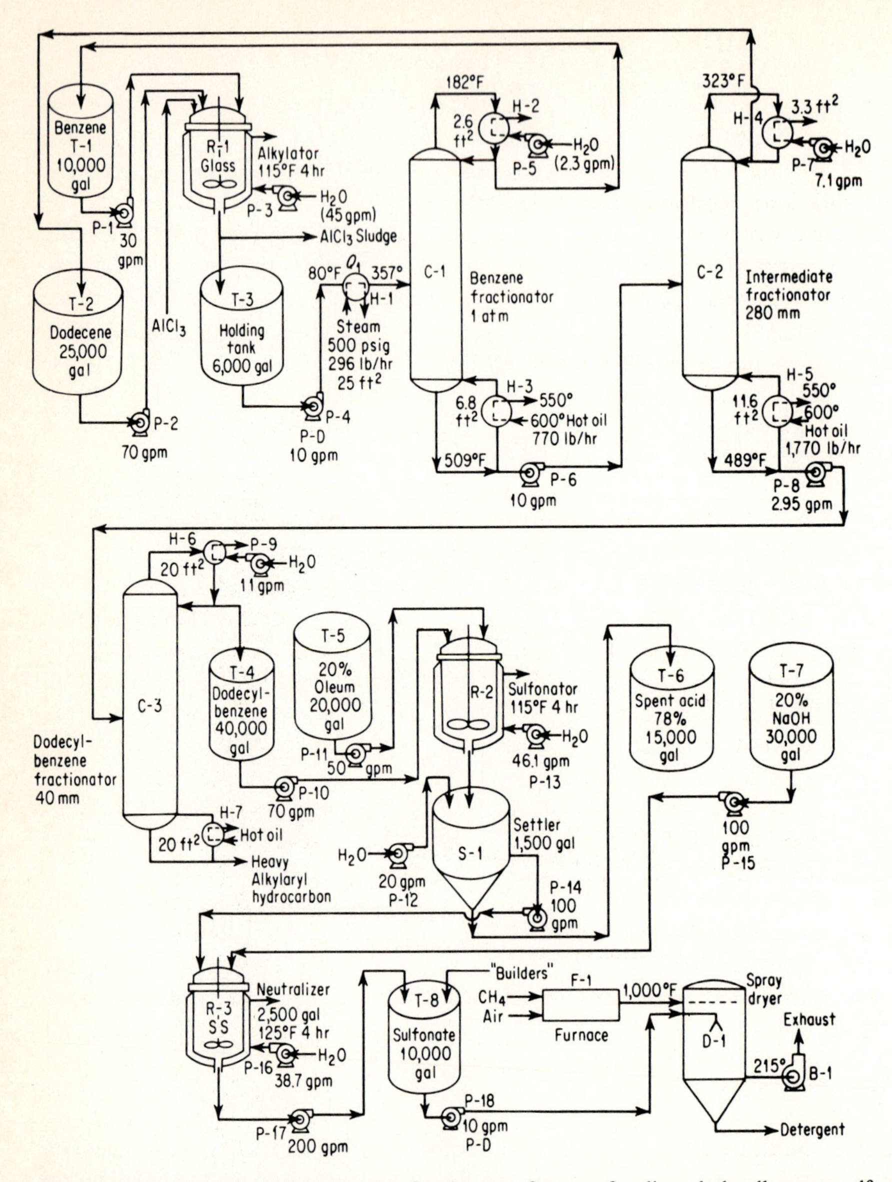


**Figure 2-3** Combined-detail flow diagram for the manufacture of sodium dodecylbenzene sulfonate. (*From M. S. Peters and K. D. Timmerhaus, "Plant Design and Economics for Chemical Engineers," 3d ed., McGraw-Hill, New York, 1980. With permission.*)

## COMPUTER-AIDED METHODS[2]

All modern engineers must be familiar with computers and should have the capability to write their own programs and make direct use of this important tool. In carrying out a full engineering project that involves applying the basic engineering and economic principles, a large number of calculations may be required, many of which are repetitive in nature and are therefore readily adaptable to computer solution using data-file programs. In recent years, effective systems have been developed that allow input of commands and data to a language interpreter for the system, which then calls for appropriate calculational subsystems from data files, makes the calculations needed, and reports the results. The application of computer-aided methods to this type of problem is referred to as *computer-aided design*. If computer-aided manufacturing is also included, the combination is referred to in abbreviated form as *CAD-CAM*.

The "first generation" of programs for computer-aided design was developed in the 1950s and early 1960s and had limited capabilities, even though the potential value was very apparent. These included programs such as the Chevron Heat and Material Balancing Program, the Kellogg Flexible Flow-Sheet, PACER, and CHESS. By the early 1970s a "second generation" of programs was well under way, allowing much more sophisticated and effective use. Included here were Monsanto's FLOWTRAN, Exxon's COPE, Union Carbide's IPES, Du Pont's CPES, and Chiyoda's CAPES programs. Because of the large expenditure the individual industries put into the development of the programs, many of the programs were highly proprietary.

In 1972, in order to investigate the possibility of getting some of the proprietary programs out into the general educational market, a committee of the Commission on Education of the National Academy of Engineering carried out an evaluation of industrial computer-aided, steady-state process design and simulation programs that might be available for use at colleges and universities. This committee had the name of CACHE (Computer Aids for Chemical Engineering Education) and made significant progress in these activities. CACHE concluded that Monsanto's FLOWTRAN was the most suitable for use at colleges and universities and published numerous articles and reports on the use of this program.[3]

While the overall development of computer-aided design is rapidly moving into a "third generation" of programs that are very sophisticated, there are many programs currently available which are extremely useful for design calculations.

[2] Adapted from M. S. Peters and K. D. Timmerhaus, *Plant Design and Economics for Chemical Engineers*, 3d ed., McGraw-Hill, New York, 1980.

[3] FLOWTRAN was first conceived in 1961 by the Applied Mathematics Department of Monsanto Company for process design and simulation and has undergone continuous development since then. In 1966, the system was put into general use at Monsanto, and from 1969 to 1973 outside companies could use FLOWTRAN through Monsanto Enviro-Chem Systems, Inc., by commercial computer networks. After 1973, FLOWTRAN was licensed and special arrangements were made by Monsanto to allow use in colleges and universities.

**Figure 2-4** Example of a graphic panel for a modern industrial plant with a computer-controlled system. (*Courtesy of C. F. Braun and Company.*)

A summary of many of these programs particularly appropriate for use in the chemical process industries has been developed through a grant to the Massachusetts Institute of Technology by the U.S. Department of Energy. The purpose of the project was to develop an advanced software computing system to meet the needs of chemical process engineers in the 1980s. The system has been named ASPEN (Advanced System for Process Engineering). A software survey to obtain sources of computer programs that would fall in the categories of commercial, university, proprietary industrial, specialty, and public was carried out, and the results have been published.

In addition to their use for design and process calculations, computers are widely used in industry for direct control of operations to give the most effective and most efficient results. Figure 2-4 is an example of a control room with computer-aided methods in use in an industrial operation. Computer technology has been moving forward at a very rapid pace, and new and better methods and equipment are regularly being developed.

## CHEMICAL ENGINEERING REFERENCES

There are many books and journals that have been published which give details related to the chemical engineering principles presented here in simplified form. New books and journals are also being produced regularly which present new ideas and approaches. Probably the best overall reference for chemical engineering is the latest edition of *Chemical Engineers' Handbook* (often referred to simply as

*Perry's Handbook*), published by the McGraw-Hill Book Company. The books in McGraw-Hill's Chemical Engineering Series listed in the front pages of this present book are basic references for obtaining additional information on any of the topics discussed here.

## PROBLEMS

**2-1** How many kilograms of carbon dioxide are produced by the complete oxidation of 24 kg of pure carbon?

**2-2** Indicate which of the following would be found in the field of unit operations and which would be found in the field of chemical technology:

(*a*) Determination of pressure drop over a packed column
(*b*) Study of the yields obtained in the production of sodium carbonate
(*c*) Investigation of the particle size obtained in a grinding operation
(*d*) Conversion of nitrogen to ammonia

**2-3** Calculate the weight in grams of 1 lb mol of sulfur dioxide.

**2-4** Convert a pressure of 32 $lb/in^2$ to grams per square centimeter.

**2-5** How many pound moles of carbon dioxide are obtained from the complete combustion of 4 kg of carbon?

**2-6** The power cost for pumping water to a plant is $10 per day if 2-in pipe is used. The cost of the installed pump and 2-in pipe is $1000. If 4-in pipe is used, the pumping cost is $5 per day and the installed pump and pipe costs $2000. If the initial investment must be written off in the first year, calculate the money saved during the first year by installing the 4-in pipe. The plant operates 365 days per year.

**2-7** Determine the Reynolds number for a fluid flowing in a circular pipe (Reynolds number $= DV\rho/\mu$ and is dimensionless) for the case where $D$, the diameter of pipe in which a fluid is flowing, is 2 ft; $V$, the velocity of fluid flow in pipe, is 30 ft/s; $\mu$, the viscosity of the fluid, is 0.50 lb/(ft)(s); and $\rho$, the density of the fluid, is 3.0 $g/cm^3$.

**2-8** A dimensionless number called the Prandtl number is defined as $N_{Pr} = c_p\mu/k$, where $c_p$ is heat capacity in units of cal/(g)(°C); $\mu$ is viscosity in units of g/(cm)(s); and $k$ is thermal conductivity. What are the net units for $k$?

**2-9** You are selling gold on the moon using an earth-calibrated spring balance to get pounds, and you are selling the gold at the regular ounce-mass price on earth of $600/oz as read on your spring balance. Local gravitational attraction ($g$) on the moon is 5 $ft/s^2$.

(*a*) Are you cheating the buyer (who will return to earth to resell the gold) by selling him the gold on the basis of the readings from your spring balance (which measures force pounds) on the moon?

(*b*) Suppose you are the purchaser, and you bought 10 oz of gold at $600/oz (pay $6000) on this basis on the moon. How much profit would you make if you sold it on earth for $600/oz?

**2-10** Mole Avogadro, high school chemistry student, complains heatedly to his father, "I don't understand this silly number you've come up with. Please explain." Father responds coolly, "It's quite simple, $6.02 \times 10^{23}$ is the number of molecules in a gram mole of any substance." The son retorts, "OK, smarty, sir, how many molecules are there in an ounce mole?" Help Mr. Avogadro by telling him how many molecules there are in an ounce mole of any substance. Also, how many ounces are there is an ounce mole of $CO_2$ (MW = 44)?

*Data*: 1 lb = 16 oz
1 lb = 454 g

**2-11** Determine the value of the *dimensionless number AB/C* involving the three variables $A$, $B$, and $C$, if $A$ is 60 $lb/ft^3$, $B$ is 100 cm/s, and $C$ is 80 when the number is expressed in cgs units (cgs means centimeter, gram, second).

**2-12** Molasses is to be pumped in January at a flow rate of 100 kg/h through a pipe 0.05 m in diameter. What is the value of the Reynolds number

$$\left(\frac{\text{diameter} \times \text{velocity} \times \text{density}}{\text{viscosity}}\right)$$

if the density of the molasses at that January temperature is 70 $lb/ft^3$ and its viscosity is 750 kg/(m)(h)?

**2-13** At his wife's insistence, John Smith, who is an astronaut, has taken on with little enthusiasm a dieting program to lose weight. At the start of his dieting program, his weight expressed as pounds-mass was 215 lb. He has a scale in his spaceship which measures his weight in pounds-force, and he is delighted to find that this scale reads his weight as 160 lb 2 weeks after he has started his diet. However, he reads the scale while circling the earth in his spaceship at a low elevation where the local acceleration due to gravity ($g$) is 25 $ft/s^2$. How many pounds-mass has he lost or gained during the first 2 weeks of his diet?

**2-14** (*This problem is intended for computer-program solution.*) A spherical tank used for chemical storage has a sight glass on one end showing the liquid level in the tank. In order to help an operator account for the material in the tank, you wish to set up a table showing cubic meters of liquid in the tank and percent of capacity versus height of liquid in the tank ($H$). Write and run a Fortran program to compute tank volume in cubic meters and percent full for readings ($H$) of 0.0, 0.1, 0.2, ... 1.9, and 2.0 times $R$, where $R$ is the inside radius of the tank. For convenience, set $R = 1.0$ m. Use the program to illustrate at least two different ways to set up a loop.

CHAPTER

# THREE

# CHEMICAL ENGINEERING STOICHIOMETRY

A chemical engineer must be adept at the mathematical manipulation of experimental data in order to obtain the maximum amount of information from any one set of results. The principles of stoichiometry indicate various methods of attack and mathematical treatments which can be applied to experimental data to give a clear picture of the overall process.

The laws of conservation of matter, conservation of energy, and combining weights in chemical reactions constitute the basis of stoichiometric considerations. The application of these laws to industrial processes is not obvious to the technical worker until he or she has had experience.

The purpose of this chapter is to acquaint the reader with the manner in which these three laws can be applied as well as to present fundamental information necessary for understanding the general stoichiometric principles.

**Table 3-1 Nomenclature for stoichiometry**

$a, b, c$ = empirical constants, dimensionless
$c_p$ = heat capacity at constant pressure, J/(kg)(K), cal/(g)(°C), or Btu/(lb)(°F)
$C_p$ = heat capacity at constant pressure, J/(g mol)(K), cal/(g mol) (°C), or Btu/(lb mol)(°F)
$c_v$ = heat capacity at constant volume, J/(kg)(K), cal/(g)(°C), or Btu/(lb)(°F)
$C_v$ = heat capacity at constant pressure, J/(g mol)(K), cal/(g mol)(°C), or Btu/(lb mol)(°F)
$n$ = number of moles (i.e., molecular weights) of a material
$p$ = absolute pressure (may refer to partial pressure), kPa, atm, or lbf/ft$^2$
$P$ = absolute total pressure (equals gage pressure plus atmospheric pressure), kPa, atm, or lbf/ft$^2$
$R$ = perfect-gas-law constant, (cm$^3$)(atm)/(g mol)(K), cal/(g mol)(K), Btu/(lb mol)(°R), etc.
$t$ = temperature, °C or °F
$T$ = absolute temperature, K (equals °C + 273) or °R (equals °F + 460)
$v$ = specific volume (may refer to pure component volume), usually cm$^3$/g or ft$^3$/lb
$V$ = total volume, cm$^3$/system or ft$^3$/system
$x$ = mole fraction (equals number of moles of one material in a system divided by the total number of moles in the system)

## MATERIAL BALANCES

As has been indicated in Chap. 2, a material balance is based on the law of conservation of matter. This law states that the total mass of all materials entering a fixed system in a given time must equal the total mass of all materials leaving plus any accumulation that occurs in the system. The application of this law may be visualized by considering an empty box into which three streams of material are flowing and from which one stream is emerging. Over any given period of time, the total weight of the three entering streams must equal the total weight of the exit stream plus any weight gained by the box during the given time interval through accumulation.

The law of conservation of matter can also be applied to the chemical elements. For all practical purposes, it can be stated that it is impossible to change one element into another. Thus, if 10 kg of carbon goes into a reactor, 10 kg of carbon must come out or accumulate in the reactor. It makes no difference what form of chemical combination the carbon may have at the entering or leaving conditions. As long as the balance is made on an element, such as carbon, hydrogen, or oxygen, the overall material balance must apply.

### Tie Substances

It frequently simplifies the mathematical work necessary for making stoichiometric calculations if a material can be found that comes into the process in just one stream and leaves unchanged in only one stream. Such a material is called a *tie substance*. An excellent example of this occurs in a continuous salt-solution concentrator. Here a dilute salt solution comes into a concentrator where water is evaporated, and the concentrated salt solution flows out. In this process, it can be assumed that no accumulation takes place in the concentrator. Since all the salt comes into the system in one stream and the same amount leaves in another stream, the salt can be taken as a tie substance.

To show the direct application of the tie substance to material-balance calculations, the case can be considered in which it is desired to determine the amount of water evaporated from 100 lb of a 10 wt% NaCl-in-water solution when it is concentrated to a 20 wt% NaCl-in-water mixture. On a basis of 100 lb of the original solution, there are 10 lb of salt and 90 lb of water before the concentration. Since no salt is lost with the evaporated water, there must be 10 lb of salt left in the final concentrated solution. Therefore, the salt may be considered as a tie substance, since all of it enters and leaves unchanged in single streams.

If $x$ represents the pounds of water left in the concentrated solution, $(10/(x + 10))(100)$ must equal the final salt concentration, or

$$\frac{10}{(x + 10)}(100) = 20$$

Solving for $x$, its value is 40. Before the concentration there was 90 lb of water per 100 lb of original solution or per 10 lb of salt. After the concentration there is

40 lb of water in the final solution per 10 lb of salt. Therefore, the amount of water evaporated per 100 lb of original solution must be 90 − 40 = 50 lb.

**Example 3-1: Material balance involving tie substance in humidification problems** In a continuous process, 100 lb of wet air per minute containing 0.02 lb of water vapor per pound of dry air enters a chamber where water vapor is added to the air. The air leaving the chamber contains 0.05 lb of water vapor per pound of dry air. Calculate the amount of water added to the initial wet air per minute.

SOLUTION The labeled diagram for the process is shown in Fig. 3-1.

**Basis**

1 min, equivalent to 100 lb of wet air entering the chamber

All the dry air entering the chamber in one stream must leave in the exit air stream; therefore, the dry air can be considered as a tie substance.

Let $x$ equal the weight of water vapor entering the chamber per minute. The weight of dry air entering the chamber must equal 100 − $x$ lb. The pounds of water vapor per pound of dry air in the entering stream equals 0.02; therefore

$$0.02 = \frac{x}{100 - x}$$

$$x = 1.96 \text{ lb/min}$$

Pounds dry air entering chamber = pounds dry air leaving chamber = 100 − 1.96 = 98.04 lb/min.

Since there is 0.05 lb water vapor per pound of dry air leaving the chamber,

$$\text{lb water vapor leaving chamber per minute} = (0.05)(98.04) = 4.90 \text{ lb/min}$$

The preceding calculations are on the basis of 1 min; therefore, the pounds of water added to the initial air per minute equals the pounds of water vapor leaving in the final air per minute minus the pounds of water vapor entering in the initial air per minute, or

$$4.90 - 1.96 = 2.94 \text{ lb water added to air per minute}$$

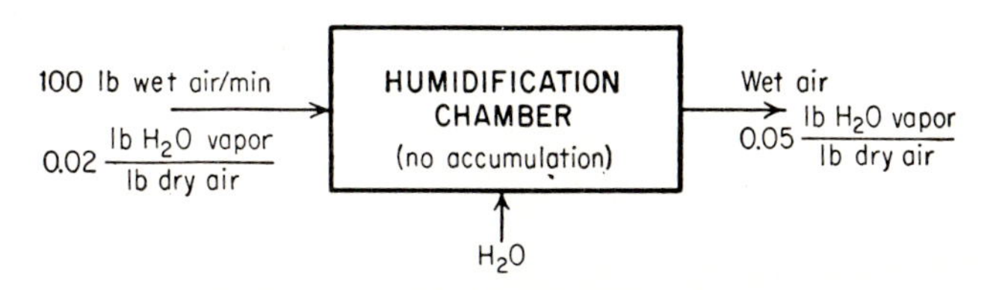


**Figure 3-1** Material balance in humidification process.

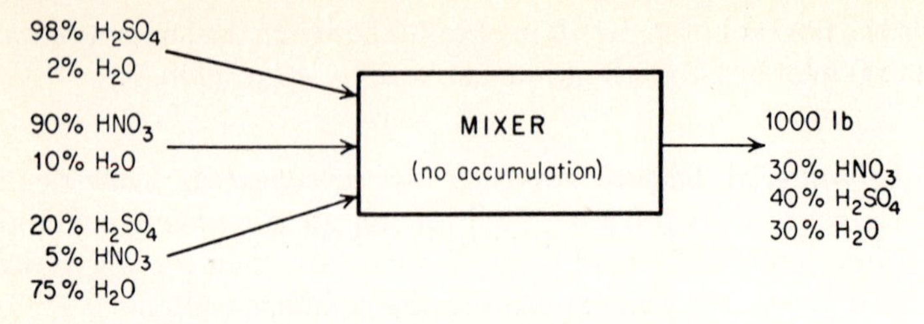


**Figure 3-2** Material balance applied to multiple streams.

**Example 3-2: Application of material balances to multiple streams** A mixture containing 30% by weight $HNO_3$, 40% by weight $H_2SO_4$, and 30% by weight water is to be made continuously by mixing concentrated $H_2SO_4$ (98 wt % $H_2SO_4$ and 2% $H_2O$), concentrated $HNO_3$ (90 wt % $HNO_3$ and 10% $H_2O$), and waste acid (20 wt % $H_2SO_4$, 75% $H_2O$, and 5% $HNO_3$). Calculate the kilograms of concentrated $H_2SO_4$, concentrated $HNO_3$, and waste acid necessary per 1000 kg of final mixture.

SOLUTION The labeled diagram for the process is shown in Fig. 3-2.

**Basis**

1000 kg of final mixture

Let $x$ = kg concentrated $H_2SO_4$ necessary
$y$ = kg concentrated $HNO_3$ necessary
$z$ = kg waste acid necessary

Overall material balance:

$$\text{Weight in} = \text{weight out}$$

$$\text{Weight in} = x + y + z$$

$$\text{Weight out} = 1000$$

$$x + y + z = 1000 \tag{A}$$

$HNO_3$ balance:

$$\text{Weight } HNO_3 \text{ in} = 0.90y + 0.05z$$

$$\text{Weight } HNO_3 \text{ out} = (1000)(0.30)$$

$$0.90y + 0.05z = 300 \tag{B}$$

$H_2SO_4$ balance:

$$\text{Weight } H_2SO_4 \text{ in} = 0.98x + 0.20z$$

$$\text{Weight } H_2SO_4 \text{ out} = (1000)(0.40)$$

$$0.98x + 0.20z = 400 \tag{C}$$

Equations (A), (B), and (C) represent three independent equations involving three unknowns. These equations can be solved simultaneously for $x$, $y$, and $z$ to give

$$x = 338 \text{ kg concentrated } H_2SO_4 \text{ necessary}$$

$$y = 313 \text{ kg concentrated } HNO_3 \text{ necessary}$$

$$z = \underline{349} \text{ kg waste acid necessary}$$

$$1000 \text{ kg total weight}$$

These results can be checked by applying a water balance to the process. Water balance:

$$\text{Weight } H_2O \text{ in} = (338)(0.02) + (313)(0.10) + (349)(0.75) = 300 \text{ kg}$$

$$\text{Weight } H_2O \text{ out} = (1000)(0.30) = 300 \text{ kg}$$

Since the weight of the water entering the system equals the weight of the water leaving the system, the water balance is satisfied, and the results check.

### Chemical Reactions and Molal Units

Where chemical reactions are involved, material balances are best handled on the basis of conservation of elements using molal units. For example, if 24 lb of carbon is burned to carbon dioxide and carbon monoxide, the amount of carbon monoxide formed can be calculated by a carbon balance when an analysis indicates that 66 lb of carbon dioxide is formed. The molecular weight of carbon dioxide is 44; therefore, using molal units, there are $\frac{66}{44} = 1.5$ lb mol of $CO_2$ formed. One pound mole of $CO_2$ contains 12 lb of carbon and 32 lb of oxygen; so 1.5 lb mol of $CO_2$ contains $(1.5)(12) = 18$ lb of carbon. Twenty-four pounds of carbon goes into the reaction, and 18 lb of this carbon is converted to carbon dioxide. The rest of the carbon, or $24 - 18 = 6$ lb, must have been converted to carbon monoxide. One pound mole of CO contains 12 lb of carbon and 16 lb of oxygen. If 6 lb of carbon is in the CO, there must be 0.5 lb mol of CO present. Since 1 lb mol of CO weighs 28 lb, the total weight of CO formed is $(0.5)(28) = 14$ lb.

## CHEMICAL LAWS OF COMBINING WEIGHTS

The relationships between the weights of reactants and products involved in a chemical reaction can be obtained from a consideration of the equation for the reaction and the molecular weights of the materials involved. The equation should be written with the reactants on the left side and the products on the right side. The molecular weight of any compound may be obtained from the formula showing its composition and a table of atomic weights.

The production of sulfur dioxide by the oxidation of iron pyrites may be taken as an example. The reactants are oxygen and pyrites, while the products are sulfur dioxide and ferric oxide. The correctly balanced equation can be written

$$4FeS_2 + 11O_2 = 2Fe_2O_3 + 8SO_2$$

The formula for the pyrites can be taken as $FeS_2$; therefore, the molecular weight of this compound equals the atomic weight of iron plus twice the atomic weight of sulfur, or the molecular weight of $FeS_2$ equals $55.85 + (2)(32.06) = 119.97$. The molecular weights of the other materials involved in the reaction can be obtained in a similar fashion.

According to the reaction equation, 4 mol of $FeS_2$ react with 11 mol of $O_2$ to form 2 mol of $Fe_2O_3$ and 8 mol of $SO_2$. This can be written in a form to indicate the weights of the materials involved in the combination.

$$\begin{array}{ccccccc} 4FeS_2 & + & 11O_2 & = & 2Fe_2O_3 & + & 8SO_2 \\ (4)(119.97) & & (11)(32.0) & & (2)(159.70) & & (8)(64.06) \\ 479.88 & & 352.00 & & 319.40 & & 512.48 \end{array}$$

Therefore, 479.88 parts by weight of $FeS_2$ reacts with 352.00 parts by weight of $O_2$ to form 319.40 parts by weight of $Fe_2O_3$ and 512.48 parts by weight of $SO_2$.

These same weight ratios must hold no matter how much of each material is present. It is possible to calculate the theoretical amounts of the other three compounds if the weight of one of the four is known. Thus, the amount of oxygen necessary to react with 100 kg of iron sulfide would equal the weight of oxygen required per kilogram of iron sulfide times 100, or $(352.00/479.88)(100) = 73.5$ kg of oxygen is required to react with 100 kg of iron sulfide. Similarly, the weight of $Fe_2O_3$ produced from 100 kg of $FeS_2$ would be $(319.40/479.88)(100) = 66.6$ kg, and the weight of $SO_2$ produced would be $(512.48/479.88)(100) = 106.9$ kg.

These results can be checked by determining whether or not an overall material balance can be made. The total weight of reactants is $100 + 73.5 = 173.5$ kg. The total weight of the products is $66.6 + 106.9 = 173.5$ kg. Since the weight of the products equals the weight of the reactants, the conditions for the overall material balance are satisfied and the results check.

## THE GAS LAWS

When applying the principles of stoichiometry to gases, the relationships among mass of material, temperature, pressure, and volume are very important. Various methods are available to indicate the effect of changes of one or more of these variables. Laws applicable to the so-called "ideal" or "perfect" gases are of fundamental importance in chemical engineering. While these laws are not precise, they do give results which are adequate for ordinary calculations as long as the gases being considered are not highly complex or under pressures in excess of 1 or 2 atm.

## The Perfect-Gas Law

Experimental investigations over a wide range of different gases and different conditions have led to the following empirically derived equation, known as the *perfect-gas law*[1]:

$$pV = nRT \tag{3-1}$$

where $n$ = number of moles of gas
$p$ = absolute pressure
$R$ = a constant proportionality factor
$T$ = absolute temperature
$V$ = volume of $n$ mol of gas

$R$ is a universal constant dependent only upon the units in which the other components of the equation are expressed. Values of this constant for various sets of units are given in App. C.

The absolute-temperature scale is based on a zero point which corresponds to the temperature where all molecular motion theoretically ceases. This point has been found to be −273.16°C, or −459.7°F. The absolute-temperature scale in kelvin corresponds to the Celsius scale, while absolute degrees Rankine corresponds to degrees Fahrenheit. The interconversion of these four methods for expressing temperature is clarified by the following conversion formulas and table (absolute zero values rounded off to −273°C and −460°F):

$$\text{Standard degrees Celsius} = {}^\circ\text{C} = \frac{{}^\circ\text{F} - 32}{1.8}$$

$$\text{Standard degrees Fahrenheit} = {}^\circ\text{F} = ({}^\circ\text{C})(1.8) + 32$$

$$\text{Absolute kelvin} = \text{K} = {}^\circ\text{C} + 273$$

$$\text{Absolute degrees Rankine} = {}^\circ\text{R} = {}^\circ\text{F} + 460$$

| | Standard-temp scale | | Absolute-temp scale | |
|---|---|---|---|---|
| | °C | °F | K | °R |
| Absolute zero | −273 | −460 | 0 | 0 |
| Freezing point of water at 1 atm pressure | 0 | 32 | 273 | 492 |
| Boiling point of water at 1 atm pressure | 100 | 212 | 373 | 672 |

The perfect-gas law can be applied by a direct mathematical treatment, or it may be used in a simpler form by a logical consideration of what should happen

[1] The $p$-$V$-$T$ relationship shown by Eq. (3-1) is a special and simplified form of a so-called "equation of state."

to a gas when its temperature, volume, or pressure is changed. For example, if a gas initially has a temperature of 100°F, a pressure of 14 $lb/in^2$, absolute, and a volume of 50 $ft^3$, the new volume of this gas when the temperature and pressure are changed to 150°F and 25 $lb/in^2$, absolute, can be found as follows: According to the perfect-gas law, the initial conditions can be expressed as $pV = nRT$, or

$$(14)(50) = nR(100 + 460) \tag{3-2}$$

Letting $V_2$ represent the volume at the new temperature and pressure, the new conditions can be expressed as

$$(25)(V_2) = nR(150 + 460) \tag{3-3}$$

Since no gas was added or taken away from the system, the number of moles of gas ($n$) must be the same in both Eqs. (3-2) and (3-3). The value of $R$ is constant for both equations since the same units were used for the corresponding terms. Dividing Eq. (3-3) by Eq. (3-2) and canceling the $n$ and $R$ yields the following value for the final volume:

$$V_2 = \frac{(150 + 460)(14)(50)}{(100 + 460)(25)} = 30.5 \text{ ft}^3$$

This same result could have been obtained by a logical consideration of what should happen to a gas when its temperature and pressure are changed. The temperature of the gas goes from 100 to 150°F. This means that the gas becomes hotter and therefore gains in energy. The increased energy causes the molecules of the gas to move around faster, resulting in more force on the walls of the container and a tendency for the volume to increase. This effect of the temperature increase on the volume can be expressed as a final volume of $(50)((150 + 460)/(100 + 460))$, where the ratio of the absolute temperatures must be greater than 1.0 to indicate that the volume tends to increase. In addition to the temperature effect, the change due to the increase in pressure from 14 to 25 $lb/in^2$, absolute, must be taken into consideration. The increased pressure tends to cause the volume to become smaller; therefore, the correction factor for the pressure change must be the ratio of the pressures in the form of a fraction less than 1.0, or $\frac{14}{25}$. The overall effect of the temperature and pressure changes can then be obtained by applying both logical corrections to the original volume of the gas to give a final volume of

$$V_2 = (50)\frac{(150 + 460)(14)}{(100 + 460)(25)} = 30.5 \text{ ft}^3$$

It can be seen that the same result is obtained whether the straight mathematical or the logical attack is used. It is usually quicker and easier to use the logical method.

The methods of using the perfect-gas law, as illustrated in the preceding paragraphs, can be applied to problems involving gases where the effect of change in temperature, pressure, volume, or the number of moles of gas is to be determined.

## Dalton's Law

In a mixture of gases, the partial pressure of any one of the component gases can be defined as the pressure that component gas would exert if it alone were present in the same volume and at the same temperature as the original mixture. Dalton's law states that the total pressure of a mixture of ideal gases equals the sum of the partial pressures of the component gases; that is,

$$P = p_a + p_b + p_c + \cdots \tag{3-4}$$

where $P$ is the total pressure of the mixture and $p_a$, $p_b$, $p_c$, etc., are the partial pressures of the individual component gases.

A very important corollary to Dalton's law states that the mole fraction of a component gas in a mixture of ideal gases equals the partial pressure of the component divided by the total pressure, or

$$x_a = \frac{n_a}{n_a + n_b + n_c + \cdots} = \frac{p_a}{P} \tag{3-5}$$

This can be proved by use of the perfect-gas law in conjunction with Dalton's law; thus, for an ideal gas,

$$p_a V = n_a RT \tag{3-6}$$

$$(p_a + p_b + \cdots)V = PV = (n_a + b_b + \cdots)RT \tag{3-7}$$

Dividing Eq. (3-6) by Eq. (3-7) gives

$$\frac{p_a}{P} = \frac{n_a}{n_a + n_b + n_c + \cdots} = x_a \tag{3-5}$$

## Amagat's Law

The pure-component volume of a component gas in a mixture of gases is defined as the volume which would be occupied by that component gas if it alone were present at the same temperature and pressure as the original mixture. Amagat's law states that the total volume occupied by a gaseous mixture equals the sum of the pure-component volumes, or

$$V = v_a + v_b + v_c + \cdots \tag{3-8}$$

where $V$ is the total volume and $v_a$, $v_b$, $v_c$, etc., are the pure-component volumes of the individual gases making up the gaseous mixture.

A corollary to Amagat's law states that the mole fraction of a component gas in a mixture of ideal gases equals the pure-component volume of that gas divided by the total volume of the mixture, or

$$x_a = \frac{v_a}{V} \tag{3-9}$$

This can be proved by the same line of reasoning used to prove the corollary to Dalton's law.

Since mole percentage equals mole fraction times 100 and volume percentage equals volume fraction times 100, the following general statement can be made: The volume percent of a component in a mixture of perfect gases always equals its mole percent in the mixture. This fact is used extensively in stoichiometric computations to simplify calculations based on volume percent analyses of gaseous mixtures.

## Standard Conditions

Since the volume of a gas depends on its temperature as well as its pressure, it has been convenient to choose a certain temperature and pressure specification by which quantities of gases can be compared when expressed in terms of volume. The temperature and pressure chosen for this specification are designated as the *standard conditions*. The values for standard conditions are 0°C and 760 mmHg pressure.[2] By Avogadro's principle, equimolal quantities of ideal gases occupy the same volume at the same conditions of temperature and pressure. At standard conditions, 1 g mol of an ideal gas occupies 22.4 L and 1 lb mol of an ideal gas occupies 359 $ft^3$.

By use of the perfect-gas law and the standard-condition volume, the volume of any ideal gas can be calculated if its temperature, pressure, and number of moles are known. As an example, the volume of 64 lb of oxygen is (64)(359)/(32) = 718 $ft^3$ at standard conditions, since the molecular weight of oxygen is 32. Using the logical application of the perfect-gas law, the volume of this same weight of oxygen at 100°F and 720 mmHg can be calculated. The absolute-temperature ratio for the temperature correction on the standard-condition volume must be a fraction greater than 1.0 since the final temperature is higher than 0°C. The pressure correction must also be a fraction greater than 1.0 because the final pressure is less than the standard-condition pressure of 760 mmHg. Applying these two corrections to the volume at standard conditions, the final volume at 100°F and 720 mmHg is

$$\frac{(718)(460 + 100)(760)}{(492)(720)} = 863 \text{ ft}^3/64 \text{ lb of oxygen}$$

A simple check on the temperature and pressure corrections applied to a standard-condition volume can be made by noting that the standard temperature of 492°R must always be in the opposite part of the fraction from the standard pressure of 760 mmHg.

The following examples illustrate the application of the gas laws to practical problems.

[2] Many engineering calculations express standard gas volumes as cubic feet at 60°F and 29.92 (or 30) inHg. These conditions are commonly used for reporting volumes of fuel gases.

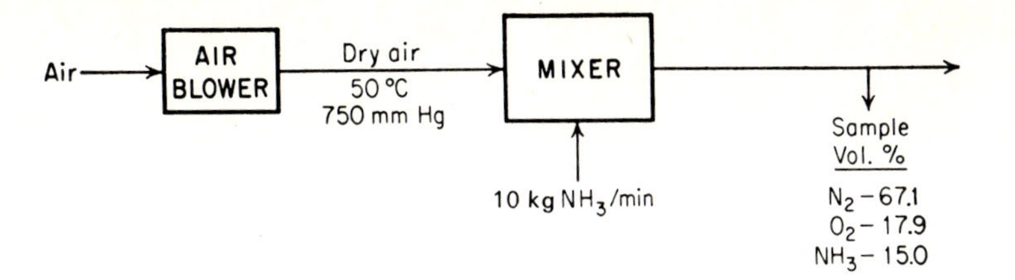


**Figure 3-3** Process for determination of air-flow rate.

**Example 3-3: Determination of gas-flow rate** The flow rate of air from a commercial blower is to be determined. The blower delivers dry air at 50°C and 750 mmHg. Ten kilograms of ammonia per minute is added to the air, and a sample of the gaseous mixture indicates 67.1% $N_2$, 17.9% $O_2$, and 15.0% $NH_3$ by volume. Calculate the rate of air delivery by the blower in cubic meters per minute. Dry air can be considered as 21% $O_2$ and 79% $N_2$ by volume.

SOLUTION The labeled diagram for the process is shown in Fig. 3-3.

**Basis**

1 min

Since 10 kg is 10,000 g and the molecular weight of ammonia is 17, gram moles of ammonia in gaseous mixture = 10,000/17 = 588 g mol. 558 g mol $NH_3$ = 15 vol % of gaseous mixture = 15 mol % of gaseous mixture

Let $x$ = total gram moles of $O_2$, $N_2$, and $NH_3$ in the mixture.

$$0.15x = 588$$

$$x = 3920 \text{ g mol}$$

Gram moles of $O_2$ and $N_2$ in the $NH_3$-air mixture = gram moles of $O_2$ and $N_2$ in the air delivered by the blower = 3920 − 588 = 3332 g mol of air delivered by the blower on the basis of 1 min.

Since 1 g mol of an ideal gas at standard conditions of 0°C and 760 mmHg pressure is equivalent to 22.4 L, or $22.4 \times 10^{-3}$ m³, of gas volume, the rate of air delivery by the blower is

| 3332 g mol | $22.4 \times 10^{-3}$ m³ at S.C. | (273 + 50) K | 760 mmHg |
|---|---|---|---|
| min | g mol | 273 K | 750 mmHg |

$$= 89.5 \text{ m}^3/\text{min}$$

**Example 3-4: Determination of average molecular weight and gas density** Calculate the average molecular weight of dry air and the density of dry air at 70°F and 730 mmHg, assuming dry air contains 21% $O_2$ and 79% $N_2$ by volume.

SOLUTION

**Basis**

1 lb mol of dry air

| Component | Volume percent | Mole fraction | Molecular weight | Pounds in 1 mol of dry air |
|---|---|---|---|---|
| $O_2$ | 21.0 | 0.21 | 32.0 | (0.21)(32) = 6.72 |
| $N_2$ | 79.0 | 0.79 | 28.0 | (0.79)(28) = 22.12 |
| Total | — | — | — | 28.84 |

1 lb mol of air weights 28.84 lb; therefore, the average molecular weight of air is 28.84.[3]

The volume of 1 lb mol of air at 70°F and 730 mmHg is

$$\frac{(359)(460 + 70)(760)}{(492)(730)} = 402 \text{ ft}^3$$

The density of dry air at 70°F and 730 mmHg is

$$\frac{28.84}{402} = 0.0718 \text{ lb/ft}^3$$

**Example 3-5: Determination of dissociation of gases** In the gaseous state, $N_2O_4$ partially dissociates according to the following equation:

$$N_2O_4 = 2NO_2$$

If it is found that 24.0 g of a gaseous mixture containing $N_2O_4$ and $NO_2$ occupies a volume of 15,100 cm$^3$ at 363 K and 97.3 kPa pressure, calculate the percentage dissociation of $N_2O_4$ to $NO_2$.

SOLUTION

**Basis**

24.0 g of $N_2O_4$ before dissociation
Molecular weight of $N_2O_4$ = 92
Gram moles of $N_2O_4$ present before dissociation = 24.0/92 = 0.261 g mol
Let $x$ = mol $N_2O_4$ dissociated.

[3] Dry air actually contains 20.99 vol % $O_2$, 78.03 % $N_2$, 0.94 % argon, and small amounts of carbon dioxide, neon, helium, krypton, and xenon. The true molecular weight of dry air is 28.96.

Since 2 mol $NO_2$ are formed per mole of $N_2O_4$ dissociated, $2x$ = mol $NO_2$ formed.

The final number of moles in the gas = mol $N_2O_4$ + mol $NO_2$ = $(0.261 - x) + 2x$.

One gram mole of an ideal gas at standard conditions of 273 K and 1 atm (1 atm is equivalent to 101.3 kPa) is equivalent to 22,400 $cm^3$ of gas volume. Therefore, by the perfect-gas law, the final number of moles of gas is

| 15,100 $cm^3$ of gas at end | 273 K | 97.3 kPa | g mol |
|---|---|---|---|
| | 363 K | 101.3 kPa | 22,400 $cm^3$ at S.C. |

= 0.487 g mol in final gas,

Therefore, $(0.261 - x) + 2x = 0.487$ and $x = 0.226$ mol $N_2O_4$ dissociated. The percentage dissociation of $N_2O_4$ to $NO_2$ is

$$\frac{0.226}{0.261}(100) = 86.7\%$$

## ENERGY BALANCES

The law of conservation of energy states that energy is indestructible and that the total amount of energy entering a given system must equal that leaving plus any accumulation within the system. A mathematical expression of this law is called an *energy balance.*

The two primary classifications of energy are potential and kinetic. *Potential energy* refers to the energy a body or substance has because of its position relative to another material or because of its components. This can be further broken down into *internal* and *external potential energy.* As an example, a lump of coal has a certain external potential energy when placed at a fixed distance from the earth's surface because of its ability to fall and strike the earth with a momentum dependent on its mass and speed. The lump of coal has internal potential energy because of its ability to give off heat when burned.

*Kinetic energy* refers to energy due to motion. Since the molecules of any gas are in constant motion, a gas has a certain internal kinetic energy. A bullet in flight contains external kinetic energy that is due to its motion.

The flow of heat from one body to another may be considered as energy in transition. When heat flows from a hot to a cold body, the internal energy of the colder body is increased at the expense of the internal energy of the body from which the flow of heat originated.

Another form of energy in transition is work. This may be defined as the energy transferred by the action of a mechanical force moving under restraint through a tangible distance. Work cannot be stored as such, but the capability to do work can be stored as potential energy or kinetic energy.

## Units of Energy

The common unit of work in the English system is the *foot-pound force* (ft · lbf). This is defined as the amount of work done by a force of 1 lb acting through a distance of 1 ft. A force of 1 lb is the standard gravitational force exerted on a mass of 1 lb at sea level.

In stoichiometric calculations, the units of heat are extensively used in energy expressions. One *calorie* (cal) is defined as the amount of heat necessary to raise the temperature of 1 g of water from 15 to 16°C, while the *British thermal unit* (Btu) is the amount of heat necessary to raise the temperature of 1 lb of water from 60 to 61°F. One *centigrade heat unit* (Chu) is the amount of heat required to raise the temperature of 1 lb of water from 15 to 16°C. In the SI system of units, *joule* (J) is the accepted unit for work and for energy in general where 1 cal (International Table) is equivalent to 4.1868 J and a joule is equivalent to a force of 1 N acting through a distance of 1 m. These four heat units are the ones most commonly used, and tables reporting heat values generally use one of them.

## Heat Capacity and Specific Heat[4]

In most chemical work concerning heat quantities, the primary item of importance is the amount of heat necessary to raise the temperature of a given body by a definite amount. This can be determined by use of the heat capacity or specific heat of the particular material. *Heat capacity* may be defined as the amount of heat required to raise the temperature of a set amount of a material by 1°. *Specific heat* is the ratio of the heat capacity of a body to the heat capacity of an equal weight of water at 15°C. Thus, if the specific heat of a material is 2.3, this means that 2.3 cal is required to raise the temperature of 1 g of the material by 1°C and 2.3 Btu is necessary to increase the temperature of 1 lb of the material by 1°F.

Heat capacities are usually expressed on the basis of unit weight or on the basis of 1 molecular weight. The molar heat capacity of water at 60°F is 18.0 Btu/(lb mol)(°F), or the heat capacity of water could be expressed as 1.0 Btu/(lb)(°F).

## Mean Heat Capacities of Gases

In stoichiometric computations, the amount of heat required to raise a gas from one temperature to another must often be calculated. The heat capacity of gases changes with temperature; therefore, knowledge of a mean (or average) heat capacity of a gas over the temperature range involved is very desirable.[5] If the mean molal heat capacity is known, the total heat can easily be calculated by multiplying this heat capacity by the temperature rise and by the number of moles of gas.

[4] Engineers often use the expressions "heat capacity" and "specific heat" interchangeably.

[5] In engineering terminology, the *mean value* of a property represents an average value which, when multiplied by the total number of divisions or parts, will give an accurate representation of the sum of the particular property.

**Table 3-2 Empirical values of constants for determination of molal heat capacities of gases $C_p = a + bT + cT^2$, where $T$ is in kelvins and $C_p$ is in cal/(g mol)(°C). Values applicable from 0 to 2000°C.**

| Gas | $a$ | $b$ | $c$ |
|---|---|---|---|
| Air | 6.27 | $2.090 \times 10^{-3}$ | $-0.459 \times 10^{-6}$ |
| CO | 6.25 | $2.091 \times 10^{-3}$ | $-0.459 \times 10^{-6}$ |
| $CO_2$ | 6.85 | $8.533 \times 10^{-3}$ | $-2.475 \times 10^{-6}$ |
| $H_2$ | 6.88 | $0.066 \times 10^{-3}$ | $+0.279 \times 10^{-6}$ |
| $H_2O$ | 6.89 | $3.283 \times 10^{-3}$ | $-0.343 \times 10^{-6}$ |
| $N_2$ | 6.30 | $1.819 \times 10^{-3}$ | $-0.345 \times 10^{-6}$ |
| NO | 6.21 | $2.436 \times 10^{-3}$ | $-0.612 \times 10^{-6}$ |
| $O_2$ | 6.26 | $2.746 \times 10^{-3}$ | $-0.770 \times 10^{-6}$ |
| $SO_2$ | 8.12 | $6.825 \times 10^{-3}$ | $-2.103 \times 10^{-6}$ |

The heat capacity of gases at any temperature can be calculated from equations of the form

$$C_p = a + bT + cT^2 \tag{3-10}$$

where $C_p$ is the heat capacity for constant pressure at the absolute temperature $T$, and $a$, $b$, and $c$ are empirical constants for each gas. Table 3-2 presents values of $a$, $b$, and $c$ for some of the common gases.

The mean heat capacity over a given temperature range can be obtained by integration of an equation similar to Eq. (3-10). However, for the common gases such as air, $O_2$, $N_2$, $H_2$, CO, $H_2O$, and $CO_2$, only a small error is introduced if the mean heat capacity is taken as the value at the average absolute temperature of the gas during the heating.

For general stoichiometric calculations, where the temperatures are higher than 0°C and the pressures are less than 5 atm, it can be assumed that the heat capacities of the common gases are not affected by pressure.

Calculations of heat contents of gases are simplified by use of tables or figures presenting mean heat capacity or sensible heat content data based on a convenient reference temperature. Table 3-3 presents data on the mean molal heat capacities at constant pressure for common gases between the reference temperature of 18°C and the final temperature $t$°C to which the gas is heated.

The use of Table 3-3 can be illustrated by determining the amount of heat necessary to heat 10 g mol of $CO_2$ from 100 to 1000°C at a total pressure of 760 mmHg. Since the heat-capacity data given in Table 3-3 are at a constant pressure of 0 atm abs, these same data will apply at a constant pressure of 760 mmHg or 1 atm pressure. From the table it can be found that the mean molal heat capacity of $CO_2$ between 18 and 100°C is 9.25 cal/(g mol)(°C). Therefore, assuming a

**Table 3-3 Mean molal heat capacities of gases between 18°C and $t$°C at constant pressure in cal/(g mol)(°C)**

$P = 0$ atm abs

| $t$°C | Air | CO | $CO_2$ | $H_2$ | $H_2O$ | $N_2$ | NO | $O_2$ | $SO_2$ |
|---|---|---|---|---|---|---|---|---|---|
| 18 | 6.94 | 6.96 | 8.70 | 6.86 | 7.99 | 6.96 | 7.16 | 7.00 | 9.38 |
| 100 | 6.96 | 6.97 | 9.25 | 6.92 | 8.04 | 6.97 | 7.16 | 7.06 | 9.81 |
| 200 | 7.01 | 7.00 | 9.73 | 6.95 | 8.13 | 7.00 | 7.17 | 7.16 | 10.22 |
| 300 | 7.06 | 7.06 | 10.14 | 6.97 | 8.23 | 7.04 | 7.22 | 7.28 | 10.59 |
| 400 | 7.14 | 7.12 | 10.48 | 6.98 | 8.35 | 7.09 | 7.31 | 7.40 | 10.91 |
| 500 | 7.21 | 7.20 | 10.83 | 6.99 | 8.49 | 7.15 | 7.39 | 7.51 | 11.18 |
| 600 | 7.28 | 7.28 | 11.11 | 7.02 | 8.62 | 7.21 | 7.47 | 7.61 | 11.42 |
| 700 | 7.35 | 7.35 | 11.35 | 7.04 | 8.76 | 7.28 | 7.55 | 7.70 | 11.62 |
| 800 | 7.43 | 7.44 | 11.57 | 7.07 | 8.91 | 7.36 | 7.63 | 7.79 | 11.79 |
| 900 | 7.50 | 7.51 | 11.76 | 7.10 | 9.06 | 7.43 | 7.71 | 7.87 | 11.95 |
| 1000 | 7.57 | 7.58 | 11.94 | 7.13 | 9.20 | 7.50 | 7.77 | 7.94 | 12.08 |
| 1100 | 7.63 | 7.65 | 12.11 | 7.16 | 9.34 | 7.57 | 7.84 | 8.00 | 12.20 |
| 1200 | 7.69 | 7.71 | 12.25 | 7.21 | 9.47 | 7.63 | 7.90 | 8.06 | 12.29 |
| 1300 | 7.74 | 7.77 | 12.37 | 7.25 | 9.60 | 7.68 | 7.95 | 8.13 | 12.39 |
| 1400 | 7.79 | 7.82 | 12.50 | 7.29 | 9.74 | 7.74 | 8.00 | 8.18 | 12.46 |
| 1500 | 7.88 | 7.85 | 12.61 | 7.33 | 9.86 | 7.79 | 8.04 | 8.22 | 12.53 |
| 1600 | 7.89 | 7.91 | 12.71 | 7.37 | 9.98 | 7.83 | 8.09 | 8.25 | 12.60 |
| 1700 | 7.93 | 7.95 | 12.80 | 7.41 | 10.11 | 7.87 | 8.13 | 8.27 | 12.65 |
| 1800 | 7.97 | 7.99 | 12.88 | 7.46 | 10.22 | 7.92 | 8.16 | 8.34 | 12.71 |
| 1900 | 8.00 | 8.03 | 12.96 | 7.50 | 10.32 | 7.95 | 8.20 | 8.39 | 12.74 |
| 2000 | 8.04 | 8.06 | 13.03 | 7.54 | 10.43 | 7.99 | 8.23 | 8.43 | 12.80 |

zero heat content at 18°C, the heat content of 10 g mol of $CO_2$ at 100°C is (9.25)(10)(100 − 18) = 7585 cal. Since the mean molal heat capacity of $CO_2$ between 18 and 1000°C is 11.94, the total heat content of the gas at 1000°C is (11.94)(10)(1000 − 18) = 117,250 cal. The total heat added to raise the temperature of the gas from 100 to 1000°C must be the difference between the heat content at 100°C and the heat content at 1000°C. The heat added, then, must be 117,250 − 7585 = 109,665 cal per 10 g mol of $CO_2$.

Most stoichiometric calculations deal with conditions at constant pressure, and the heat capacities discussed in the preceding paragraphs have been at constant pressure. When a gas is heated at constant volume, less heat must be added for a given temperature rise than would be necessary at constant pressure, because no energy is required for work of expansion. It has been found that the molal heat capacity of the ideal gases at constant volume equals their molal heat capacity at constant pressure minus the perfect-gas-law constant $R$ in consistent units, or

$$C_v = C_p - R \tag{3-11}$$

Thus, if the molal heat capacity at constant pressure of $CO_2$ equals 9.25 cal/(g mol)(°C), the molal heat capacity of this gas at constant volume equals

9.25 − 1.99 = 7.26 cal/(g mol)(°C). In making this conversion, the units of $R$ must be the same as the heat-capacity units. In the preceding example, $R$ was 1.99 cal/(g mol)(°C); so the unit requirements were satisfied.

## Heats of Reaction

Many chemical reactions evolve or absorb heat. When considering energy balances in stoichiometric calculations, the heats of the reactions are always indicated in molal units so they can be applied directly to the equation showing the chemical change. For example, if 1 g mol of CaO at 18°C reacts with 1 g mol of $CO_2$ at 18°C to produce 1 g mol of $CaCO_3$ at 18°C, 43,600 cal (182,500 J) of heat is evolved. This can be expressed as

$$CaO + CO_2 = CaCO_3 + 43{,}600 \text{ cal } (182{,}500 \text{ J})$$

If this reaction is reversed, 43,600 cal, or 182,500 J, of heat must be supplied before 1 g mol of $CaCO_3$ at 18°C can decompose into 1 g mol of CaO and 1 g mol of $CO_2$ (both at 18°C).

In general, values of heats of reaction are indicated as the amount of heat absorbed when the reaction occurs (considering both reactants and products as in their normal states at 18°C and 1 atm pressure).

A consideration of these heats of reaction along with the sensible heat content of the materials entering and leaving a system constitutes the basis of most stoichiometric energy balances. To simplify the calculations, an appropriate temperature should always be chosen and considered as the zero energy-content level.

**Example 3-6: Application of heat-energy balance to combustion** One hundred pounds of pure carbon is burned to $CO_2$, using the amount of dry air (79% by volume $N_2$ and 21% by volume $O_2$) theoretically necessary to supply the oxygen for the combustion. The air and carbon enter the burner at 18°C, and the product gases leave at 2000°C. The burner is operated at a constant pressure of 760 mmHg. Calculate the amount of heat unaccounted for. When 1 lb mol of carbon combines with 1 lb mol of oxygen to form 1 lb mol of $CO_2$ at atmospheric pressure and 18°C, 169,250 Btu is evolved.

SOLUTION

**Basis**

100 lb of pure carbon and 18°C as the energy level

The complete reaction is

$$C + O_2 = CO_2 + 169{,}250 \text{ Btu}$$

Preliminary calculations:

Pound moles carbon $= \frac{100}{12} = 8.34$

Pound moles $O_2$ necessary from air = 8.34

$$\text{Pound moles air necessary} = \frac{(8.34)\,|\,(100)}{\quad\;|\,(21)} = 39.7$$

$$\frac{(\text{mol } O_2)\,|\,(\text{mol air})}{\quad\;|\,(\text{mol } O_2)} = \text{mol air}$$

Pound moles $CO_2$ in product gas = 8.34

Pound moles $N_2$ in product gas = (39.7)(0.79) = 31.36

Heat-energy balance:

Input:

Sensible heat in air = 0 (since air comes in at temperature of zero energy level)

Sensible heat in carbon = 0 (since carbon comes in at temperature of zero energy level)

Potential heat-energy content of carbon in conversion to $CO_2$ = (8.34)(169,250) = 1,410,000 Btu

Total heat-energy input = 1,410,000 Btu

Output:

Mean molal heat capacity of $CO_2$ at constant pressure between 18 and 2000°C = 13.03 cal/(g mol)(°C), and

$$\frac{(13.03)\,|\,(454)\,|\quad}{\quad|\quad|\,(252)} = 23.45 \text{ Btu/(lb mol)(°C)}$$

$$\frac{(\text{cal})\,|\,(\text{g mol})\,|\,(\text{Btu})}{(\text{g mol °C})\,|\,(\text{lb mol})\,|\,(\text{cal})} = \text{Btu/(lb mol)(°C)}$$

Mean molal heat capacity of $N_2$ at constant pressure between 18 and 2000°C = 7.99 cal/(g mol)(°C) = 14.38 Btu/(lb mol)(°C)

Total sensible heat out in gases = (8.34)(23.45)(2000 − 18) + (31.36)(14.38)(2000 − 18) = 1,280,000 Btu

Potential heat-energy content of carbon in gases = 0 (since all carbon is in form of $CO_2$)

Total heat-energy output = 1,280,000 + unaccounted-for losses

Applying overall heat-energy balance:

Total heat-energy input = total heat-energy output

Unaccounted-for losses = 1,410,000 − 1,280,000 = 130,000 Btu

The unaccounted-for losses would mainly be found in heat losses from the surface of the outside of the burner.

## FUELS AND COMBUSTION

The application of stoichiometric principles to problems involving fuels and combustion is important in modern industrial work. Methods have been developed which permit the rapid calculation of results that would otherwise require expensive and cumbersome equipment for their determination.

### Classes of Fuels

Fuels can be divided into three general classes: solid, liquid, and gaseous. The stoichiometric treatments of problems involving the different classes of fuels are similar. Consequently, in the study of the methods for manipulation of the data supplied, the actual form of the fuel is not of great importance.

### Heating Values

The heating value of a fuel is one of its most important properties. This value represents the amount of heat evolved in the complete combustion of a given quantity of the fuel.

The major materials produced by the complete combustion of a fuel are carbon dioxide and water. The carbon dioxide comes from the oxidation of the carbon in the fuel, and the water comes from the oxidation of any hydrogen in the fuel plus any water originally present as such in the fuel. Because of the presence of varying amounts of water in different fuels, two methods for expressing heating value are in common use. The *gross heating value* is defined as the amount of heat evolved in the complete combustion of a unit weight of the fuel under constant pressure at 18°C when all the water originally present and formed is condensed to liquid water at 18°C. The *net heating value* is defined as the amount of heat evolved in the complete combustion of a unit weight of the fuel under constant pressure at 18°C when all the products, including the water, are in the gaseous state at 18°C.

The gross heating value is greater than the net heating value by the latent heat of vaporization at 18°C of the total amount of water originally present in the fuel and formed by oxidation of the hydrogen in the fuel.

Table 3-4 presents the heating values of some of the more important materials occurring as components of fuels.

### Excess Air

In most combustion processes, air is the source of the oxygen required for the oxidation. To ensure an adequate supply of oxygen, it is a common practice to introduce more oxygen (or air) than is actually needed for the process. This means that some oxygen comes out unchanged in the product gases.

*Excess air* is defined as the amount of air supplied above that theoretically necessary for *complete* oxidation of the combustible materials in the fuel. The percent excess air is 100 multiplied by the ratio of the amount of excess air to the amount of air theoretically required.

**Table 3-4 Molal heats of combustion***

| Fuel | State $g$ = gas $s$ = solid $l$ = liquid | Combustion products ($CO_2$, CO, and $SO_2$ are in gaseous state) | Gross heating value (water condensed) | Net heating value (water uncondensed) |
|---|---|---|---|---|
| Carbon (graphite) (C) | $s$ | $CO_2$ | 169,250 | 169,250 |
| Carbon (graphite) (C) | $s$ | CO | 47,550 | 47,550 |
| Carbon monoxide (CO) | $g$ | $CO_2$ | 121,700 | 121,770 |
| Hydrogen ($H_2$) | $g$ | $H_2O$ | 122,970 | 104,040 |
| Methane ($CH_4$) | $g$ | $CO_2$, $H_2O$ | 383,040 | 345,170 |
| Ethane ($C_2H_6$) | $g$ | $CO_2$, $H_2O$ | 671,080 | 614,270 |
| Propane ($C_3H_8$) | $g$ | $CO_2$, $H_2O$ | 955,090 | 879,350 |
| $n$-Butane ($C_4H_{10}$) | $g$ | $CO_2$, $H_2O$ | 1,238,370 | 1,143,690 |
| Ethylene ($C_2H_4$) | $g$ | $CO_2$, $H_2O$ | 607,020 | 569,150 |
| Acetylene ($C_2H_2$) | $g$ | $CO_2$, $H_2O$ | 559,110 | 540,170 |
| Propylene ($C_3H_6$) | $g$ | $CO_2$, $H_2O$ | 885,580 | 828,770 |
| Benzene ($C_6H_6$) | $l$ | $CO_2$, $H_2O$ | 1,405,760 | 1,348,960 |
| Sulfur (S) | $s$ | $SO_2$ | 127,660 | 127,660 |

* Reference conditions: 18°C and 1 atm pressure.
Heating values expressed as Btu evolved per pound mol.
Divide heating value by molecular weight to obtain Btu/lb.
Divide heating value by 7.54 to obtain J/(g mol) or by 1.8 to obtain cal/(g mol)
All values for carbon compounds are based on the graphite form.

There are several alternate ways to determine the percent excess air as indicated in the following:

$$\text{Percent excess air} = \frac{\text{excess air}}{\text{theoretically necessary air}}(100)$$

$$\text{Percent excess air} = \frac{\text{total air} - \text{theoretically necessary air}}{\text{theoretically necessary air}}(100)$$

$$\text{Percent excess air} = \frac{\text{excess air}}{\text{total air} - \text{excess air}}(100)$$

The determination of percent excess air can be illustrated by an example wherein the combustion of 100 lb of a coal containing 80% by weight carbon, 10% by weight hydrogen, and 10% by weight ash gives a gas containing 60 lb mol of nitrogen plus CO, $CO_2$, $O_2$, and $H_2O$. On the basis of 100 lb of coal, the amount of air theoretically necessary is the amount needed to burn 80 lb of carbon to $CO_2$, plus the amount needed to burn 10 lb of hydrogen to $H_2O$. Writing the reactions involved,

$$C + O_2 = CO_2$$

$$H_2 + \tfrac{1}{2}O_2 = H_2O$$

it can be seen that lb mol $O_2$ theoretically necessary from air equals

$$\frac{(80)(1)}{(12)} + \frac{(10)(1)}{(2.02)(2)} = 9.14 \text{ mol } O_2$$

Considering air as 21 mol % $O_2$ and 79 mol % $N_2$, there is 100 mol of air supplied for every 21 mol of $O_2$ supplied, or

$$\text{Moles air theoretically necessary} = \frac{(9.14)(100)}{(21)} = 43.5 \text{ mol air}$$

Retaining the basis of 100 lb of coal, there is 60 mol of nitrogen in the product gas. Since the nitrogen can be considered as a tie substance, 60 mol of nitrogen must have been supplied with the original air, or the total moles of air actually supplied = (60)(100)/(79) = 76.0 mol.

$$\text{Percent excess air} = \frac{\text{total} - \text{theoretically necessary}}{\text{theoretically necessary}}(100)$$

$$= \frac{(76.0 - 43.5)(100)}{(43.5)} = 74.7\%$$

The fact that some of the carbon in the coal is not completely oxidized does not affect the calculation of the percent excess air in any way. It is important to note that calculations of percent excess air involve only considerations of what *theoretically* could have happened to the coal supplied.

## Analysis of Solid Fuels

Solid fuels consist of free carbon, moisture, hydrocarbons, oxygen mostly in the form of oxygenated hydrocarbons, small amounts of sulfur and nitrogen, and nonvolatile noncombustible materials designated as ash. Two types of analyses are in common use for expressing the composition of solid fuels. These two types are known as the ultimate analysis and the proximate analysis.

In an *ultimate analysis*, determination is made of the carbon, moisture, ash, nitrogen, sulfur, net hydrogen, and combined water in the fuel. "Moisture" represents the weight lost on heating the finely divided fuel at 105°C in an open container for 1 h. Since the hydrogen content is always in excess of the amount needed to form water with all the oxygen present, it is assumed that the oxygen is all united with the available hydrogen, and this combination is reported as "combined water." The hydrogen in excess of that necessary to combine with the oxygen in the fuel is termed "net hydrogen."

The common approximate method for expressing the composition of a solid fuel is to report it as moisture, volatile combustible matter, fixed carbon, and ash. This type of analysis is known as a *proximate analysis*. The "moisture" in the proximate analysis is the same as in the ultimate analysis. The "volatile combustible matter" consists of a large amount of the carbon in the fuel, which is lost as volatile hydrocarbons, plus the hydrogen and combined water which are given off on ignition in a covered crucible under specified conditions of temperature and time. The "fixed carbon" is the carbon left in the fuel after the volatile combustible matter has been removed.

Following are the ultimate and proximate analyses of a typical Pennsylvania coal:

| | Ultimate % by wt | | Proximate % by wt |
|---|---|---|---|
| Moisture | 4.61 | Moisture | 4.61 |
| Carbon | 78.25 | V.C.M. | 22.01 |
| Net hydrogen | 3.44 | Fixed carbon | 67.70 |
| Combined water | 6.18 | | |
| Sulfur | 0.64 | | |
| Nitrogen | 1.20 | | |
| Ash | 5.68 | Ash | 5.68 |
| Total | 100.00 | Total | 100.00 |

The difference between these two analyses should be thoroughly understood. The 78.25% carbon in the ultimate analysis appears in the proximate analysis partly as the 67.70% fixed carbon and partly in the 22.01% volatile combustible matter.

## Liquid and Gaseous Fuels

Liquid fuels consist, essentially, of carbon and hydrogen in the form of various hydrocarbons. Some of the common components of gaseous fuels are methane, ethane, propane, carbon monoxide, hydrogen, oxygen, nitrogen, and carbon dioxide.

The analyses of solid and liquid fuels are almost always reported as weight percent. The analyses of gaseous fuels are usually reported on a volume percent basis.

**Example 3-7: Application of fuel and flue-gas analyses** A coal containing 7.1% by weight moisture and 72.0% by weight carbon is burned in a furnace using dry air. The flue-gas analysis indicates 15.00% by volume $CO_2$, 1.67% CO, 2.77% $O_2$, and 80.56% $N_2$ on a dry basis. The total pressure of the flue gases including the water vapor is 740 mmHg. The partial pressure of the water vapor in the flue gas is 55 mmHg. Assuming the original coal contains no sulfur or nitrogen and no combustible material left in the residue, what is the ultimate analysis of the coal?

SOLUTION By choosing a basis of 100 kg mol of dry flue gas and tabulating the flue-gas data, the ultimate analysis can be obtained as follows:

**Basis**

100 kg mol of dry flue gas

| Fuel-gas component | Moles | Atoms carbon | Kilograms carbon | Moles $O_2$ |
|---|---|---|---|---|
| $CO_2$ | 15.00 | 15.00 | 180.0 | 15.00 |
| CO | 1.67 | 1.67 | 20.0 | 0.83 |
| $O_2$ | 2.77 | ..... | ..... | 2.77 |
| $N_2$ | 80.56 | | | |
| Total | 100.00 | 16.67 | 200.0 | 18.60 |

There is 72 kg of carbon per 100 kg of original coal. Since all the carbon goes into the flue gas, the total kilograms of coal originally charged to the furnace to give 200.0 kg of carbon in the flue gas must have been

$$\frac{(200.0)(100)}{(72)} = 277.8 \text{ kg}$$

By Dalton's law, the mole fraction of water in the wet flue gas = $p_{H_2O}$/total pressure = 55/740, and the mole fraction of dry gas in the wet flue gas = 685/740. Therefore, the moles of water vapor per mole of dry gas = 55/685, and the moles of water vapor per 100 mol of dry flue gas = (55/685)(100) = 8.03.

The kilograms of water vapor per 100 kg mol of dry flue gas = (8.03)(18.0) = 144.5.

The original 277.8 kg of coal contained 7.1% moisture, which equals (7.1/100)(277.8) = 19.7 kg of water in the flue gas from the moisture in the coal.

The rest of the water in the flue gas = 144.5 − 19.7 = 124.8 kg, which must have come from the combined water and net hydrogen in the coal.

The tabulation of the flue-gas data shows that 18.6 mol of oxygen was accounted for in the dry flue gas. Using the nitrogen in the flue gas as a tie substance, the amount of oxygen supplied by the entering air = (80.56)(21/79) = 21.4 mol. This means that 21.4 − 18.6 = 2.8 mol of the oxygen that came in with the air was used to burn the net hydrogen in the coal to water.

Since 4.0 kg of hydrogen is needed for every mol of oxygen to form water, the total weight of net hydrogen in the coal per 100 kg mol of dry flue gas is (4.0)(2.8) = 11.2 kg.

The weight of water produced by the net hydrogen is (18.0)(11.2/2.0) = 100.8 kg.

Retaining the basis of 100 mol of dry flue gas, the weight of combined water in the original coal is 124.8 − 100.8 = 24.0 kg.

These results can be summarized to give the ultimate analysis of the coal.

**Basis**

100 kg mol of dry flue gas

| Component of coal | Weight, kg | Ultimate analysis, % by weight |
|---|---|---|
| Moisture | 19.7 | 7.1 |
| Carbon | 200.0 | 72.0 |
| Net hydrogen | 11.2 | 4.0 |
| Combined water | 24.0 | 8.6 |
| Ash (by difference) | 22.9 | 8.3 |
| Total | 277.8 | 100.0 |

## Unburned Combustible in Refuse

In the preceding examples, it has been assumed that all the carbon in the fuel was burned to carbon monoxide or carbon dioxide. In general practice, this is not completely true, since some of the carbon in the original fuel is usually left in the final residue. This carbon loss can be determined from an analysis of the residue employing the ash content in the fuel and in the residue as a tie substance. Thus, if 100 lb of a coal containing 12% ash gives a residue analyzing 91% ash and 9% carbon, the amount of carbon lost in the residue per 100 lb of coal can be calculated as follows: On the basis of 100 lb of coal, 12 lb of ash is charged to the furnace. This same 12 lb of ash leaves the furnace as a part of the residue. Since there is 9 lb of carbon in the residue for every 91 lb of ash in the residue, 9/91 must represent the pounds of carbon in the residue per pound of ash in the residue. On the basis of 100 lb of original coal, there is 12 lb of ash in the residue, or (9/91)(12) = 1.19 lb of unburned carbon is lost in the final residue per 100 lb of original coal.

**Example 3-8: Excess air in burning of coal if all carbon and hydrogen is burned to $CO_2$ and $H_2O$** Coal containing 72% C, 12% $H_2$, and 16% ash by weight is burned with 20% excess air to give a gas product of $CO_2$, $H_2O$, $N_2$, and $O_2$. How many liters of product gas at 740 mmHg and 200°C are obtained if *all* the carbon and hydrogen are burned to $CO_2$ and $H_2O$?

SOLUTION

**Basis**

$$1000 \text{ g coal} = 720 \text{ g C and } 120 \text{ g } H_2$$

$$\text{MW carbon} = 12.0 \qquad \text{MW hydrogen} = 2.0$$

$$C + O_2 \longrightarrow CO_2 \qquad 2H_2 + O_2 \longrightarrow 2H_2O$$

Gram moles $O_2$ required with 20% excess air is

$$\left(\frac{720 \text{ g C} \mid 1 \text{ g mol } O_2}{\mid 12.0 \text{ g C}} + \frac{120 \text{ g } H_2 \mid 1 \text{ g mol } O_2}{\mid 2 \times 2.0 \text{ g } H_2}\right)\frac{120}{100}$$

$= (60.0 \text{ g mol for } CO_2 + 30.0 \text{ g mol for } H_2O)(1.2)$
$= 108.0 \text{ g mol } O_2 \text{ required} = 108.0 \text{ g mol } O_2 \text{ supplied with air}$

Since air is 79 mol % nitrogen and 21 mol % oxygen, the total moles of $N_2$ in = total moles of $N_2$ out is

$$\frac{108.0 \text{ g mol } O_2 \mid 79 \text{ g mol } N_2}{\mid 21 \text{ g mol } O_2} = 406.3 \text{ mol } N_2 \text{ in and out}$$

Total moles of final gas = 60.0 g mol $CO_2$ + 30.0 × 2 g mol $H_2O$ from burned $H_2$ + (108.0 − 90.0) g mol excess $O_2$ + 406.3 g mol $N_2$ = 544.3 g mol. 544.3 g mol of final gas at 740 mmHg and 200 °C is

$$\frac{544.3 \text{ g mol} \mid 22.4 \text{ L} \mid 760 \text{ mmHg} \mid (200 + 273) \text{ K}}{\mid \text{g mol at S.C.} \mid 740 \text{ mmHg} \mid 273 \text{ K}} = 21{,}700 \text{ L}$$

of final gas at 740 mmHg and 200°C

**Example 3-9: Excess air in burning of coal if all carbon and hydrogen is *not* burned to $CO_2$ and $H_2O$** Repeat Example 3-8, except that the refuse from the burner contains 20% carbon, 5% hydrogen, and 75% ash by weight, with the rest of the carbon and hydrogen in the coal being burned to $CO_2$ and $H_2O$.

SOLUTION

**Basis**

1000 g coal = 720 g C, 120 g $H_2$, and 160 g ash

MW Carbon = 12.0 MW hydrogen = 2.0

$$C + O_2 \longrightarrow CO_2 \qquad 2H_2 + O_2 \longrightarrow 2H_2O$$

The amount of oxygen *theoretically* required to burn all of the carbon and hydrogen in the coal to $CO_2$ and to $H_2O$, as in Example 3-8, would be

$$\frac{720 \text{ g C} \mid 1 \text{ g mol } O_2}{\mid 12.0 \text{ g C}} + \frac{120 \text{ g } H_2 \mid 1 \text{ g mol } O_2}{\mid 2 \times 2.0 \text{ g } H_2}$$

$$= 60.0 \text{ g mol for } CO_2 + 30.0 \text{ g mol for } H_2O = 90 \text{ g mol } O_2$$

theoretically necessary. If 20% excess air above that *theoretically* necessary is provided, then the total moles of oxygen provided = (90)(120/100) = 108.0 g mol, exactly as in Example 3-8.

Since air is 79 mol % $N_2$ and 21 mol % $O_2$, the total moles of $N_2$ in = total moles of $N_2$ out is

$$\frac{108.0 \text{ g mol } O_2 \mid 79 \text{ g mol } N_2}{\mid 21 \text{ g mol } O_2} = 406.3 \text{ mol } N_2 \text{ in and out}$$

as in Example 3-8. Ash in = ash out = 160 g. With incomplete combustion, carbon lost in ash is

$$\frac{160 \text{ g ash} \mid 0.2 \text{ g C in ash}}{\mid 0.75 \text{ g ash}} = 42.7 \text{ g C in ash}$$

Hydrogen lost in ash is

$$\frac{160 \text{ g ash} \mid 0.05 \text{ g } H_2 \text{ in ash}}{\mid 0.75 \text{ g ash}} = 10.7 \text{ g } H_2 \text{ in ash}$$

Carbon burned to $CO_2$ = 720 − 42.7 = 673.3 g C = 673.3/12.0 = 56.4 g mol $CO_2$ formed and going to exit gas.

Hydrogen burned to

$$H_2O = 120 - 10.7 = 109.3 \text{ g } H_2 = \frac{109.3 \mid 2}{4.0 \mid} = 54.7 \text{ g mol } H_2O$$

formed and going to exit gas.

Unused $O_2$ = 108 g mol total $O_2$ in − 56.4 g mol $O_2$ used for $CO_2$ − (54.7/2) g mol $O_2$ used for $H_2O$ from burned hydrogen = 24.3 g mol unused $O_2$ going to exit gas. Total moles of exit gas = 56.4 g mol $CO_2$ + 54.7 g mol $H_2O$ + 24.3 g mol unused $O_2$ + 406.3 g mol $N_2$ = 541.7 g mol.

541.7 g mol of final gas at 740 mmHg and 200°C

$$= \frac{541.7 \text{ g mol} \mid 22.4 \text{ L} \mid 760 \text{ mmHg} \mid (200 + 273)\text{ K}}{\mid \text{g mol at S.C.} \mid 740 \text{ mmHg} \mid 273 \text{ K}} = 21{,}600 \text{ L}$$

of final gas at 740 mmHg and 200°C.

## LIMEKILN PERFORMANCE

Calcium oxide is commonly obtained by heating calcium carbonate in limekilns. The operation of the kiln can be considered stoichiometrically to give a mathematical indication of the quantities of the materials involved in the process.

Limestone, containing calcium carbonate, is charged to the kiln, where heat is supplied by the combustion of a fuel such as coal, coke, producer gas, or oil. The heat causes the calcium carbonate to decompose according to the following reaction:

$$CaCO_3 = CaO + CO_2$$

The gaseous products from the kiln contain the $CO_2$ from the limestone plus $CO_2$, CO, and $H_2O$ from the combustion of the fuel, along with nitrogen and excess oxygen from the air.

The pounds of calcium oxide produced per pound of fuel used is designated as the *fuel ratio*. It is possible to calculate this fuel ratio from analyses of the product gases and the fuel.

**Example 3-10: Calculation of limekiln fuel ratio from product-gas and fuel analyses** A limestone containing $CaCO_3$ and inert material is burned with a coke containing 80% by weight carbon and 20% by weight ash. The gases produced have a composition by volume of 25.0% $CO_2$, 5.0% $O_2$, and 70.0% $N_2$. If all the $CaCO_3$ is decomposed to CaO and $CO_2$ and all the carbon in the coke is burned to $CO_2$ with air, calculate the fuel ratio.

SOLUTION

**Basis**

100 lb mol of gas produced

There is 25 lb mol of $CO_2$ in the gas = 25 atoms of carbon = 25 mol of oxygen.

The total moles of oxygen accounted for in the product gases = 25 + 5 = 30.

The moles of oxygen supplied by the air, based on the nitrogen in the product gases = (70)(21/79) = 18.6.

The difference between the moles of oxygen appearing in the final gases and the moles of oxygen supplied by the air must be the moles of oxygen coming from $CO_2$ produced by $CaCO_3$ decomposition; therefore, the pound moles of oxygen in the $CO_2$ from $CaCO_3$ decomposition = 30.0 − 18.6 = 11.4.

For every mole of oxygen appearing as $CO_2$ in the decomposition reaction $CaCO_3 = CaO + CO_2$, 1 mol of $CO_2$ is formed and also 1 mol of CaO is formed. Therefore, the total moles of CaO formed = 11.4 lb mol = (11.4)(56) = 638 lb of CaO formed.

Since there was a total of 25 mol of $CO_2$ in the product gas, 11.4 mol of which was from decomposition of $CaCO_3$, the balance, or 25.0 − 11.4 = 13.6 lb mol of $CO_2$, must have come from the combustion of the coke. This means that (13.6)(12) = 163 lb of carbon was supplied by the original coke per 100 lb mol of product gas. There was 100 lb of original coke per 80 lb of carbon in the coke; therefore, the total pounds of coke supplied = (163)(100/80) = 204 lb.

$$\text{Fuel ratio} = \frac{\text{pounds CaO formed}}{\text{pounds fuel used}} = \frac{638}{204} = 3.13$$

## PROBLEMS

**3-1** A moist paper containing 20% water by weight goes into a drier in a continuous process. The paper leaves the drier containing 2% water by weight. Calculate the weight of water removed from the paper per 100 lb of the original moist paper.

**3-2** One hundred pounds of wet air containing 0.10 lb of water vapor per pound of dry air is mixed with 50 lb of another wet air containing 0.02 lb of water vapor per pound of dry air. Calculate the pounds of water vapor per pound of dry air in the final mixture.

**3-3** Twenty pounds of pure carbon is burned with air to give a gaseous product that contains 16% by weight $CO_2$ and 4% by weight CO. Calculate the pounds of $CO_2$ formed.

**3-4** Assuming the reaction $Na_2CO_3 + Ca(OH)_2 \rightarrow CaCO_3 + 2NaOH$ goes to completion, calculate the following:

(*a*) Pounds of $Ca(OH)_2$ to react with 100 lb of $Na_2CO_3$

(*b*) Pounds of $CaCO_3$ produced from 100 lb of $Na_2CO_3$

(*c*) Grams of $Ca(OH)_2$ necessary to produce 100 g of NaOH

(*d*) Pounds of $Na_2CO_3$ necessary to produce 400 g of $CaCO_3$

**3-5** How many kilograms of oxygen are needed for complete reaction with iron pyrites to give 100 kg of sulfur dioxide?

**3-6** We are going to purify seawater by freezing pure $H_2O$ out of it as ice, and we want to get 2000 kg/h of pure water in the form of ice. If the seawater entering the freezing unit is 2% by weight NaCl and 98% by weight $H_2O$, how many kilograms of entering seawater are needed per hour if the exit liquid brine concentration is 10% by weight NaCl and 90% by weight $H_2O$? The process is continuous and steady state, i.e., there is no accumulation.

**3-7** Determine the pounds of water per pound of dry sand in the final mixture resulting by mixing 100 lb of a wet sand *A* containing 0.8 lb $H_2O$/1 lb dry sand with 50 lb of another wet sand *B* containing 0.2 lb $H_2O$/1 lb dry sand.

**3-8** A mill produces wet paper containing 15% water by weight with the rest being dry paper. This wet paper is fed in a continuous steady-state operation through a drier where the water content of the paper is reduced to 6% by weight. If the heating cost is 5 cents for every pound of water removed from the paper in the drying operation, what is the heating cost per 100 lb of the initial wet paper?

**3-9** Moist air at a total pressure of 720 mmHg and a temperature of 35°C containing water vapor at a partial pressure of 30 mmHg is passed through a dehumidifier at an entering flow rate of 100 $ft^3$/h. If the air comes out of the dehumidifier at 30°C and a total pressure of 720 mmHg containing water vapor at a partial pressure of 20 mmHg, how many kilograms of water are removed from the air per hour?

**3-10** We have developed a system to produce ice to sell by freezing pure water out of seawater in a continuous and steady-state process. The input to the process is 5000 kg of seawater per day containing 3% by weight NaCl and 97% by weight $H_2O$. If the exit liquid brine from the system is 12% by weight NaCl and 88% by weight $H_2O$ and the selling price of ice is $2/kgram, how much income per day can we get from selling all the ice produced?

**3-11** A paper mill has a drying operation in which wet paper is fed in a continuous steady-state operation through a drier where the water content of the paper is reduced to 10% by weight. The wet paper entering the drier contains 20% water by weight with the rest being dry paper. If 100 kg of water is removed from the paper per hour, how many kilograms of the dried paper (10% by weight water) are produced per hour?

**3-12** Wet sand containing 20% water by weight and 80% sand by weight is being dried in a continuous steady-state operation to yield a product sand ready for sale containing 8% water by weight with the rest being sand. The cost for heating has been set at 5 cents for every *pound* of water removed from the wet sand. Under these conditions, what is the heating cost as *cents per kilogram* of the initial wet sand?

**3-13** Calculate the volume of 4 lb of hydrogen in cubic feet (*a*) at standard conditions; (*b*) at 20°C and 740 mmHg absolute pressure; (*c*) at 80°F and 14.5 lb/$in^2$, absolute.

**3-14** Calculate the pounds of water vapor per pound of dry air in an air-water vapor mixture at a total pressure of 750 mmHg when the partial pressure of the water vapor in the mixture is 40 mmHg.

**3-15** Calculate the density in pounds per cubic foot at standard conditions of a gas analyzing 25% by volume $CO_2$, 10% by volume CO, 5% by volume $O_2$, and 60% by volume $N_2$

**3-16** The vapor pressure of water vapor is 200 mmHg in a mixture of air and water vapor. The total pressure is 740 mmHg and the temperature is 160°F. What is the density of the gaseous mixture in pounds per cubic foot?

**3-17** Under conditions of constant pressure, 1010 cal of heat is added to 22.4 L of an ideal gas initially at 0°C and 760 mmHg pressure. What is the final temperature of this gas if the mean heat capacity at constant pressure over the temperature range involved is 7.2 cal/(g mol) (°C)?

**3-18** One gram mole of a gas containing 30 vol % $CO_2$, 60 vol % $N_2$, and 10 vol % $O_2$ at 100°C and 760 mmHg is heated from 100 to 500°C at constant pressure. Calculate the heat in calories added to this gas during this temperature change.

**3-19** If 1 lb of pure carbon (graphite) at 18°C is oxidized to pure CO at 2000°C when the theoretical amount of pure oxygen necessary is supplied at 18°C, how many Btu of heat are unaccounted for?

**3-20** The gases leaving a furnace analyze by volume 15% $CO_2$, 5% $O_2$, and 80% $N_2$. The coal charged to the furnace has an ultimate analysis of 80% carbon, 4% net hydrogen, and 16% ash. The residue left in the furnace contains 80% ash and 20% pure carbon by weight. Calculate the weight of $CO_2$ produced per 100 lb of coal charged to the furnace.

**3-21** With a platinum catalyst and high temperature, $NH_3$ is oxidized to NO by the following reaction:

$$4NH_3 + 5O_2 = 4NO + 6H_2O$$

How many pounds of oxygen must be supplied to produce 20 lb of NO using 35% excess oxygen? Assume the above oxidation reaction goes to completion.

**3-22** A wood drier takes into it wood containing 8% water by weight. The wood comes out containing 3% water by weight. Under these conditions, a company can afford to pay 10 cents per 100 lb of initial wet wood for the heating. What is the maximum amount the company can pay for the heating expressed as dollars per 100 lb of water removed?

**3-23** A mixture of dry air and ammonia enters an absorption tower at a rate of 100 $ft^3$/min. The total pressure of the mixture is 740 mmHg, and the vapor pressure of the ammonia in the mixture is 50 mmHg. The temperature of the entering gas is 80°F. Water enters the tower at a rate of 200 lb/min and absorbs part of the ammonia. Assume no water is vaporized in the tower. The gases leave the tower at a total pressure of 730 mmHg and a temperature of 60°F and contain 0.2% $NH_3$ by volume. Calculate the pound moles of ammonia in the liquid leaving the tower per 1000 lb of water.

**3-24** Fifty pounds of a NaCl solution (40% NaCl and 60% $H_2O$), 100 lb of a sugar solution (20% sugar and 80% $H_2O$), and 40 lb of a waste solution (10% NaCl, 5% sugar, and 85% $H_2O$) are mixed together and heated. Some of the water is lost by evaporation. If the final mixture contains 15% NaCl, what percent of the total water charged was lost through evaporation? (All percents are weight percents.)

**3-25** One hundred pounds of a limestone containing 80% $CaCO_3$ and 20% inerts is burned with 100 lb of pure carbon using 20% excess dry air. If all the $CaCO_3$ decomposes to CaO and $CO_2$ and all the carbon is burned to $CO_2$, calculate the volume percent composition of the gaseous mixture produced.

**3-26** One hundred kilograms of pure carbon (graphite) per hour is burned with 30% excess air in a continuous process to yield a gas containing $CO_2$, $O_2$, and $N_2$. The carbon and the air enter the burner at 18°C and 1 atm pressure. Assuming all of the carbon is converted to $CO_2$ and negligible heat losses from the burner unit, what will be the final temperature of the exit gas if the mean heat capacity of the outlet gas mixture is taken as 7.5 cal/(g mol)(°C)? The heat of combustion of carbon (graphite) based on 18°C and 1 atm and $CO_2$ product is 94,030 cal of heat released per gram mole of carbon burned.

**3-27** How many calories of heat energy must be added to 5 $m^3$ of air initially at a pressure of 640 mmHg and a temperature of 18°C to heat the air to 572°F with the pressure remaining constant?

**3-28** One hundred pounds of a limestone containing 80 wt % $CaCO_3$ and 20 wt % inerts is burned with 100 pounds of pure carbon using 20% excess dry air. If all the $CaCO_3$ decomposes to CaO and $CO_2$ and the residue from the burner contains 40 wt % unburned carbon and 60 wt % inerts, what is the volume percent composition of $CO_2$ in the gaseous mixture of $CO_2$, $O_2$, and $N_2$ produced?

**3-29** Pure carbon is being burned with 30% excess air, with the carbon and the air entering the burner at 18°C and 1 atm pressure. The operation is continuous and at steady state, with 90 percent of the entering carbon being converted to $CO_2$ and the rest of the carbon going to CO. Thus, there is no residue and the leaving gases contain CO, $CO_2$, $O_2$, and $N_2$. Assuming negligible heat losses from the burner unit, what will be the *final temperature* of the exit gas if the mean heat capacity of the outlet gas mixture is assumed to be 7.4 cal/(g mol)(°C)? The heat of combustion of carbon (graphite) based on 18°C and 1 atm and $CO_2$ product is 94,030 cal of heat released per gram mole of carbon burned. The heat of combustion of carbon (graphite) based on 18°C and 1 atm and CO product is 26,400 cal of heat released per gram mole of carbon burned. The pure carbon being burned can be considered as graphite. (*Note*: This differs from Prob. 3-26 by the CO production, so that carbon combustion is not actually complete. Remember the definition of excess air.)

**3-30** Coal containing 70% carbon, 10% hydrogen, and 20% ash by weight is burned with 20% excess air to give a gas that contains only $CO_2$, $O_2$, $N_2$, and $H_2O$. If the residue from the bottom of the burner contains 83% ash, 14% carbon, and 3% hydrogen by weight, what is the volume percent of water in the exit gas? The process is steady state and all the carbon and hydrogen that does not come out in the residue is burned to $CO_2$ and $H_2O$.

**3-31** (*This problem is intended for computer-program solution.*) The following are heat-capacity equations for a gas, liquid, and solid:

1. Carbon dioxide

$$C_p = 18.036 - 4.474 \times 10^{-5}T - 158.08/\sqrt{T}$$

$C_p$ in cal/(g mol)(K), $T$ in K

2. Acetone

$$C_p = 17.20 + 4.805 \times 10^{-2}T - 3.056 \times 10^{-5}T^2 + 8.307 \times 10^{-9}T^3$$

$C_p$ in cal/(g mol)(°C), $T$ in °C

3. Ferric oxide

$$C_p = 24.72 + 1.604 \times 10^{-2}T - 4.234 \times 10^5/T^2$$

$C_p$ in cal/(g mol)(K), $T$ in K

(*a*) Calculate the heat capacity in kJ/(kg)(K) from 200 to 700 K in 10 K increments.

(*b*) Present the results of (*a*) in a neat table.

(*c*) Find the largest value of all of the heat capacities computed in (*a*). Determine all other values as percentages of the largest value. Again, present the percentages in a neat table.

CHAPTER

# FOUR

# INDUSTRIAL CHEMICAL ENGINEERING EQUIPMENT

In addition to understanding the basic principles of chemical engineering, the practical engineer should have a visual concept of the equipment used in industrial processes. The true value of a chemical engineer becomes apparent when he or she demonstrates the ability to apply the theoretical principles to actual operations. This can only be accomplished by a clear understanding of the basic principles combined with a practical knowledge of the equipment used to perform the operations.

A chemical engineer should have some knowledge of equipment *before* starting study of the underlying theories. The purpose of this chapter is to acquaint the reader with general types of equipment used in chemical engineering processes. Detailed descriptions are not given, but sufficient information is included to give a background for a practical understanding of the material presented in the following chapters.

The basic types of equipment commonly used in industrial chemical engineering operations are presented and discussed. The actual sizes and special modifications must be determined for each particular job on the basis of practical experience, theoretical principles, and modern catalog information available from individual manufacturers or suppliers.

## TRANSPORTATION OF MATERIALS

### Pipes

Pipes, valves, fittings, and pumps are used in every industry to convey fluids from one point to another. Pipes of circular cross section are used almost exclusively, since pipes of this shape have the maximum strength per unit weight of constructional material and also give the maximum cross-sectional area per unit of wall-surface area. The most common material of construction for pipes is steel, although copper, brass, wrought iron, cast iron, stainless steel, or other materials may also be used.

**Specifications.** Steel pipes were originally classified according to the wall thickness as standard, extra strong, and double extra strong. Modern industrial demands for more exact specifications have made these three classifications obsolete. Pipes are now specified according to wall thickness by a standard formula for "schedule number" as designated by the Americal Standards Association. *Schedule number* is defined as the approximate value of

$$1000\left(\frac{\text{internal working pressure}}{\text{allowable fiber stress under the operating conditions}}\right)$$

where both the internal working pressure and the allowable fiber stress are expressed in the same units.

Ten schedule numbers are in use at the present time. These are 10, 20, 30, 40, 60, 80, 100, 120, 140, and 160. For pipe diameters up to 10 in, Schedule 40 corresponds to the former "standard" pipe and Schedule 80 corresponds to the former "extra strong" pipe.

Pipe sizes are ordinarily based on the approximate diameter and are reported as nominal pipe sizes. For example, all steel pipes of nominal 2-in diameter have an outside diameter of 2.375 in. The wall thickness and inside diameter are determined by the schedule number. The inside diameter for a nominal 2-in diameter pipe of Schedule 40 is 2.067 in. If the schedule number were 80, the inside diameter would be 1.939 in. The outside diameter is kept constant to permit the use of standard fittings on pipes of different schedule numbers. A table showing outside diameters, inside diameters, and wall thicknesses for pipes of different sizes and schedule numbers is presented in App. C.

## Pipe Fittings

Pipes are purchased in standard sizes of diameter and length, and pieces of pipes may be connected together by means of fittings. Figure 4-1 shows some standard fittings used for joining pipes or changing the direction in which a fluid is flowing.

*Straight couplings* are used for joining two sections of pipe with no change in pipe diameter or direction of flow. *Reducing couplings* permit two pipes of different diameters to be connected together. A *union* is used to join two pieces of pipe and differs from a coupling in that it permits the junction to be broken by merely unscrewing half of the union. *Bushings* combine male and female threads to permit a reduction in pipe diameter. A 90° *ell* permits a right-angle change in the direction in which the contained fluid is flowing. *Tees* and *crosses* allow three or four sections of pipe to be connected at the same point. A *nipple* is merely a short section of pipe threaded on both ends. *Caps* and *plugs* are used for closing off an end of a pipe.

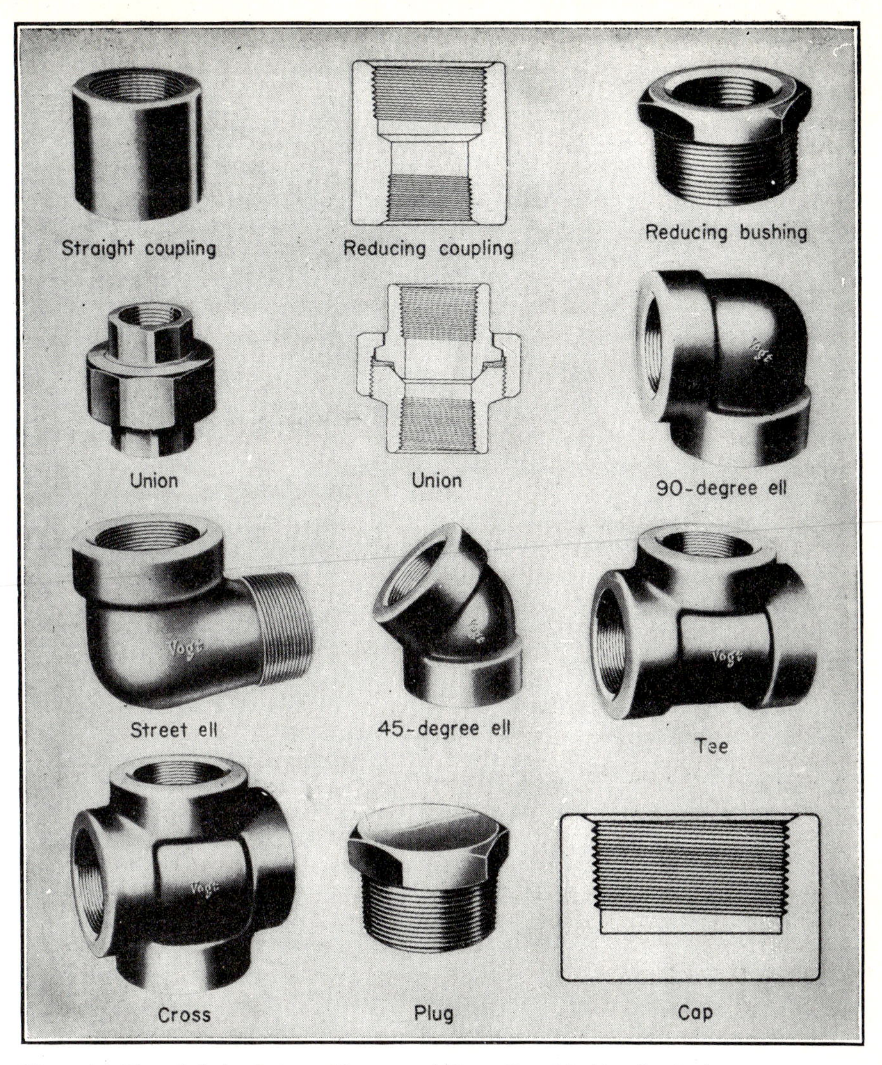


**Figure 4-1** Threaded pipe fittings. (*Courtesy of Henry Vogt Machine Co., Inc.*)

## Valves

Valves are used for shutting off or regulating the flow of a fluid. Many types of valves are available, but the two most common types are the gate valve and the globe valve.

A *gate valve* is shown in Fig. 4-2. When the valve is open, the fluid flows straight through the opening and there is little pressure drop caused by the presence of the valve. As the valve is closed, a disk-shaped face moves perpendicularly across the flowing fluid. When the face hits the bottom seat, the liquid flow is shut off. The face moving across the fluid acts as a closing gate; thus the name "gate valve." This

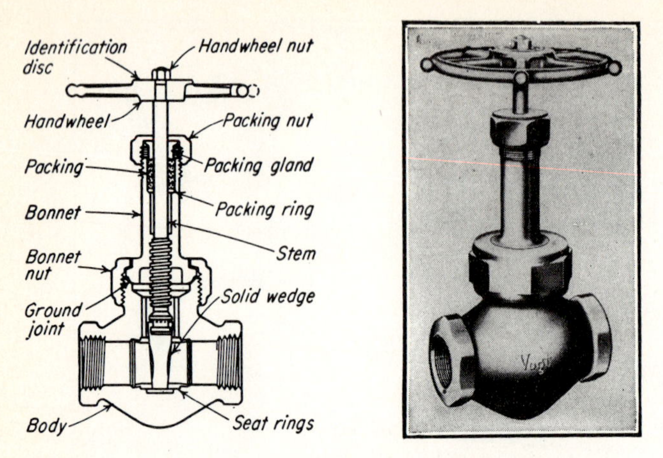


**Figure 4-2** Gate valve. (*Courtesy of Henry Vogt Machine Co., Inc.*)

type of valve does not give accurate regulation of the amount of fluid flowing and is ordinarily used in an open or closed position.

A *globe valve* is shown in Fig. 4-3. As liquid flows through the open valve, the fluid must change its direction of flow as it passes through the seat opening and then turns back to the original direction. This type of valve is useful for regulating the rate of flow, but the friction brought about by the changes in flow direction causes an appreciable pressure drop.

## Pumps

A pump does work on a fluid. This work may cause the fluid to flow from one point to another, or it may merely increase the pressure of the fluid. Power must be supplied to the pump from some outside source. Thus, electrical or steam energy

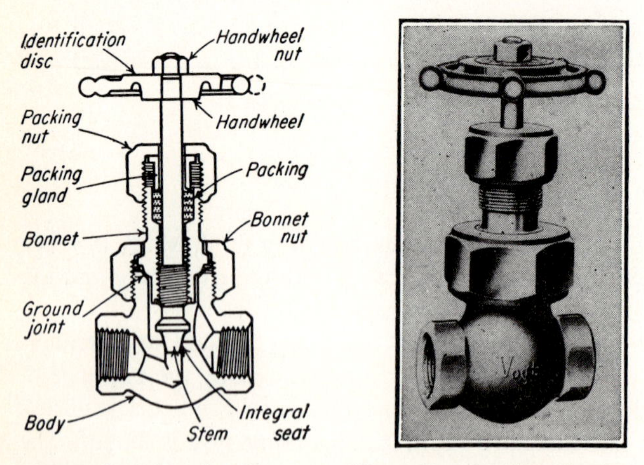


**Figure 4-3** Globe valve. (*Courtesy of Henry Vogt Machine Co., Inc.*)

**Figure 4-4** Reciprocating single-piston pump. (*Courtesy of Worthington Pump and Machinery Corporation.*)

may be transformed into mechanical power, which is used to drive the pump. The efficiency of a pumping apparatus is usually designated as the amount of mechanical energy supplied to the fluid divided by the net amount of energy actually supplied to the power machine.

A *reciprocating pump*, as shown in Figs. 4-4 and 4-5, delivers power to a flowing fluid by means of a piston acting through a cylinder. Steam is often used as the source of power in this type of pump. As the steam flows into the power unit, it forces a piston through a cylinder containing the flowing fluid. This piston compresses the fluid and forces it out of the cylinder. By a system of opening and closing valves, the piston is forced to reciprocate, delivering energy to the fluid with every stroke.

A *rotary positive pump* combines a rotary motion with a positive displacement of the fluid. Pumps of this type deliver fluid at a constant rate and are capable of discharging materials at high pressures. An external gear pump, as shown in Fig. 4-6, is a common type of rotary pump. The two intermeshing gears are fitted into an outside casing with a sufficiently close spacing to seal off effectively each separate tooth space. As the gears rotate in opposite directions, fluid is picked up by each tooth space and is delivered to the exit side of the pump. Thus a constant rate of delivery is obtained, and the fluid may be delivered at high pressures.

Rotary positive pumps require no priming and are well adapted for pumping highly viscous fluids. Because of the small clearance that must be maintained between the gear teeth and the casing, this type of pump should not be used with nonlubricating fluids or with fluids containing solid particles.

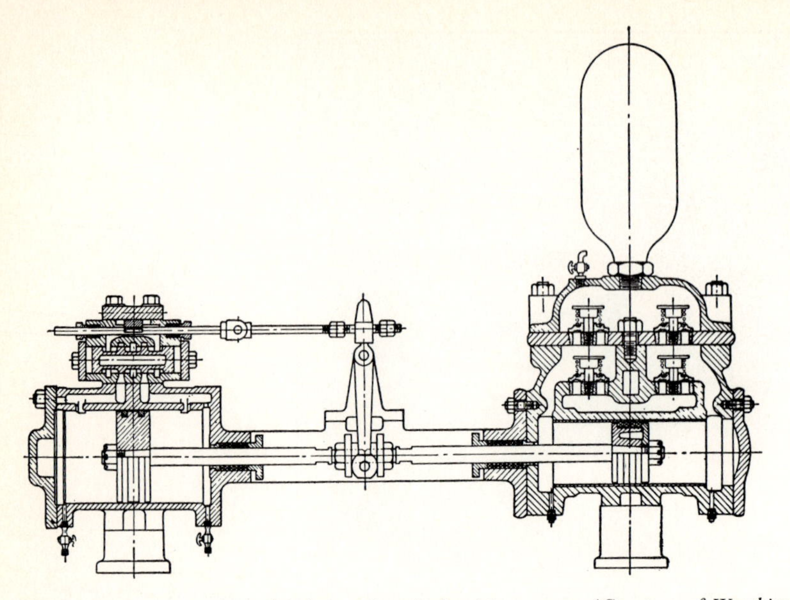

**Figure 4-5** Cutaway view of reciprocating single-piston pump. (*Courtesy of Worthington Pump and Machinery Corporation.*)

*Centrifugal pumps* operate on the principle of a rotating impeller throwing a fluid to the periphery by means of centrifugal force. In most types of centrifugal pumps, the fluid enters at the axis (or center) of a rotating impeller and is discharged at the outer edge of the impeller through an outlet in the pump casing.

A typical centrifugal pump is shown in Fig. 4-7. These pumps are satisfactory for use with fluids containing suspended solids. They may be used where the rate of delivery must be varied, since the delivery pressure does not increase greatly when the outlet line is partly closed or even shut off entirely. Centrifugal pumps must be primed, and they cannot pump fluids against high pressures.

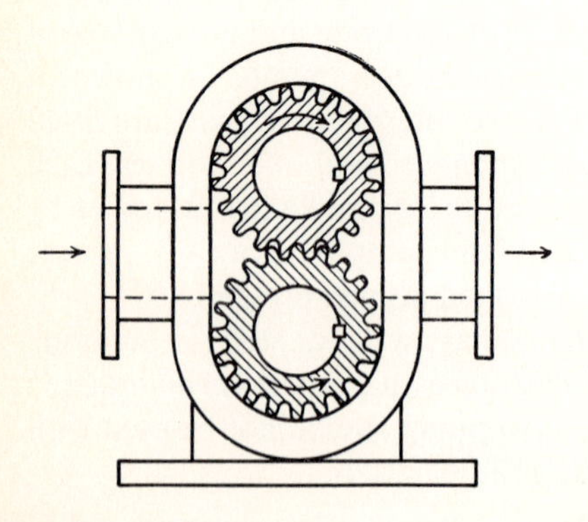

**Figure 4-6** Cutaway view of external gear (rotary) pump.

**Figure 4-7** Centrifugal pump and motor. (*Courtesy of Worthington Pump and Machinery Corporation.*)

## Steam Traps

Steam traps are used to remove condensate from live steam. Figure 4-8 illustrates a typical bucket-type steam trap. The trap is attached to a steam chamber through the bottom inlet opening. When condensate enters the steam trap, the liquid fills the entire body of the trap and passes out the exit opening in the top. A small hole in the top of the inverted bucket permits trapped air to escape.

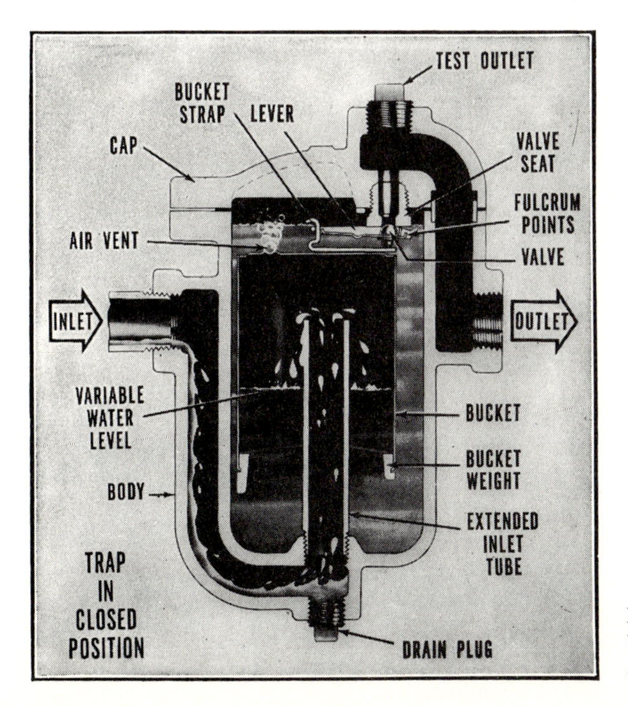


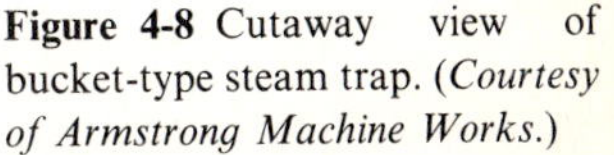

**Figure 4-8** Cutaway view of bucket-type steam trap. (*Courtesy of Armstrong Machine Works.*)

**Figure 4-9** Belt conveyor. (*Courtesy of C. O. Bartlett and Snow Co.*)

When the condensate has been removed, live steam enters the trap and displaces liquid from the inverted bucket. The buoyant effect of the steam lifts the bucket and shuts off the outlet opening at the top of the trap. As long as live steam remains in the bucket, the outlet remains closed. As soon as sufficient condensate enters the trap, the bucket drops down and the liquid is discharged. Thus, the trap discharges intermittently during the entire time it is in use.

### Conveyors

Many different types of open conveyors are used to transport solid materials. Helical screw, belt, chain, and bucket are examples of common types of conveyors. Figure 4-9 shows a belt conveyor.

## HEAT TRANSFER

Modern heat exchangers, as equipment for use in transferring heat from one medium to another, range from simple concentric-pipe exchangers to complex surface condensers with thousands of square feet of heating area. Between these two extremes are found the conventional shell-and-tube exchangers, coil heaters, bayonet heaters, extended-surface finned exchangers, plate exchangers, furnaces, and many other varieties of equipment. Exchangers of the shell-and-tube type are used extensively in industry, and they are often named specifically for distinguishing design features. For example, U-tube, finned-tube, fixed-tubesheet, and floating-

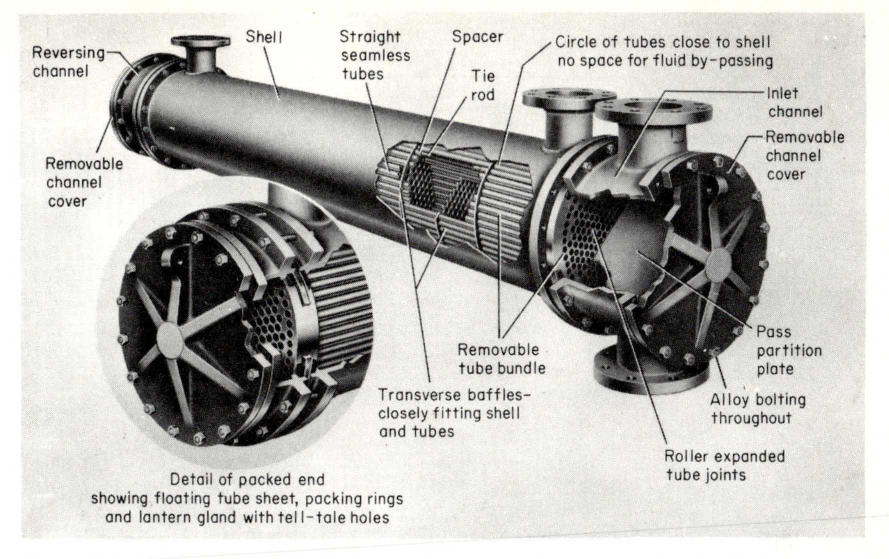


**Figure 4-10** Two-pass shell-and-tube heat exchanger showing construction details. (*Courtesy of Ross Heat Exchanger Division of American Standard.*)

head exchangers are common types of shell-and-tube exchangers. Figure 4-10 shows design details of a conventional two-pass exchanger of the shell-and-tube type.

Because heating and cooling of the various metal parts of a heat exchanger can cause thermal stresses due to variable expansion, it is often desirable to provide some means for the tube bundle and the shell to expand independently to reduce thermal strains. Figure 4-11 shows a heat exchanger with fixed tubesheets so that such differential expansion of the tube bundle and the shell is not possible, while Figs. 4-12 and 4-13 show the use of an internal floating head and an external floating head that allow the differential thermal expansion to occur without any stresses on the equipment.

Baffles are commonly used in the shell side of a heat exchanger to get better mixing and increased turbulence for the shell-side fluid despite the increased pressure drop such baffles can cause. The distance between baffles is known as the

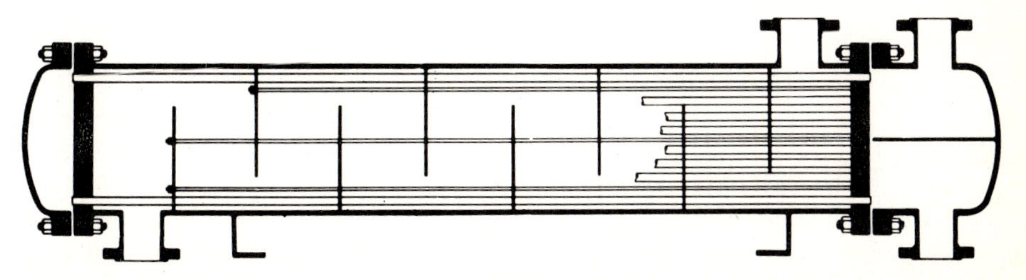

**Figure 4-11** Heat exchanger with fixed tubesheets, two tube passes, and one shell pass. (*Courtesy of Struthers-Wells Corporation.*)

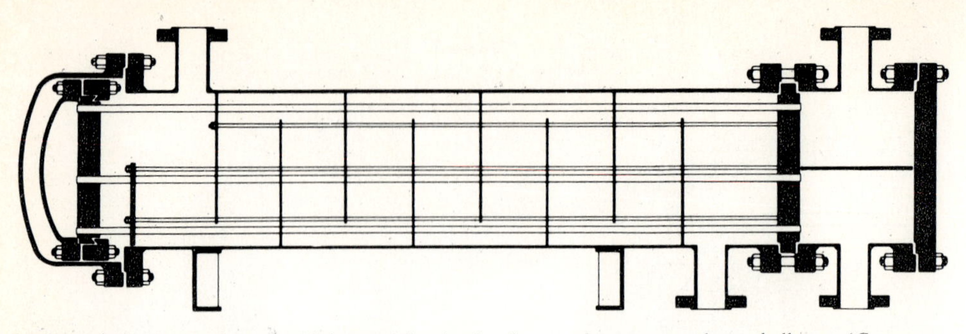

**Figure 4-12** Heat exchanger with internal floating head, two tube passes, and one shell pass. (*Courtesy of Struthers-Wells Corporation.*)

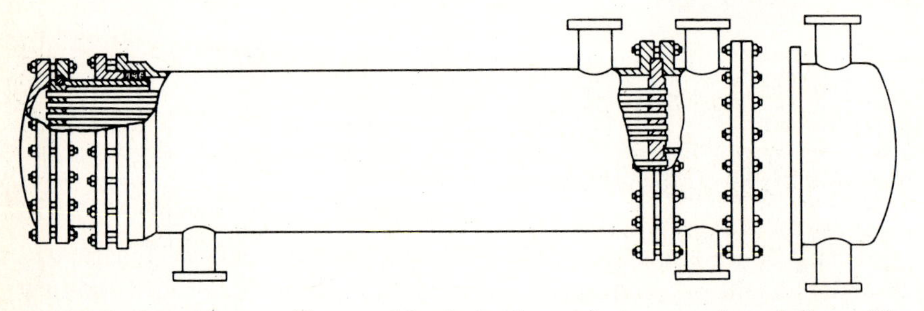

**Figure 4-13** Heat exchanger with external floating head, two tube passes, and one shell pass. (*Courtesy Swenson Evaporator Company.*)

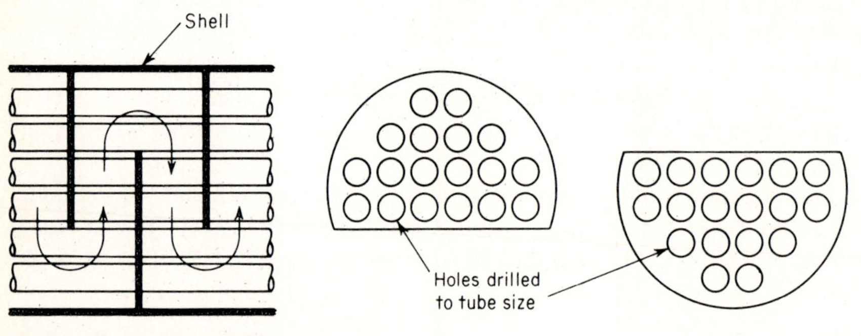


**Figure 4-14** Segmental baffles.

*baffle spacing.* Baffle spacing usually is not greater than the diameter of the shell or less than one-fifth of the shell diameter. The most common type of baffle used in heat exchangers is the segmental baffle shown in Fig. 4-14. Many segmental baffles have a baffle height that is 75 percent of the inside diameter of the shell. There are also other types of baffles used, such as the disk-and-doughnut baffle illustrated in Fig. 14-15.

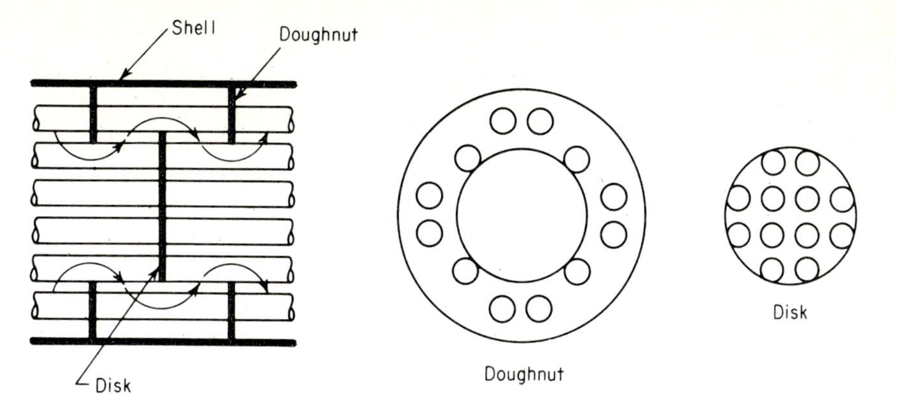


**Figure 4-15** Disk-and-doughnut baffles.

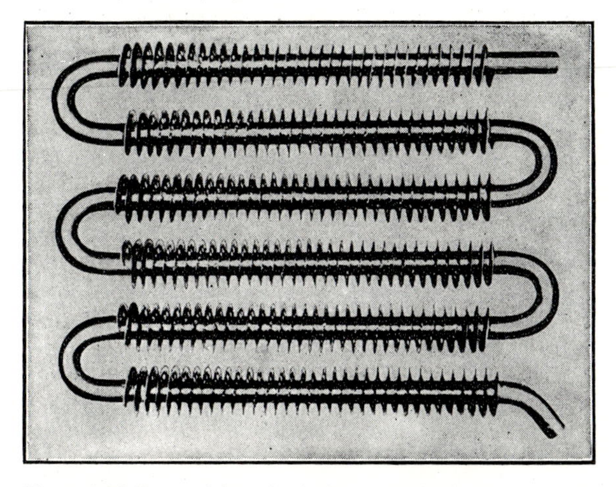

**Figure 4-16** Exposed finned-tube heat exchanger. (*Courtesy of Schutte & Koerting Co.*)

Added heat-transfer surface may be obtained by the use of finned tubes as illustrated in Fig. 4-16. The exposed area of the fins, as well as the exposed area of the tubes, can be used for transferring heat. The additional area of the fins results in a great increase in the amount of heat that can be transferred as compared to an unfinned tube.

## SEPARATION BY MASS TRANSFER

The unit operations of evaporation, distillation, absorption, extraction, humidification, dehumidification, and drying involve mass transfer combined with thermal effects. Certain basic types of equipment have been developed for obtaining effective mass transfer in these operations. A general description of the basic equipment for mass-transfer operations is presented in this section.

## Evaporators

A *basket evaporator* contains an enclosed steam chest with annular tubes passing through the chest, as shown in Fig. 4-17. The feed material passes through the annular tubes where some of the water in the feed is evaporated. The concentrated product is removed at the base of the evaporator. The vapors evolved in the course of the evaporation pass out the top of the evaporator.

A long-tube evaporator using forced circulation is pictured in Fig. 4-18. A pump forces the liquid up through the inside of tubes surrounded with steam. Part of the liquid is evaporated as the liquid passes through the tubes. The vapors go out the top of the evaporator, while the remaining solution may be withdrawn as product or recirculated for additional evaporation.

## Stagewise Contactors

In order to accomplish effective mass transfer between vapor and liquid phases in operations such as distillation and absorption, it is necessary to use equipment that provides intimate contact between the two phases. One common way of accomplishing this is by using a series of interconnected individual units as the physical equipment. In a distillation process, for example, the vapors rise through the distillation column into each stage in series where devices are used to disperse the rising vapors into the descending liquid with the maximum amount of surface contact. This intimate contact enhances the rate of mass transfer between phases and allows enriching of the rising vapors to give a desired purified product at the top of the column.

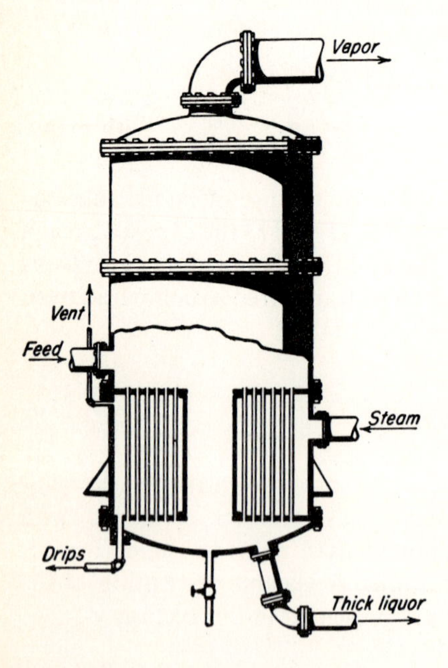


**Figure 4-17** Basket-type evaporator. (*Courtesy of Swenson Evaporator Company.*)

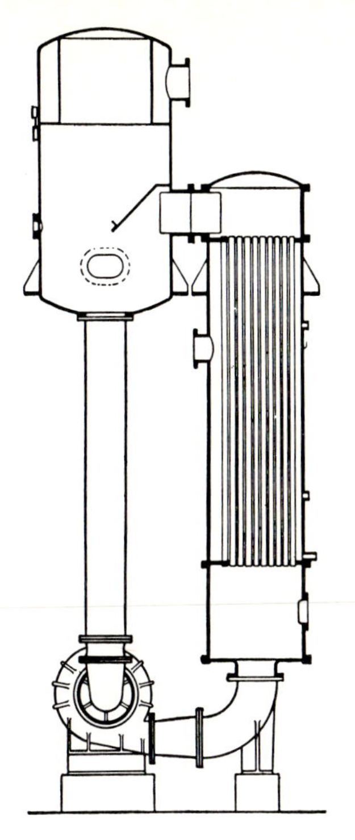

**Figure 4-18** Cross-sectional view of long-tube forced-circulation evaporator. (*Courtesy of Zaremba Company.*)

The most common devices used to accomplish this type of two-phase contact in stagewise operations are sieve-tray, valve-tray, or bubble-cap-tray contactors. Figure 4-19 shows the general operating characteristics of a tray-type stagewise contactor, with examples of sieve, valve, and bubble-cap contactors.

The *sieve-tray contactor* as shown in Fig. 4-19 is presented in a form known as a *crossflow plate contactor*. The tray consists of a flat plate perforated with many small holes that are drilled or punched in a size range of 1/8- to 1/2-in diameter. The rising vapors pass through the holes and bubble directly into the liquid where they are dispersed and rise through the liquid on each tray. This results in a large amount of interfacial area between the vapor and liquid phases, thereby permitting effective mass transfer. Liquid flows across the tray, as shown in Fig. 4-19, through the froth or spray that develops and passes over a weir into the downcomer leading to the tray below. The upward flow of the vapors keeps the liquid from flowing through the holes, and the overall operation of the tray is basically the same as that of a valve tray or a bubble-cap tray. If the flow of gas is low, some or all of the liquid may drain down through the perforations so that some of the contacting area may be bypassed. If the entire transfer of the liquid from one tray to another is by this so-called "weeping" action with no downcomer being used, the type of unit is designated as a *counterflow plate contactor*.

Because best results are normally obtained with full crossflow-plate operation for a sieve tray, units may be designed with a lift valve over the hole in the tray or

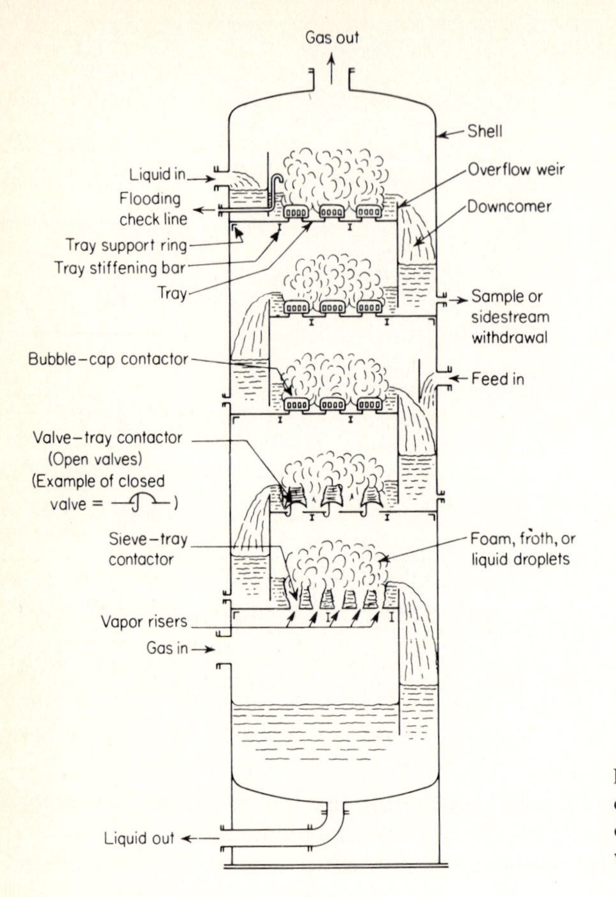


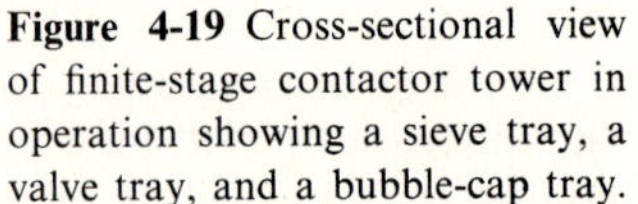
**Figure 4-19** Cross-sectional view of finite-stage contactor tower in operation showing a sieve tray, a valve tray, and a bubble-cap tray.

over a riser from the tray so that the rising vapors lift this valve to allow the vapors to pass horizontally into the liquid, as is illustrated in Fig. 4-19. The liquid cannot easily flow back down the holes in the tray when the gas flow is low because the valve tends to close with the reduced vapor flow. This type of stagewise contactor is designated as a *valve-tray contactor*.

The *bubble-cap tray* consists of a large number of individual bubble-cap units distributed over the tray. A bubble cap consists of a cup-shaped metallic cap with slots around the edge. The cap is inverted over a vapor chimney or riser. The riser is merely a hollow cylinder passing through the supporting plate and extending into the inverted cap. The riser extends sufficiently far into the cap to keep the liquid from running down inside the riser. The vapors rise upward through the vapor riser into the bubble cap. When the vapors hit the inside of the inverted cap, the direction of flow is reversed, and the vapors pass out the slots at the base of the cap. Liquid is maintained on the plate at a depth sufficient to keep the slots submerged so that the vapors must bubble through the liquid after passing outward through the slots.

**Figure 4-20** Bubble caps and risers. (*Courtesy of Badger Manufacturing Company.*)

Figure 4-20 shows various types of bubble caps and risers, while Fig. 4-21 presents a cross-sectional view of a bubble-cap tray showing details of the flow of vapors through the cap and flow of liquid on one tray. Bubble caps may be obtained in a variety of shapes and sizes. Circular caps of the type shown in Fig. 4-20, with outside diameters ranging from 1 to 6 in, are in common use, although most new stagewise contactors are designed with sieve trays because they combine lower cost and about equally good operating characteristics.

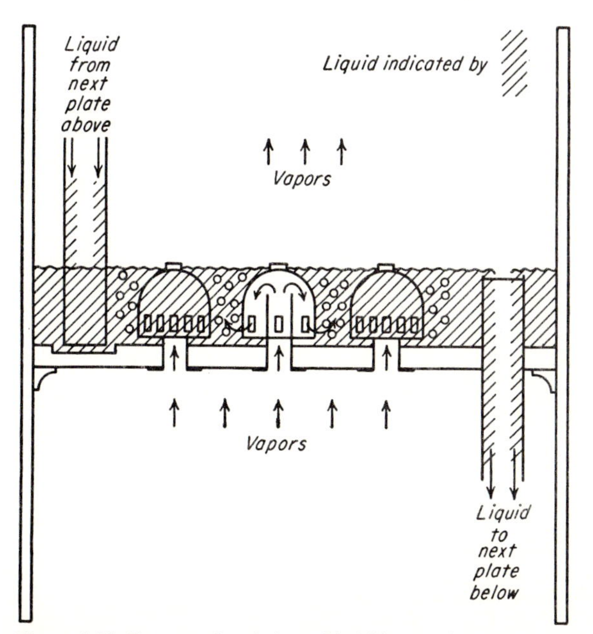


**Figure 4-21** Cross-sectional view of bubble-cap plate tower (one plate shown).

## Packed Column Contactors

A packed column achieves the necessary intimate contact between vapor and liquid phases for effective mass transfer by having the liquid flow down the column over some sort of distributive solid particles (or packing) countercurrent to the rising vapors. Many different types of packing are available; they are normally made up of a large number of small pieces of solid material. These pieces are "dumped" or "stacked" into a containing cylindrical column. When vapors and liquid are passed countercurrently through the packed column, the packing causes the vapors and liquid to mix together, resulting in efficient mass-transfer operations. Figure 4-22 shows a cross-sectional view of a packed tower in operation, with characteristics of the liquid flow indicated.

Although many different types of packing are available for obtaining efficient contact between two fluid phases, the types can generally be classified as random or stacked. A *random packing* is one that is merely dumped into a containing shell, and the individual pieces are not arranged in any particular pattern. Pall rings, Intalox saddles, Raschig rings, and Berl saddles, as shown in Fig. 4-23, are the most common of the random packings used in industrial operations. Pall rings and Intalox saddles are generally replacing the older Raschig rings and Berl

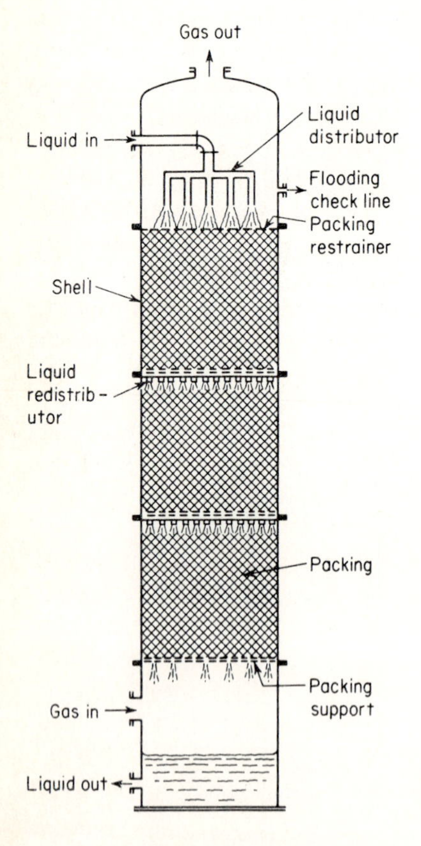


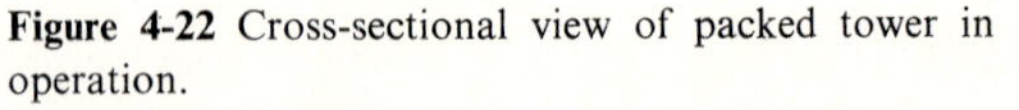
**Figure 4-22** Cross-sectional view of packed tower in operation.

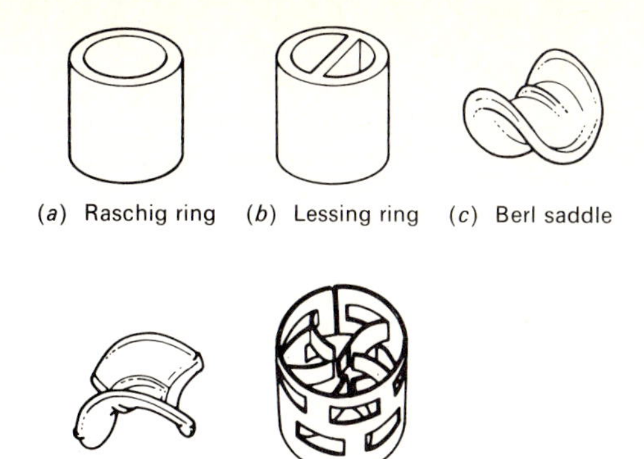


**Figure 4-23** Single pieces of typical random packings.

saddles because, in most cases, the Pall rings and Intalox saddles permit a more economical tower design than the other packings. Pall rings, which are made of metal or plastic, have the same general form as Raschig rings, being open cylinders with the height equal to the diameter. However, Pall rings are stamped during the forming operation so that part of the original cylinder wall is cut and the projections are bent inward, leaving holes in the wall. The projections nearly touch at the center, and the result is an opening of the ring and utilization of the interior of the ring to give improved vapor-liquid contact. They are available in sizes ranging from 5/8 in to 3 in or more. Other forms of packing, such as Flexipac and Koch-Sulzer, are also in common use.

Saddle-shaped packings, such as Intalox saddles and Berl saddles, are available in sizes from $\frac{1}{4}$ to 2 in. These packings are formed from chemical stoneware, plastics, or any other material that can be shaped by punch dies. They form an interlocking structure that gives less side thrust and more active surface area per unit volume than Raschig rings.

Raschig rings, as illustrated in Fig. 4-23, are simply hollow cylinders with the outside diameter equal to the height. They are usually made of inert materials that are cheap and light, such as porcelain, chemical stoneware, or carbon. Other materials of construction, such as clay, plastic, steel, and metal alloys, are also used. Raschig rings are available in sizes ranging from $\frac{1}{4}$ to 3 in or more. Because breakage of fragile packing may occur when the pieces are dropped into an open shell, the initial packing charge is sometimes made by filling the empty tower with water and then dumping the packing slowly into the water.

Additional active surface can be provided by adding a single web or cross web on the inside of a Raschig ring. When a single web is present, the packing is known as Lessing rings. With a solid cross web, the packing is known as cross-partition rings; these are normally available in sizes ranging from 3 to 6 in and are almost always used as a stacked packing.

*Stacked packings*, in general, give lower pressure drops for equivalent fluid capacities than random packings. However, this advantage is gained at the expense of higher initial costs because of the extra installation labor. The ring packings

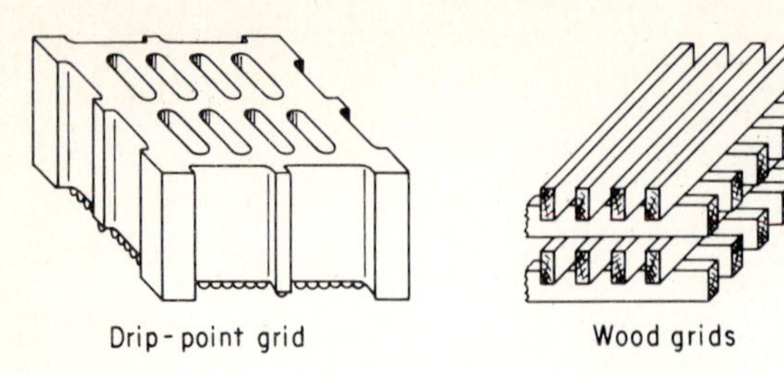


**Figure 4-24** Stacked grid packings.

of nominal sizes 3 in or larger are often used as a stacked packing. Other examples of stacked packings are shown in Fig. 4-24.

## Driers

Figure 4-25 shows a set of trays ready to be introduced into a tunnel-type tray drier. The material to be dried is put on the trays, and the tray rack passes slowly through the "tunnel." Hot gases are blown over the wet material as it passes through the drier, and the moisture is removed to yield a dry, solid product.

A *rotary drier*, as shown in Fig. 4-26, consists of a horizontal cylindrical shell arranged so that the shell rotates about its horizontal axis. One end of the cylinder is elevated slightly above the other end to permit the solid feed to pass continuously through the unit by gravity pull. Wet feed is introduced at the higher end of the rotating shell, and hot air or gas is introduced at the lower end. The feed and the hot gases pass countercurrently through the inside of the shell.

The rotating action lifts the wet material and drops it through the hot gases, resulting in a rapid rate of drying. After picking up the moisture from the wet feed, the gases pass out an exhaust stack. A stationary blade scraper may be located inside the cylindrical shell to keep the feed material from forming large lumps and to prevent the formation of a thick cake of solid on the inside walls. The pitch of the blades is set to force the solids toward the exit end of the drier.

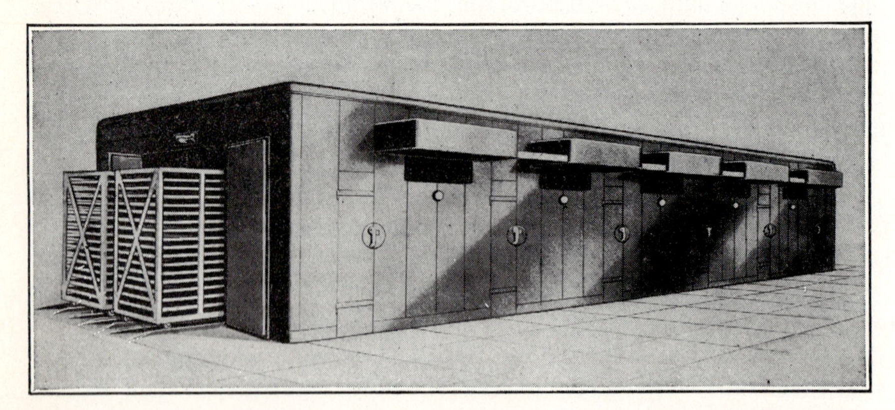

**Figure 4-25** Tunnel-type tray drier. (*Courtesy of National Drying Machinery Company.*)

**Figure 4-26** Rotary drier. (*Courtesy of Denver Equipment Company.*)

## MECHANICAL SEPARATION

### Filtration

A slurry, composed of solids suspended in a liquid, may be separated by filtration. When the mixture is forced into a cloth or fine-mesh wire screen, the liquid passes through the cloth or screen and the solid material collects as a cake on the filtering medium. The separated liquid is known as the *filtrate*. It is often necessary to wash the final cake by passing water through it. The resultant liquid, obtained from the filter, is called *wash water*.

A partly assembled *plate-and-frame filter press*, along with one plate and one frame, is shown in Fig. 4-27. The plates are solid, with cross corrugations on the surface in the central part of the plate and a smooth surface around the edge to form a seal with the adjoining frame. A cloth filtering medium covers both faces of the plate. The frames have an open space in the middle, as shown in Fig. 4-27. The entire assembly is held in place by a screw-driven mechanical force which may be applied manually by means of a handwheel.

The slurry enters the filtering unit through the series of holes in one corner of the plates and frames and passes into the open part of the frames. The filtrate then passes through the cloth, leaving the cake deposited on the cloth. The filtrate follows the plate corrugations to one of the outlet holes in the corner of the plate and is withdrawn from the unit. Small holes drilled through the frame walls permit the

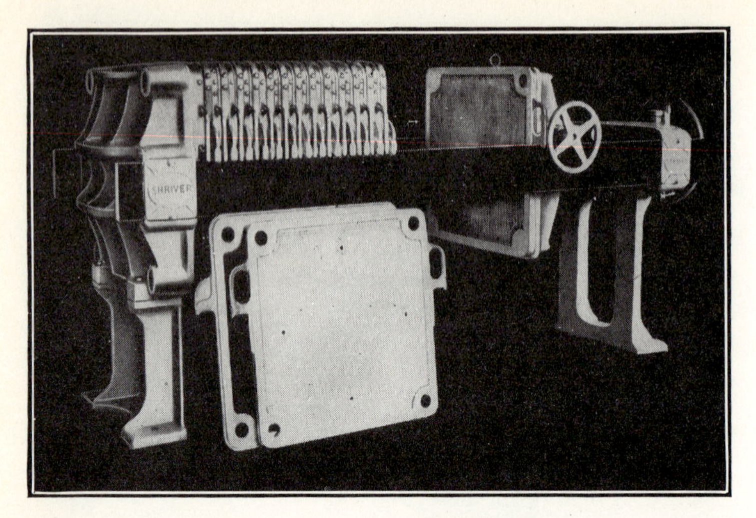


**Figure 4-27** Plate-and-frame filter press showing a separate plate and separate frame. (*Courtesy of T. Shriver and Company.*)

**Figure 4-28** Rotary vacuum filter (complete unit). (*Courtesy of the Eimco Corporation.*)

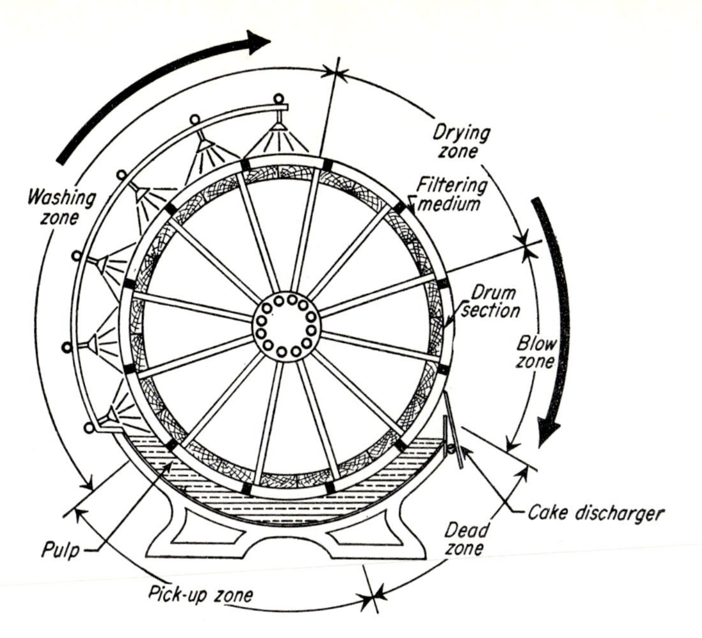


**Figure 4-29** Cross-sectional view of rotary vacuum filter. (*Courtesy of the Eimco Corporation.*)

slurry to pass from the main feed line into the open frame space. Similar holes deliver the filtrate from the corrugated plate face into the filtrate delivery line. The liquid inlets and outlets for the plates and the frames may be arranged to permit backward or forward washing of the filter cake.

A *rotary vacuum filter* is shown in Figs. 4-28 and 4-29. A rotary vacuum filter consists essentially of a perforated drum revolving about its cylindrical axis. A filter cloth covers the outside surface of the drum. A vacuum on the inside of the revolving drum sucks filtrate, wash water, or air through the filter cloth and cake. With units of this type, an internal segmented valve permits delivery of filtrate and wash water in separate streams. In the "pickup" zone, filtrate is pulled through the cloth and cake, and the cake is deposited on the cloth. In the "washing" and "drying" zones, wash water and some air are pulled through the cloth and cake. After passing the drying zone, the cake is removed from the drum by means of an angled scraper (or cake discharger).

Many rotary filters have an "air blow" connection which blows air out against the cake just before it is to be removed by the scraper. This "air blow" loosens the cake from the filter cloth and permits easy and complete cake removal.

## Screening

Dry solids may be separated into the various sizes of particles by the use of a series of screens. Figure 4-30 shows a *vibrating screen separator*. The solid particles are introduced on the top screen, and the entire assembly is subjected to a vibrating

**Figure 4-30** Vibrating screen separator. (*Courtesy of Denver Equipment Company.*)

motion. The individual screens have different mesh sizes, with the largest hole size at the top and the smallest at the bottom. The vibrating screens separate the various sizes of solids, ending up with the coarsest particles on the top screen and the finest particles on the bottom screen or base catch pan.

## SIZE REDUCTION

Many different types of equipment are available for reducing the size of solid particles. Most of the machines accomplish the size reduction by crushing, grinding, or cutting. Jaw crushers, ball mills, roller crushers, gyratory crushers, hammer mills, rod mills, cone crushers, and attrition mills are common examples of equipment used for the reduction of solid-particle size.

Figure 4-31 shows a typical *jaw crusher*. One jaw is mounted in a fixed position, and the other jaw pivots about its base by the action of an eccentric axis. The movable jaw moves up and down in such a manner as to crush the solids entering the top and deliver the product out the bottom. The space separating the two jaws may be varied to give different product sizes.

A *roller crusher* consists of two solid cylinders mounted with their axes parallel. The cylinders rotate in opposite directions. Solid material is fed into the space between the rollers. The solid is crushed to a size approximating the distance separating the roller surfaces, and the crushed products drop out below the two rollers.

A *ball mill* can be used to obtain a finely divided product. As indicated in Fig. 4-32, a ball mill consists of a horizontal cylinder or cone approximately half

**Figure 4-31** Jaw crusher with one side removed. (*Courtesy of Denver Equipment Company.*)

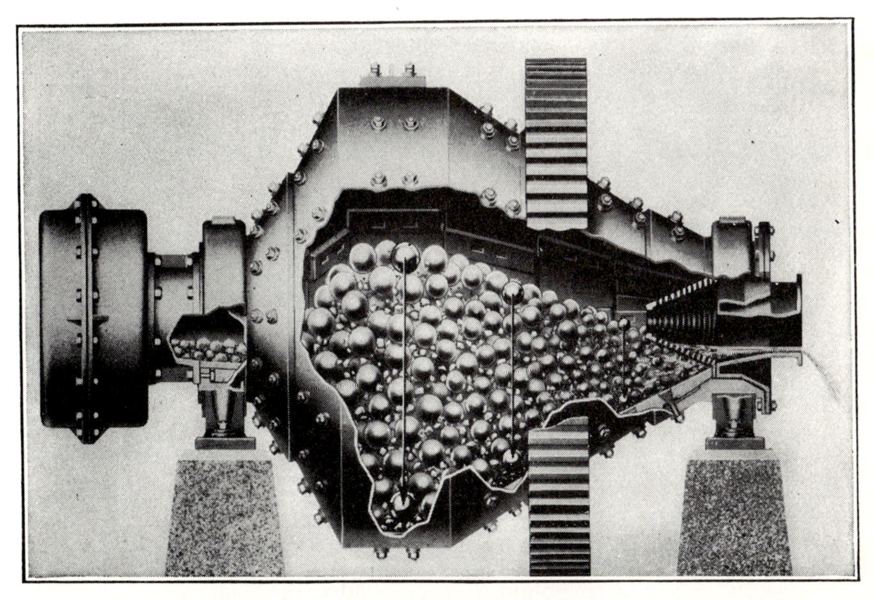

**Figure 4-32** Cutaway view of conical ball mill. (*Courtesy of Hardinge Company, Inc.*)

full of steel balls or flint stones. The cylinder rotates about its axis, and the balls are lifted by the rotating chamber and then fall back into the bottom of the cylinder. As the balls drop down, they hit the material which is being ground. The impact breaks up the solid particles, thus resulting in a size reduction. Ball mills may be operated continuously, with feed entering one end and product discharging from the other end.

## MATERIALS OF CONSTRUCTION

When specifying or designing equipment for use in chemical engineering processes, it is essential to choose materials of construction which will be chemically resistant as well as physically resistant to the conditions of operation. For example, copper would not be a suitable material of construction for a unit producing nitric acid, since copper would react chemically with the nitric acid. While glass does not react chemically with nitric acid, it would, nevertheless, be a poor choice for a material of construction. It is very difficult to fabricate glass in the large sizes needed for a nitric acid unit, and the glass could not withstand the large stresses that would be put on it. A good material of construction for a nitric acid plant would be stainless steel. Stainless steel is chemically inert toward nitric acid, and it is physically strong enough to withstand large stresses.

The chemical engineer should realize the importance of selecting the correct material of construction. If there is any doubt as to materials which are suitable for the construction of equipment, reference should be made to the literature, or laboratory tests should be carried out with the proposed material under conditions simulating the actual conditions of operation.

Table 4-1 presents information on the chemical resistance of certain materials of construction to common industrial chemicals.

### Galvanic Action Between Two Dissimilar Metals

When two dissimilar metals are used in the construction of equipment containing a conducting fluid in contact with both metals, an electric potential may be set up between the two metals. The resulting galvanic action can cause one of the metals to dissolve into the conducting fluid and deposit on the other metal. As an example, if a piece of copper equipment containing a solution of sodium chloride in water is connected to an iron pipe, electrolysis can occur between the iron and copper, causing high rates of corrosion. As indicated in Table 4-2, iron is higher in the electromotive series than copper, and the iron pipe will gradually dissolve and deposit on the copper. The farther apart the two metals are in the electromotive series, the greater is the possible extent of corrosion due to electrolysis.

## Table 4-1 Corrosion resistance of constructional materials

*Code designation for corrosion resistance*

A = acceptable, can be used successfully
C = caution, resistance varies widely depending on conditions; used when some corrosion is permissible
X = unsuitable
Blank = information lacking

*Code designation for gasket materials*

a = asbestos, white (compressed or woven)
b = asbestos, blue (compressed or woven)
c = asbestos (compressed and rubber-bonded)
d = asbestos (woven and rubber-frictioned)
e = GR-S or natural rubber
f = Teflon

| | Metals | | | | | | | | Nonmetals | | | | | |
|---|---|---|---|---|---|---|---|---|---|---|---|---|---|---|
| | | | Stainless steel | | | | | | | | | | | |
| Chemical | Iron and steel | Cast iron (Ni-resist) | 18-8 | 18-8 Mo | Nickel | Monel | Red brass | Aluminum | Industrial glass | Carbon (Karbate) | Phenolic resins (Haveg) | Acrylic resins (Lucite) | Vinylidene chloride (Saran) | Acceptable nonmetallic gasket materials |
| Acetic acid, crude | C | C | C | C | C | C | C | A | A | A | A | A | C | b, c, d, f |
| Acetic acid, pure | X | X | C | A | C | A | X | A | A | A | A | X | X | b, c, d, f |
| Acetic anhydride | C | C | A | A | A | A | X | A | A | A | A | X | C | b, c, d, f |
| Acetone | A | A | A | A | A | A | A | A | A | A | C | X | C | a, e, f |
| Aluminum chloride | X | C | X | X | C | C | A | A | A | A | A | ... | A | a, c, e, f |
| Aluminum sulfate | X | C | C | A | C | C | X | A | A | A | A | A | A | a, c, d, e, f |
| Alums | X | C | C | A | C | A | X | A | A | A | A | A | A | a, c, d, e, f |
| Ammonia (gas) | A | A | C | A | A | A | X | C | A | ... | A | ... | C | a, f |
| Ammonium chloride | C | A | C | C | A | A | C | C | A | A | A | A | A | b, c, d, e, f |
| Ammonium hydroxide | A | A | A | A | C | C | X | C | A | ... | A | A | C | a, c, d, f |
| Ammonium phosphate (monobasic) | X | C | A | A | ... | C | X | X | A | A | A | ... | ... | b, c, d, e, f |
| Ammonium phosphate (dibasic) | C | A | A | A | ... | A | C | C | A | A | A | ... | ... | a, c, d, e, f |
| Ammonium phosphate (tribasic) | A | A | A | A | A | A | X | C | A | A | A | ... | ... | a, c, d, e, f |
| Ammonium sulfate | C | A | C | C | A | A | C | A | A | A | A | A | A | b, c, d, e, f |
| Aniline | A | A | A | A | ... | A | X | ... | A | A | C | ... | C | a, f |

**Table 4-1 Corrosion resistance of constructional materials** (*Continued*)

| Chemical | Metals | | | | | | | | Nonmetals | | | | | |
|---|---|---|---|---|---|---|---|---|---|---|---|---|---|---|
| | Iron and steel | Cast iron (Ni-resist) | Stainless steel 18-8 | Stainless steel 18-8 Mo | Nickel | Monel | Red brass | Aluminum | Industrial glass | Carbon (Karbate) | Phenolic resins (Haveg) | Acrylic resins (Lucite) | Vinylidene chloride (Saran) | Acceptable nonmetallic gasket materials |
| Benzene, benzol | A | A | A | A | A | A | A | A | A | A | A | ... | C | a, f |
| Boric acid | X | C | A | A | A | A | C | A | A | A | A | ... | A | a, c, d, e, f |
| Bromine | X | C | C | C | C | C | C | ... | A | C | X | ... | X | b, f |
| Calcium chloride | C | A | C | C | A | A | C | C | A | A | A | A | A | b, c, d, e, f |
| Calcium hydroxide | A | A | A | A | ... | A | C | ... | ... | ... | ... | ... | ... | a, c, d, e, f |
| Calcium hypochlorite | X | C | C | A | C | C | C | C | A | A | C | ... | C | b, c, d, f |
| Carbon tetrachloride | C | C | C | A | A | A | C | C | A | A | A | A | C | a, f |
| Carbonic acid | C | C | A | A | A | A | C | A | A | A | A | A | A | a, e, f |
| Chloracetic acid | X | ... | X | X | C | C | X | C | A | ... | A | ... | ... | b, f |
| Chlorine, dry | A | A | C | A | A | A | A | A | A | A | A | ... | X | b, e, f |
| Chlorine, wet | X | X | X | X | X | X | X | X | A | C | A | ... | X | b, e, f |
| Chromic acid | C | C | C | C | C | C | X | C | A | X | X | X | A | b, f |
| Citric acid | X | C | C | A | C | A | C | A | A | A | A | A | A | b, c, d, e, f |
| Copper sulfate | X | C | A | A | C | C | X | X | A | A | A | X | ... | b, c, d, e, f |
| Ethanol | A | A | A | A | A | A | A | A | A | A | A | ... | A | a, c, e, f |
| Ethylene glycol | A | A | A | A | A | A | A | A | A | A | A | A | C | a, c, e, f |
| Fatty acids | C | C | A | A | A | A | C | A | A | A | A | ... | A | a, e, f |
| Ferric chloride | X | X | X | C | X | X | X | X | A | C | A | ... | A | b, e, f |
| Ferric sulfate | X | X | C | A | C | C | X | C | A | C | A | ... | A | b, c, e, f |
| Ferrous sulfate | C | A | A | A | A | A | C | C | A | A | A | ... | C | |
| Formaldehyde | C | C | A | A | A | A | C | A | A | A | A | ... | A | a, c, e, f |
| Formic acid | X | ... | C | C | C | C | X | X | A | A | A | ... | A | b, c, e, f |
| Glycerol | A | A | ... | A | A | A | A | A | A | A | C | A | C | a, c, e, f |
| Hydrocarbons (aliphatic) | A | A | A | A | A | A | A | A | A | A | A | A | C | a, c, d, f |
| Hydrochloric acid | X | X | X | X | C | C | X | X | A | A | A | A | C | b, c, d, f |
| Hydrofluoric acid | C | X | X | X | C | C | X | X | X | A | C | ... | C | b, f |

| | | | | | | | | | | | | | | |
|---|---|---|---|---|---|---|---|---|---|---|---|---|---|---|
| Hydrogen peroxide | C | . . . | C | C | C | C | C | A | A | A | A | A | C | a, e, f |
| Lactic acid | X | C | C | A | C | C | A | C | A | A | A | . . . | . . . | a, b, c, d, e, f |
| Magnesium chloride | C | C | C | C | A | A | C | C | A | A | A | . . . | A | b, c, e, f |
| Magnesium sulfate | A | A | A | A | A | A | A | A | A | . . . | A | . . . | A | b, c, e, f |
| Methanol | A | A | A | A | A | A | A | A | A | A | A | . . . | A | a, c, e, f |
| Nitric acid | X | C | C | C | X | X | X | C | A | C | C | . . . | C | b, f |
| Oleic acid | C | C | A | A | A | A | C | A | A | A | A | . . . | A | a, e, f |
| Oxalic acid | C | C | C | C | C | A | C | C | A | A | A | . . . | . . . | b, c, d, e, f |
| Phenol (carbolic acid) | C | A | C | A | A | A | C | A | A | A | C | A | C | a, f |
| Phosphoric acid | C | C | C | A | C | C | X | X | C | A | A | . . . | A | b, c, f |
| Potassium hydroxide | C | C | A | A | A | A | X | X | . . . | . . . | . . . | . . . | C | a, e, f |
| Sodium bisulfate | X | C | A | A | A | A | C | C | A | A | A | . . . | A | b, c, d, e, f |
| Sodium carbonate | A | A | A | A | A | A | C | C | C | A | A | X | . . . | a, c, d, e, f |
| Sodium chloride | A | A | C | C | A | A | C | C | A | A | A | . . . | . . . | a, c, d, e, f |
| Sodium hydroxide | A | A | A | A | A | A | C | X | C | A | A | A | C | a, c, d, f |
| Sodium hypochlorite | X | C | C | A | C | C | C | X | A | C | X | . . . | A | b, c, d, f |
| Sodium nitrate | A | A | A | A | A | A | C | A | A | A | A | . . . | . . . | b, c, d, e, f |
| Sodium sulfate | A | A | A | C | A | A | A | A | A | A | A | . . . | A | a, c, d, e, f |
| Sodium sulfide | A | A | C | A | A | A | X | X | C | A | A | . . . | . . . | a, e, f |
| Sodium sulfite | A | A | A | A | A | A | C | C | A | A | A | | | |
| Sodium thiosulfate | C | . . . | A | A | A | A | C | C | A | A | A | . . . | A | a, c, d, e, f |
| Stearic acid | C | A | A | A | A | A | C | A | A | A | A | . . . | . . . | a, e, f |
| Sulfur | A | C | C | C | C | C | C | C | A | A | A | . . . | . . . | a, e, f |
| Sulfur dioxide | C | C | C | C | C | C | C | A | A | A | A | . . . | A | a, f |
| Sulfuric acid (98% to fuming) | A | C | X | C | X | X | X | C | A | X | X | X | C | b, f |
| Sulfuric acid (75–95%) | A | C | X | X | X | C | X | X | A | C | X | X | C | b, f |
| Sulfuric acid (10–75%) | X | C | X | X | C | C | X | X | A | A | C | C | A | b, f |
| Sulfuric acid (<10%) | X | C | X | C | C | C | C | C | A | A | C | A | A | a, b, c, e, f |
| Sulfurous acid | X | . . . | C | A | X | X | C | C | A | A | A | . . . | C | b, c, d, e, f |
| Trichloroethylene | C | A | C | A | A | A | C | C | A | A | A | . . . | C | a, f |
| Zinc chloride | C | C | C | X | A | A | X | C | . . . | . . . | A | A | . . . | b, c, d, e, f |
| Zinc sulfate | C | A | A | A | A | A | C | C | . . . | . . . | . . . | . . . | . . . | b, c, d, e, f |

*Source*: From M. S. Peters and K. D. Timmerhaus, *Plant Design and Economics for Chemical Engineers*, 3d ed., McGraw-Hill, New York, 1980.

## Table 4-2 Electromotive series of metals

List of metals arranged in decreasing order of their tendencies to pass into ionic form by losing electrons

| Metal | Ion | Standard electrode potential at 25°C |
|---|---|---|
| Lithium | $Li^{+}$ | +2.96 |
| Potassium | $K^{+}$ | 2.92 |
| Calcium | $Ca^{++}$ | 2.87 |
| Sodium | $Na^{+}$ | 2.71 |
| Magnesium | $Mg^{++}$ | 2.40 |
| Aluminum | $Al^{3+}$ | 1.70 |
| Manganese | $Mn^{++}$ | 1.10 |
| Zinc | $Zn^{++}$ | 0.76 |
| Chromium | $Cr^{++}$ | 0.56 |
| Gallium | $Ga^{3+}$ | 0.50 |
| Iron | $Fe^{++}$ | 0.44 |
| Cadmium | $Cd^{++}$ | 0.40 |
| Cobalt | $Co^{++}$ | 0.28 |
| Nickel | $Ni^{++}$ | 0.23 |
| Tin | $Sn^{++}$ | 0.14 |
| Lead | $Pb^{++}$ | 0.12 |
| Iron | $Fe^{3+}$ | 0.045 |
| Hydrogen | $H^{+}$ | 0.0000 |
| Antimony | $Sb^{3+}$ | −0.10 |
| Bismuth | $Bi^{3+}$ | −0.23 |
| Arsenic | $As^{3+}$ | −0.30 |
| Copper | $Cu^{++}$ | −0.34 |
| Copper | $Cu^{+}$ | −0.47 |
| Silver | $Ag^{+}$ | −0.80 |
| Lead | $Pb^{4+}$ | −0.80 |
| Platinum | $Pt^{4+}$ | −0.86 |
| Gold | $Au^{3+}$ | −1.36 |
| Gold | $Au^{+}$ | −1.50 |

CHAPTER

# FIVE

# FLUID FLOW

The flow of liquids and gases in industrial processes is an important unit operation, and the principles involved should be thoroughly understood by all chemical engineers. The expression "fluid flow" means the movement of materials such as liquids, gases, or dispersed solids through certain bounded regions.

Energy balances and material balances, along with the laws of fluid friction, constitute the basis of the principles of fluid flow. Application of these fundamentals gives methods for determining relationships between rates of flow and pressure drops in a given system. These relationships may be used to determine power requirements or flow rates for different types of equipment.

The extensive applications of fluid-flow principles make it necessary for all types of engineers to have an understanding of at least the elementary laws involved. A basic knowledge of this subject indicates the answers to such practical questions as why a gate valve may be preferable to a globe valve in lines containing flowing fluids, what horsepower motor would be required on a pump to force water from a well into an existing overhead tank, or what diameter pipe should be used to handle a flow of 100,000 gal of water per hour.

The basic principles of fluid flow are presented in this chapter along with some practical applications of these principles in normal plant operation. Steady mass rate of flow and absence of chemical reaction will be assumed in the following discussion.

## TYPES OF STEADY FLOW

When a fluid passes through a pipe at a steady mass rate of flow, the mass of fluid entering one end of the pipe in unit time must equal the mass of fluid leaving the other end of the pipe in the same unit time. This is merely a statement of the law of conservation of matter expressed for the conditions of no accumulation or depletion. Similarly, on the basis of unit time, the mass of fluid passing *any* total

**Table 5-1 Nomenclature for fluid flow**

$A$ = area of one layer of fluid parallel to another layer, $m^2$ or $ft^2$
$C$ = coefficient of discharge, dimensionless
$d$ = prefix indicating differential, dimensionless
$D$ = diameter (inside) of circular pipe, m or ft
$D_o$ = diameter of orifice opening, m or ft
$f$ = Fanning friction factor, dimensionless
$F, \Sigma F$ = mechanical energy loss due to friction, J/kg or ft · lbf/lbm
$F_c$ = mechanical energy loss due to sudden contraction, J/kg or ft · lbf/lbm
$F_e$ = mechanical energy loss due to sudden expansion, J/kg or ft · lbf/lbm
$g$ = local acceleration due to gravity, usually taken as 9.8 $m/s^2$ or 32.17 $ft/s^2$
$g_c$ = Universal gravitational constant, conversion factor in Newton's law of motion, 32.17 ft · lbm/($s^2$)(lbf)
$H_v$ = difference in static head (equals $v(p_1 - p_2)$), J/kg or ft · lbf/lbm
$i$ = enthalpy, J/kg or ft · lbf/lbm
$K_c$ = coefficient in the contraction-loss equation, dimensionless
$L$ = length of straight pipe, m or ft
$L_e$ = fictitious length of straight pipe equivalent to the resistance of a pipe fitting of same nominal diameter as pipe, m or ft
$N_{Re}$ = Reynolds number equals $DV\rho/\mu$, dimensionless
$p$ = absolute pressure, Pa, $N/m^2$, or $lbf/ft^2$
$q$ = volumetric rate of flow, $m^3/s$ or $ft^3/s$
$Q$ = net heat energy added to a system from an outside source, J/kg or ft · lbf/lbm
$R_H$ = hydraulic radius (equals $S/\psi$), m or ft
$S_1$ = cross-sectional area normal to path of fluid flow, upstream section, $m^2$ or $ft^2$
$S_2$ = cross-sectional area normal to path of fluid flow, downstream section, $m^2$ or $ft^2$
$S_o$ = cross-sectional area of orifice opening, $m^2$ or $ft^2$
$S_p$ = maximum cross-sectional area of plummet in a rotameter, $m^2$ or $ft^2$
$u$ = internal energy, J/kg or ft · lbf/lbm
$v$ = specific volume of fluid (equals $1/\rho$), $m^3/kg$ or $ft^3/lbm$
$V$ = average linear velocity, m/sec or ft/s
$V_i$ = instantaneous or point linear velocity, m/s or ft/s
$V_P$ = volume of plummet in a rotameter, $m^3$ or $ft^3$
$W$ = net external work done on a flow system, J/kg or ft · lbf/lbm
$W_o$ = mechanical work imparted to fluid from outside source, J/kg or ft · lbf/lbm
$x$ = distance between layers of fluid, m or ft
$Z$ = vertical distance above an arbitrarily chosen datum plant, m or ft

Greek symbols

$\alpha$ = alpha, correction coefficient for streamline or turbulent flow, dimensionless
$\epsilon$ = epsilon, equivalent roughness of pipe surface, m or ft
$\mu$ = mu, absolute viscosity of fluid, kg/(s)(m) or lbm/(s)(ft)
$\rho$ = rho, density of fluid, $kg/m^3$ or $lbm/ft^3$
$\rho_p$ = rho, density of plummet in a rotameter, $kg/m^3$ or $lbm/ft^3$
$\psi$ = psi, wetted perimeter normal to direction of fluid flow, m or ft

cross-sectional area of the pipe must equal the mass of fluid flowing past *any other* total cross-sectional area of the pipe.

The flow of a fluid through a pipe can be divided into two general classes, streamline flow or turbulent flow, depending upon the type of path followed by the individual particles of the fluid. When the flow of all the fluid particles is essentially along lines parallel to the axis of the pipe, the flow is called *streamline* (also *viscous* or *laminar*). When the course followed by the individual particles of the fluid deviates greatly from a straight line so that vortices and eddies are formed in the fluid, the flow is called *turbulent*.

The distinction between streamline flow and turbulent flow can be shown clearly by means of a simple experiment. The experiment is carried out by injecting a small stream of colored liquid into a fluid flowing inside a glass tube. If the fluid is moving at a sufficiently low velocity, the colored liquid will flow through the system in a straight line. No appreciable mixing of the two fluids will occur, and the straight-line path of the colored liquid can be observed visually. Under these conditions, streamline flow exists. If the velocity of the main stream is increased steadily, a velocity will finally be reached where the colored liquid no longer flows in a straight line. It now starts to mix with the main body of the fluid, and eddies and whirls can actually be observed through the glass walls of the tube. As the main-stream velocity is further increased, the mixing effect becomes more noticeable until the colored liquid is finally dispersed at random throughout the entire body of the main fluid. Under these conditions, turbulent flow exists. The particles of the fluid are no longer moving in smooth, straight lines but are moving in irregular directions throughout the tube.

There is a large difference between the characteristics of streamline flow and those of turbulent flow, and many of the important chemical engineering relationships apply only to one particular type of flow. Therefore, it is important to be able to distinguish between the different types of flow.

## Reynolds Number

The type of flow, whether streamline or turbulent, has been experimentally shown to depend on the inside diameter of the tube ($D$), the velocity of the flowing fluid ($V$), the density of the fluid ($\rho$), and the viscosity of the fluid ($\mu$). The numerical value of a dimensionless[1] grouping of these four variables serves to indicate whether the flow is streamline or turbulent. This dimensionless group is known as the *Reynolds number* and is expressed as follows:

$$\text{Reynolds number} = N_{Re} = \frac{DV\rho}{\mu}$$

When the Reynolds number exceeds 2100, turbulent flow may exist, while streamline flow exists at Reynolds numbers less than about 2100.

Streamline flow may occur at Reynolds numbers higher than 2100 if the flow condition is obtained by gradually increasing the Reynolds number toward and past the value of 2100. However, if the flow is orginally turbulent, it will stay turbulent until the Reynolds number drops below about 2100.

[1] The expression "dimensionless" means there are no physical units for the particular symbol or group. The units of the individual components of the Reynolds number cancel out, resulting in a "dimensionless" group.

## VELOCITY DISTRIBUTION IN PIPES

Because of the resistance encountered by a flowing fluid at the pipe-wall surface, the fluid particles at the wall surface may be considered to have no net forward velocity. This means there is no slippage at the wall and that the velocity of the fluid is zero at the wall surface. The particles further from the wall are less affected by this frictional resistance, and the maximum velocity of the fluid particles occurs at the center of the pipe. Thus, a cross-sectional view of the velocity distribution of a fluid flowing in a long straight pipe would show the maximum velocity at the center of the pipe, with the velocity gradually decreasing to zero as any portion of the wall is approached.

The average linear velocity of flow through a pipe is taken as the flow, expressed as volume per unit time, divided by the cross-sectional area of the pipe. For example, if a fluid is flowing at the rate of 6 ft$^3$/min through a pipe having an inside diameter of 2 in, the average linear velocity of the fluid is

$$\frac{6\ \text{ft}^3 \,\big|\, \,\big|\, (12)^2\ \text{in}^2 \,\big|\, \text{min}}{\text{min} \,\big|\, (2/2)^2\ \pi\ \text{in}^2\ \text{cross-sectional area} \,\big|\, \text{ft}^2 \,\big|\, 60\ \text{s}} = 4.59\ \text{ft/s}$$

If this liquid has a density of 30 lb/ft$^3$ and a viscosity of 0.002 lb/(s)(ft), the Reynolds number is

$$N_{Re} = \frac{DV\rho}{\mu} = \frac{2\ \text{in} \,\big|\, \text{ft} \,\big|\, 4.59\ \text{ft} \,\big|\, (\text{s})(\text{ft}) \,\big|\, 30\ \text{lb}}{\,\big|\, 12\ \text{in} \,\big|\, \text{s} \,\big|\, 0.002\ \text{lb} \,\big|\, \text{ft}^3} = 11{,}500$$

Working the same example in SI units, 6 ft$^3$/min is 0.170 m$^3$/min, and an inside diameter of 2 in is 0.0508 = m inside diameter. Therefore, the average linear velocity of the fluid is

$$\frac{0.170\ \text{m}^3 \,\big|\, \,\big|\, \text{min}}{\text{min} \,\big|\, (0.0508/2)^2\ \pi\ \text{m}^2\ \text{cross-sectional area} \,\big|\, 60\ \text{s}} = 1.40\ \text{m/s}$$

A density of 30 lb/ft$^3$ is 481 kg/m$^3$, and a viscosity of 0.002 lb/(s)(ft) is 0.00298 kg/(s)(m). Therefore, the Reynolds number is

$$N_{Re} = \frac{DV\rho}{\mu} = \frac{0.0508\ \text{m} \,\big|\, 1.40\ \text{m} \,\big|\, (\text{s})(\text{m}) \,\big|\, 481\ \text{kg}}{\,\big|\, \text{s} \,\big|\, 0.00298\ \text{kg} \,\big|\, \text{m}^3} = 11{,}500$$

The same final result is obtained for the Reynolds number no matter which units system is used because the Reynolds number is dimensionless. In this example, the Reynolds number is greater than 2100, so the flow may be considered as turbulent.

## TYPES OF PRESSURE

**Static pressure.** The pressure exerted on a plane parallel to the direction of flow of a moving fluid is called the *static pressure*. The static pressure is commonly measured at the inner surface of the pipe wall where the liquid velocity is negligible, and it is often simply called the "pressure."

**Impact pressure.** When the pressure is measured on a plane perpendicular to the direction of the fluid flow, it is called the *impact pressure*.

**Velocity pressure.** The difference between the impact pressure and the static pressure, when both are measured at the same point in the fluid, is known as the *velocity pressure*. For a stationary fluid, the pressure is the same in all directions; therefore, in this case, the static pressure equals the impact pressure, and the velocity pressure is zero.

**Fluid head.** If a vertical tube, open to the atmosphere at one end, is attached to a pipe containing a fluid under any pressure greater than atmospheric, the fluid will rise in the tube. The fluid will continue to rise until its weight in the tube produces enough pressure at the bottom to balance the difference between the pressure in the pipe and the atmospheric pressure. This height of fluid is termed the *fluid head*. The actual pressure in the pipe can be obtained by calculating the pressure due to the weight of the fluid in the tube, or due to the fluid head, and adding this pressure to the atmospheric pressure exerted at the top, open end of the tube. Thus, if an open vertical tube attached to a water main shows water in the tube to a height of 6 ft (1.83 m), the total pressure of the water in the main in American engineering units of pounds-force per square foot is

$$\frac{6 \text{ ft fluid head} \mid 62.4 \text{ lbm} \mid g = 32.17 \text{ ft/s}^2}{\phantom{6 \text{ ft fluid head}} \mid \text{ft}^3 \mid g_c = 32.17 \text{ ft} \cdot \text{lbm/(s}^2\text{)(lbf)}} + 2117 \frac{\text{lbf}}{\text{ft}^2} = 2491 \frac{\text{lbf}}{\text{ft}^2}$$

assuming the density of water as 62.4 lbm/ft$^3$ (999.6 kg/m$^3$) and atmospheric pressure as 2117 lbf/ft$^2$ (101.3 kPa). Using SI units for the calculation,[2] the total pressure of the water in the main in units of kilopascals is

$$\frac{1.83 \text{ m fluid head} \mid 999.6 \text{ kg} \mid g = 9.8 \text{ m/s}^2}{\phantom{1.83 \text{ m fluid head}} \mid \text{m}^3 \mid} + 101{,}300 \text{ Pa} = 17{,}930 \frac{\text{(m)(kg)}}{\text{(s}^2\text{)(m}^2\text{)}}$$

$$+ 101{,}300 \text{ Pa} = 17{,}930 \text{ N/m}^2 + 101{,}300 \text{ Pa} = 17{,}930 \text{ Pa} + 101{,}300 \text{ Pa}$$

$$= 119{,}230 \text{ Pa} = 119.23 \text{ kPa (same as 2491 lbf/ft}^2\text{)}$$

In the preceding example, it may be assumed that the tube is connected to the water main with the plane of the connection opening parallel to the direction of fluid flow. In this case, the fluid head is due to the static pressure (or pressure). In presenting data concerning fluid heads, the angle between the plane of the opening to the indicating tube and the direction of fluid flow should be specified.

[2] See App. A for conversions. One newton is 1 (m)(kg)/s$^2$.

## VISCOSITY

The viscosity of a fluid is a property of the material by virtue of which it resists shearing forces. Fluids having low viscosities, such as water, offer less resistance to a shearing force than fluids having high viscosities, such as oils. A thin-bladed knife will easily cut through water, while much more force would be required to force the knife through a heavy oil at the same speed. The force exerted on the knife as it cuts through the fluid is a shearing force, and the difference in resistance offered by the two fluids is due to the difference in their viscosities.

The concept of viscosity is particularly important in the consideration of the flow of fluids, since the magnitude of the viscosity affects the resistance to flow offered by the fluid. A fluid having a low viscosity flows more freely than a fluid having a high viscosity.

The expression for viscosity may be obtained by considering two small parallel layers of fluid, each with an area of $A$ $m^2$ or $ft^2$ and a differential distance of $dx$ m or ft apart. Under these conditions, a certain shearing force (expressed in poundals or newtons) must be exerted on the top layer to cause it to move parallel to the other layer at a relative differential velocity of $dV$ m/s or ft/s. It has been observed experimentally that this force is directly proportional to the velocity ($dV$) and to the area ($A$) of the layers and is inversely proportional to the distance ($dx$) between the layers. This may be expressed in equation form as

$$\text{Force} = \mu \frac{dV}{dx} A \tag{5-1}$$

where $\mu$ is the proportionality constant.

For gases and most liquids, the value of $\mu$ is constant it the temperature and pressure are fixed. Fluids of this type are called *Newtonian fluids*. The proportionality constant $\mu$ is defined as the *absolute viscosity* for all Newtonian fluids. Equation (5-1) may be rearranged as follows:

$$\text{Absolute viscosity} = \mu = \frac{(\text{force})(dx)}{(A)(dV)} = \frac{\text{force}}{(A)(dV/dx)} \tag{5-2}$$

From Eq. (5-2), the absolute viscosity of a fluid may be defined as the ratio of the shearing force per unit of parallel area to the resultant velocity gradient ($dV/dx$) perpendicular to the direction of the shear force.

### Units of Viscosity

The units of force (poundals) may be expressed as pounds-mass times acceleration, or $(\text{lbm})(\text{ft})/\text{s}^2$;[3] therefore, the units of viscosity in the American engineering system are

$$\frac{(\text{lbm})(\text{ft})}{\text{s}^2}\bigg|\frac{}{\text{ft}^2}\bigg|\frac{\text{s}}{\text{ft}}\bigg| = \frac{\text{lbm}}{(\text{ft})(\text{s})}$$

[3] See section on units of force in Chap. 2.

The units of viscosity in the cgs (or metric) system are g/(cm)(s), and a viscosity of 1 g/(cm)(s) is designated as 1 *poise*. Viscosities are commonly expressed as *centipoises* (cP), where 1 cP equals 0.01 poise. Centipoises may be converted to pounds per foot per second by multiplying the number of centipoises by 0.000672. The English unit of viscosity as pounds per foot per hour may be obtained by multiplying the number of centipoises by 2.42. The SI unit of viscosity as kilogram per meter per second is gotten by multiplying 0.001 times the number of centipoises.

The viscosity of air at room temperatures is approximately 0.02 cP, while the viscosity of water at ordinary temperatures is about 1 cP. Oils may have viscosities ranging from 10 to 5000 cP, depending on the temperature and type of oil.

The absolute viscosity of a fluid divided by its density is defined as *kinematic viscosity*. When a capillary viscometer is used for measuring viscosity, the kinematic viscosity is the value actually obtained. This value times the density of the fluid gives the *absolute viscosity*. The common unit of kinematic viscosity is the *centistoke* (cSt). One centistoke equals 0.01 cm$^2$/s.

## TOTAL ENERGY BALANCE

A fluid flowing through any type of conduit, such as a pipe, contains energy in three fundamental forms:

1. The fluid has a certain potential energy that is due to its position relative to a reference plane and is caused by the force exerted by the local gravitational field. Letting $Z$ represent the vertical distance from the fluid center of gravity to the reference plane and choosing a basis of unit mass of the flowing fluid, the potential energy due to the gravitational field per unit of body mass is $Z$ times the local gravitational acceleration times the appropriate conversion factor for units, as shown by Eq. (2-1). With American engineering units, the potential energy is expressed as $Z(g/g_c)$. The units of this expression are

$$\frac{\text{ft}\,\big|\,\text{ft}\,\big|\,(\text{s}^2)(\text{lbf})}{\quad\big|\,\text{s}^2\,\big|\,(\text{ft})(\text{lbm})} = \frac{\text{ft}\cdot\text{lbf}}{\text{lbm}}$$

With SI units,[4] the $g_c$ conversion factor becomes unity, and the potential-energy expression per kilogram of mass becomes $Zg$ with units of

$$\frac{\text{m}\,\big|\,\text{m}}{\quad\big|\,\text{s}^2} = \frac{(\text{m}^2)(\text{kg})}{(\text{s}^2)(\text{kg})} = \left[\frac{(\text{m})(\text{kg})}{\text{s}^2}\right]\frac{\text{m}}{\text{kg}} = \frac{(\text{N})(\text{m})}{\text{kg}} = \frac{\text{J}}{\text{kg}}$$

2. The fluid contains kinetic energy that is due to the velocity of flow. Letting $V_i$ represent the instantaneous velocity of the fluid at any point, the kinetic

[4] By App. A, 1 N is 1 (m)(kg)/s$^2$ and 1 J is 1 (N)(m).

energy per unit mass may be expressed as $V_i^2/2g_c$ for American engineering units and $V_i^2/2$ for SI units. The net units for the American system are

$$\frac{ft^2}{s^2}\left|\frac{(s^2)(lbf)}{(ft)(lbm)}\right. = \frac{ft \cdot lbf}{lbm}$$

and for SI are $m^2/s^2 = J/kg$.

3. The fluid contains a certain amount of internal energy that is due to the temperature level and the state. The internal energy may be expressed by the symbol $u$ having the units of foot-pound force per pound-mass or joule per kilogram[5].

The total energy per pound of fluid at any point in the system may be expressed as[6]

$$Z\frac{g}{g_c} + \frac{V_i^2}{2g^c} + u$$

According to the law of conservation of energy, the total energy put into a system in unit time must equal the amount of energy leaving the system in unit time, plus any accumulation. Consider a fluid flowing through a pipe at a steady rate with no accumulation or depletion of energy or material. The fluid entering the pipe contains potential, kinetic, and internal energy, while the fluid leaving the pipe contains the same three fundamental forms of energy. As the fluid flows through the system, the total energy of the fluid can only be altered by having external work done on or by the fluid or by allowing heat energy to enter or leave the flow system.

The following energy balance can be made around the entire system, employing a basis of unit mass of the flowing fluid and designating the fluid entrance conditions by subscript 1 and the exit conditions by subscript 2:

$$Z_1\frac{g}{g_c} + \frac{V_{i_1}^2}{2g_c} + u_1 + Q + W = Z_2\frac{g}{g_c} + \frac{V_{i_2}^2}{2g_c} + u_2 \tag{5-3}$$

where $W$ = net external work done on the system, J/kg or ft · lbf/lbm

$Q$ = net heat energy added to the system from an outside source, J/kg or ft · lbf/lbm

A certain amount of external work is done on the fluid as it enters the system. This work is supplied by the pushing force exerted by the rest of the fluid flowing toward the entrance point. The amount of this external work may be expressed as $p_1v_1$, where $p$ is the static pressure and $v$ is the specific volume of the fluid. Similarly, the fluid leaving the system does an amount of work on the surroundings

[5] The symbol $E$ is sometimes used to designate energy in place of $u$.

[6] In all cases, $g_c$ is included in the equations when American engineering units are involved, and $g_c$ becomes 1 when SI units are involved.

(or on the fluid already past the final point) equal to $p_2 v_2$. Therefore, the net work done on the system per pound of flowing fluid can be expressed as

$$W = W_o + p_1 v_1 - p_2 v_2 \tag{5-4}$$

where $W_o$ is the work imparted to the system from outside mechanical sources such as a pump or a turbine.

When a fluid flows through a pipe, the point velocity varies over the cross-sectional area of the pipe. When the velocity distribution is approximately uniform across the pipe, as in turbulent flow, only small errors are introduced in Eq. (5-3) if the kinetic energy term is expressed as $V^2/2g_c$, where $V$ is the average linear velocity. When the flow is streamline, the velocity distribution over the cross-sectional area of the pipe is parabolic, and the kinetic energy term may be expressed as $V^2/g_c$. Since the average linear velocity $V$ of a fluid flowing in a conduit can be determined easily, it is customary to use the following expression for the kinetic-energy terms in Eq. (5-3):

$$\frac{V_i^2}{2g_c} = \frac{V^2}{2\alpha g_c} \tag{5-5}$$

where $\alpha = 1.0$ if the flow is turbulent

$\alpha = 0.5$ if the flow is streamline

Substituting Eqs. (5-4) and (5-5) in Eq. (5-3) gives the following common form for the *total energy balance*:

$$Z_1 \frac{g}{g_c} + p_1 v_1 + \frac{V_1^2}{2\alpha_1 g_c} + u_1 + Q + W_o = Z_2 \frac{g}{g_c} + p_2 v_2 + \frac{V_2^2}{2\alpha_2 g_c} + u_2 \tag{5-6}$$

The preceding equation is very important, and it should be thoroughly understood. It takes all types of energy changes into consideration, including surface effects, chemical effects, and frictional effects. The equation is based on material and energy balances, and it is completely valid for the assumed conditions. For all practical purposes, $g$ and $g_c$ are numerically equal everywhere on the earth's surface. Therefore, with American engineering units, the potential-energy term is often simply expressed as $Z$, with the $g$ and $g_c$ numerically canceled. With SI units, the $g_c$ values become unity and all other terms are unchanged.

The entrance and exit potential energies $Z_1$ and $Z_2$ are based on some convenient reference plane so that the change in height, or $Z_2 - Z_1$, represents the external potential-energy change over the system. In a similar manner, the entrance and exit internal energies are based on a standard reference state. Consequently, the absolute values of the internal energies have no particular significance, but the difference, or $u_2 - u_1$, gives the actual change in internal energy from the entrance to the exit of the system.

The symbol $Q$ represents net heat added to the system from some external source. This heat is usually added through the walls enclosing the system, as, for example, heat added to a flowing gas by means of a direct flame applied to the pipe containing the flowing gas.

Work added to the system from an external source is represented by the symbol $W_o$. This work is almost always added by means of some mechanical device such as a pump operating on the flowing material.[7]

## Enthalpy

The internal energy and the external work term $pv$ are often added together and treated as a single function called *enthalpy*. This is done merely as a matter of convenience for calculations. The change in enthalpy is identical with the sum of the changes in the internal energy and the $pv$ external work terms, since the enthalpy, internal energy, and $pv$ expression are all point functions. Thus the enthalpy change from point 1 to point 2, designated by $i_2 - i_1$,[8] equals $u_2 + p_2 v_2 - u_1 - p_1 v_1$.

The change in enthalpy of a perfect gas equals the mean heat capacity of the gas at constant pressure times the temperature change. The change in internal energy of a perfect gas equals the mean heat capacity of the gas at constant volume times the temperature change.

**Example 5-1: Application of the total energy balance to a flowing gas** One pound mole (0.4536 kg mol) of dry air per second, flowing at an average linear velocity of 200 ft/s (60.96 m/s), at an absolute pressure of 200 $lb/in^2$ (1379 kPa), and at a temperature of 200°F (366.3 K) enters a horizontal baffled pipe. The air leaves the pipe at an absolute pressure of 100 $lb/in^2$ (689.3 kPa) and a temperature of 100°F (310.8 K). The average linear velocity of the air leaving the pipe is 550 ft/s (167.6 m/s). The change in enthalpy of the air may be taken as the mean heat capacity of the air at constant pressure [7.0 Btu/(lb mol)(°F) or 29,310 J/(kg mol)(K)] times the change in temperature. Flow in the pipe is turbulent so that $\alpha$ in Eq. (5-6) may be taken as 1.0. Calculate the amount of heat lost from the walls of the pipe (*a*) as Btu per second using American engineering units and (*b*) as joules per second using SI units.

SOLUTION

(*a*) With American engineering units:

**Basis**

1 lb of dry air (molecular weight of dry air = 29)

Since the pipe is horizontal, $Z_1 = Z_2$, and these terms disappear from the total energy balance.

[7] Since $W_o$ represents the actual work imparted to the system from outside mechanical sources, the loss of mechanical energy due to pump friction must be included in $Q$.

[8] The symbol $h$ is sometimes used to denote enthalpy in place of $i$.

The change in enthalpy is

$$i_2 - i_1 = u_2 + p_2 v_2 - u_1 - p_1 v_1 = \frac{7.0\ \text{Btu}}{(\text{lb mol})(^\circ\text{F})}\left|\frac{(100 - 200)^\circ\text{F}}{}\right|\frac{\text{lb mol}}{29.0\ \text{lbm}}$$

$$= -24.1\ \text{Btu/lbm}$$

Since there is 778 ft · lbf/Btu, the enthalpy change is

$$(-24.1)(778) = -18{,}750\ \text{ft} \cdot \text{lbf/lbm}$$

No external work is added to the system; so $W_o = 0$.

$$V_1 = 200\ \text{ft/s} \qquad V_2 = 550\ \text{ft/s} \qquad g_c = 32.17\ \text{ft} \cdot \text{lbm/(s}^2\text{)(lbf)}$$

From the total energy balance,

$$Q = u_2 + p_2 v_2 - u_1 - p_1 v_1 + \frac{V_2^2}{2g_c} - \frac{V_1^2}{2g_c}$$

$$Q = -18{,}750 + \frac{(550)^2 - (200)^2}{(2)(32.17)} = -14{,}670\ \text{ft} \cdot \text{lbf/lbm}$$

$$Q = -\frac{14{,}670}{778} = -18.85\ \text{Btu/lbm}$$

The sign of the calculated $Q$ is negative. This means heat is lost from the system in the amount of 18.85 Btu/lb of air flowing.

Since 1 lb mol of air, or 29 lb of air, passes through the system per second, the total amount of heat lost from the walls of the pipe per second is

$$(18.85)(29) = 547\ \text{Btu/s}$$

(*b*) With SI units:

**Basis**

1 kg of dry air (molecular weight of dry air = 29)

Since the pipe is horizontal, $Z_1 = Z_2$, and these terms disappear from the total energy balance.

The change in enthalpy is

$$i_2 - i_1 = u_2 + p_2 v_2 - u_1 - p_1 v_1 = \frac{29{,}310\ \text{J}}{(\text{kg mol})(\text{K})}\left|\frac{(310.8 - 366.3)\text{K}}{}\right|\frac{\text{kg mol}}{29.0\ \text{kg}}$$

$$= -56{,}090\ \text{J/kg}$$

No external work is added to the system; so $W_o = 0$.

$$V_1 = 60.96\ \text{m/s} \qquad V_2 = 167.6\ \text{m/s}$$

From the total energy balance written for SI units,

$$Q = u_2 + p_2 v_2 - u_1 - p_1 v_1 + \frac{V_2^2}{2} - \frac{V_1^2}{2}$$

$$Q = -56{,}090 + \frac{(167.6)^2 - (60.96)^2}{2} = -43{,}900 \text{ J/kg}$$

The sign of the calculated $Q$ is negative. This means heat is lost from the system in the amount of 43,900 J/kg of air flowing.

Since 0.4536 kg mol of air, or (29)(0.4536) = 13.15 kg of air, passes through the system per second, the total amount of heat lost from the walls of the pipe per second is (43,900)(13.15) = 577,000 J/s.

Check by comparing answer (*a*) to answer (*b*). Since 1 Btu is 1055 J by App. A, 547 Btu/s [answer (*a*)] × 1055 J/Btu = 577,000 J/s [answer (*b*)], and answer (*b*) is gotten by conversion from answer (*a*).

## TOTAL MECHANICAL-ENERGY BALANCE

The total energy balance as presented in the preceding section can be written in a form involving only mechanical energies and, as such, is of considerable value to engineers.[9] A fluid flowing through a pipe will ordinarily not undergo a chemical change, and minor energy changes such as surface effects can usually be neglected. Making these assumptions, a mechanical-energy balance can be set up for a system containing a fluid flowing at a steady rate.

If a unit weight of fluid entering a steady-flow system is assumed as a basis, the total mechanical-energy balance between the entrance and exit points of the system may be written as

$$Z_1 \frac{g}{g_c} + p_1 v_1 + \frac{V_1^2}{2\alpha_1 g_c} + \int_1^2 p\,dv + W_o = Z_2 \frac{g}{g_c} + \frac{V_2^2}{2\alpha_2 g_c} + p_2 v_2 + \Sigma F \quad (5\text{-}7)$$

Equation (5-7) is written in a form for use with American engineering units, with the units of each of the terms in the equation being foot-pounds force per pound-mass, just as in the total energy balance. When SI units are used, Eq. (5-7) applies, except that all of the $g_c$ conversion factors become 1.0 and the units of the individual terms in the equation are joules per kilogram.

The flow of a fluid through any system is always accompanied by friction. The friction causes loss in mechanical energy which could have done work but

[9] See App. B for a direct derivation of the total mechanical-energy balance from the total energy balance.

was lost as a result of irreversible changes occurring in the flowing fluid. The term $\Sigma F$ in Eq. (5-7) represents the loss in mechanical energy caused by frictional effects and is best defined as the amount of mechanical energy necessary to balance the output energy against the input energy in a total mechanical-energy balance.

Mechanical energy may enter and leave a system in the form of kinetic energy ($V^2/2\alpha g_c$) and potential energy ($Z(g/g_c)$). Mechanical energy may be added to the system in the form of work from an outside source ($W_o$). The fluid entering the system adds mechanical energy because of the pushing force $pv$ exerted by the adjacent fluid, and, in a similar manner, mechanical energy is delivered to the surroundings at the point where the fluid leaves the system.

When a compressible fluid such as a gas is involved, there is usually a change in volume per unit mass as the fluid flows from one point to another. This volume change is due to changes in temperature and pressure. When the volume of a fluid changes, a certain amount of work must be supplied or withdrawn. This work of expansion or contraction between points 1 and 2 in a system can be expressed by $\int_1^2 p\,dv$. Much engineering work deals with essentially noncompressible fluids such as water. With fluids of this type, the expansion work term $\int_1^2 p\,dv$ is practically zero and can be neglected.

**Example 5-2: Application of total mechanical-energy balance** Water flows at a constant mass rate through a long section of uniform-diameter pipe. The density of the water is 62.3 lb/ft$^3$ (998.0 kg/m$^3$). Because water may be considered as a noncompressible fluid, the velocity is constant throughout the pipe. The section exit is 10 ft (3.05 m) higher than the entrance. The entrance pressure is 15 lb/in$^2$, absolute (103.4 kPa), and the exit pressure is 25 lb/in$^2$, absolute (172.4 kPa). A pump supplies 50 ft · lb (67.8 J) to every pound of water flowing. Water density in the section may be assumed as constant, and the Reynolds number is higher than 2100 throughout the system. Calculate the mechanical-energy frictional loss in the pipe section as (*a*) foot-pounds force per pound of flowing water using American engineering units and (*b*) joules per kilogram of flowing water using SI units.

SOLUTION

(*a*) With American engineering units:

**Basis**

1 lbm of water

Assuming the entrance to the section as the reference level, $Z_1 = 0$ and $Z_2 = 10$. Since the Reynolds number is higher than 2100, the flow may be considered as turbulent, and $\alpha$ in the kinetic energy expressions of Eq. (5-7) is unity. The frictional loss in the pipe may be determined by applying the total mechanical-energy balance as follows:

$$V_1 = V_2$$

Density of fluid = 62.3 lb/ft$^3$

$v_1 = v_2 = 1/62.3 \text{ ft}^3/\text{lb}$

$$p_1 v_1 = \frac{15 \text{ lbf} \mid 144 \text{ in}^2 \mid \text{ft}^3}{\text{in}^2 \mid \text{ft}^2 \mid 62.3 \text{ lbm}} = 34.7 \text{ ft} \cdot \text{lbf/lbm}$$

$$p_2 v_2 = \frac{(25)(144)}{62.3} = 57.8 \text{ ft} \cdot \text{lbf/lbm}$$

$$\int_1^2 p\, dv = 0 \text{ (since water is a noncompressible fluid)}$$

$W_o = 50 \text{ ft} \cdot \text{lbf/lbm}$

$$\Sigma F = \text{frictional loss} = Z_1 \frac{g}{g_c} - Z_2 \frac{g}{g_c} + \frac{V_1^2}{2g_c} - \frac{V_2^2}{2g_c} + p_1 v_1 - p_2 v_2 + W_o$$

$$\Sigma F = 0 - 10 + 34.7 - 57.8 + 50 = 16.9 \text{ ft} \cdot \text{lbf/lbm}$$

(*b*) With SI units:

**Basis**

1 kg of water

Assuming the entrance to the section as the reference level, $Z_1 = 0$ and $Z_2 = 3.05$ m. Because the Reynolds number is higher than 2100, the flow may be considered as turbulent, and $\alpha$ in the kinetic energy expressions of Eq. (5-7) is unity. The frictional loss in the pipe may be determined by applying the total mechanical-energy balance equation as follows, noting that 1 N is 1 (m)(kg)/s$^2$ and 1 (N)(m) is 1 J.

$$Z_1 = 0 \qquad Z_1 g = 0 \qquad Z_2 = 3.05 \text{ m}$$

$$Z_2 g = \frac{3.05 \text{ m} \mid 9.80 \text{ m}}{\mid \text{s}^2} = 29.9 \frac{\text{m}^2}{\text{s}^2} = 29.9 \left[\frac{(\text{m})(\text{kg})}{\text{s}^2}\right] \frac{(\text{m})}{\text{kg}} = 29.9 \text{ (N)(m)/kg}$$

$= 29.9$ J/kg

$V_1 = V_2$

Density of fluid = 998.0 kg/m$^3$

$v_1 = v_2 = 1/998.0 \text{ m}^3/\text{kg}$

$p_1 = 103.4 \text{ kPa} = 103{,}400 \text{ Pa} = 103{,}400 \text{ N/m}^2$

$p_2 = 172.4 \text{ kPa} = 172{,}400 \text{ Pa} = 172{,}400 \text{ N/m}^2$

$$p_1 v_1 = \frac{103{,}400 \text{ N} \mid \text{m}^3}{\text{m}^2 \mid 998.0 \text{ kg}} = 103.6 \text{ (N)(m)/kg} = 103.6 \text{ J/kg}$$

$$p_2 v_2 = \frac{172{,}400 \text{ N} \mid \text{m}^3}{\text{m}^2 \mid 998.0 \text{ kg}} = 172.7 \text{ (N)(m)/kg} = 172.7 \text{ J/kg}$$

$$\int_1^2 p\, dv = 0 \text{ since water is a noncompressible fluid}$$

$$W_o = \frac{67.8 \text{ J} \mid 2.2046 \text{ lb}}{\text{lb} \mid \text{kg}} = 149.5 \text{ J/kg}$$

By Eq. (5-7) with all $g_c$ values at 1.0 and $\alpha$ at 1.0,

$$\Sigma F = \text{frictional loss} = Z_1 g - Z_2 g + \frac{V_1^2}{2} - \frac{V_2^2}{2} + p_1 v_1 - p_2 v_2 + W_o$$

$$\Sigma F = 0 - 29.9 + 103.6 - 172.7 + 149.5 = 50.5 \text{ J/kg}$$

Check by comparing answer (*a*) to answer (*b*). Since 1 ft · lbf is 1.3558 J and 1 kg is 2.2046 lb, 16.9 ft · lbf/lbm [answer (*a*)] × 1.3558 J/ft · lbf × 2.2046 lb/kg = 50.5 J/kg [answer (*b*)], and answer (*b*) is gotten by conversion from answer (*a*).

# FRICTION

## Fanning Equation

When a fluid flows through a conduit, the amount of energy lost because of friction depends on the properties of the flowing fluid and the extent of the conduit system. For the case of steady flow through long straight pipes of uniform diameter, the variables affecting the amount of frictional losses are the velocity at which the fluid is flowing, the density of the fluid, the viscosity of the fluid, the diameter of the pipe, the length of the pipe, and the roughness of the pipe.

By applying the method of dimensional analysis[10] to these variables, the following expression can be obtained for the frictional loss in the system, where the mechanical energy lost as friction ($F$) is equivalent to the pressure drop over the system due to friction divided by the density of the flowing fluid:

$$F = 2f \frac{V^2 L}{g_c D} \tag{5-8}$$

This expression is known as the *Fanning equation* and is strictly applicable only to point conditions or to a system in which the fluid density, viscosity, and linear velocity are constant. The conversion factor $g_c$ is used with American engineering units, while the conversion factor $g_c$ becomes 1.0 with SI units.

The friction factor $f$ is based on experimental data and has been found to be a function of the Reynolds number and the relative roughness of the pipe. The equivalent pipe roughness is designated by the symbol $\epsilon$ and represents the average roughness or depth of the surface irregularities. The relative roughness is defined as the dimensionless ratio of the equivalent pipe roughness to the pipe diameter or $\epsilon/D$, where $\epsilon$ and $D$ are expressed in the same units.

The effect of pipe-surface characteristics (or relative roughness) was neglected by early workers in the field of fluid flow. The original data of these workers showed very poor correlations (especially at high Reynolds numbers) until the effect of relative roughness was correctly included.

[10] See App. B for a discussion of the methods of dimensional analysis and a derivation of Eq. (5-8) by dimensional analysis.

Figure 5-1 presents a plot of the friction factor versus the Reynolds number in straight pipes. In the streamline region, the friction factor is not affected by the relative roughness of the pipe. Therefore, only one line is shown in Fig. 5-1 for Reynolds numbers up to about 2100. In the turbulent region, the relative roughness of the pipe has a large effect on the friction factor. Curves with different parameters of $\epsilon/D$ are presented in Fig. 5-1 for values of Reynolds numbers greater than 2100. A table on the plot indicates values of $\epsilon$ for various pipe-construction materials. The methods for determining $f$ do not permit high accuracy;[11] therefore, the value of $f$ should only be determined to two significant figures, and Fig. 5-1 gives adequate accuracy for determining the numerical values of the friction factor.

Curves similar to Fig. 5-1 are sometimes presented in the literature with a different defining value for $f$. For example, mechanical engineers usually define the friction factor so that it is exactly four times the friction factor given in Eq. (5-8).

The Reynolds number range between 2100 and 4000 is commonly designated as the *critical region.* In this range, under ordinary conditions, there is considerable doubt as to whether the flow is streamline or turbulent. For design purposes, it is the safest practice to assume that turbulent flow exists at all Reynolds numbers greater than 2100. However, it should be realized that this assumption may result in overdesign if the flow is in the critical region.

A mathematical expression for the friction factor can be obtained from the equation for the straight line in the streamline-flow region of Fig. 5-1. Thus, at Reynolds numbers below 2100,

$$f = \frac{16\mu}{DV\rho} = \frac{16}{N_{Re}} \tag{5-9}$$

**Example 5-3: Calculation of friction term in the total mechanical-energy balance** A liquid flows through a straight steel pipe at a velocity of 15 ft/s (4.57 m/s). The inside diameter of the pipe is 2.067 in (0.0525 m). The density of the liquid is 40 lb/ft$^3$ (641 kg/m$^3$), and its viscosity is 0.003 lb/(ft)(s) [0.00446 kg/(m)(s)]. If the pipe is 60 ft (18.3 m) long, calculate the mechanical-energy loss due to friction as (*a*) foot-pound force per pound-mass using American engineering units and (*b*) joules per kilogram using SI units.

SOLUTION (*a*) With American engineering units:

$$D = \frac{2.067}{12}\ \text{ft} \qquad V = 15\ \text{ft/s} \qquad \rho = 40\ \text{lb/ft}^3$$

$$\mu = 0.003\ \text{lb/(s)(ft)} \qquad L = 60\ \text{ft} \qquad \epsilon = 0.00015\ \text{ft by Fig. 5-1}$$

$$\frac{\epsilon}{D} = \frac{(0.00015)(12)}{2.067} = 0.00087$$

[11] See App. B for a discussion of the experimental methods for determining the values of the Fanning friction factor ($f$) and Example 5-4 for a method of calculating the value of the Fanning friction factor from experimental data.

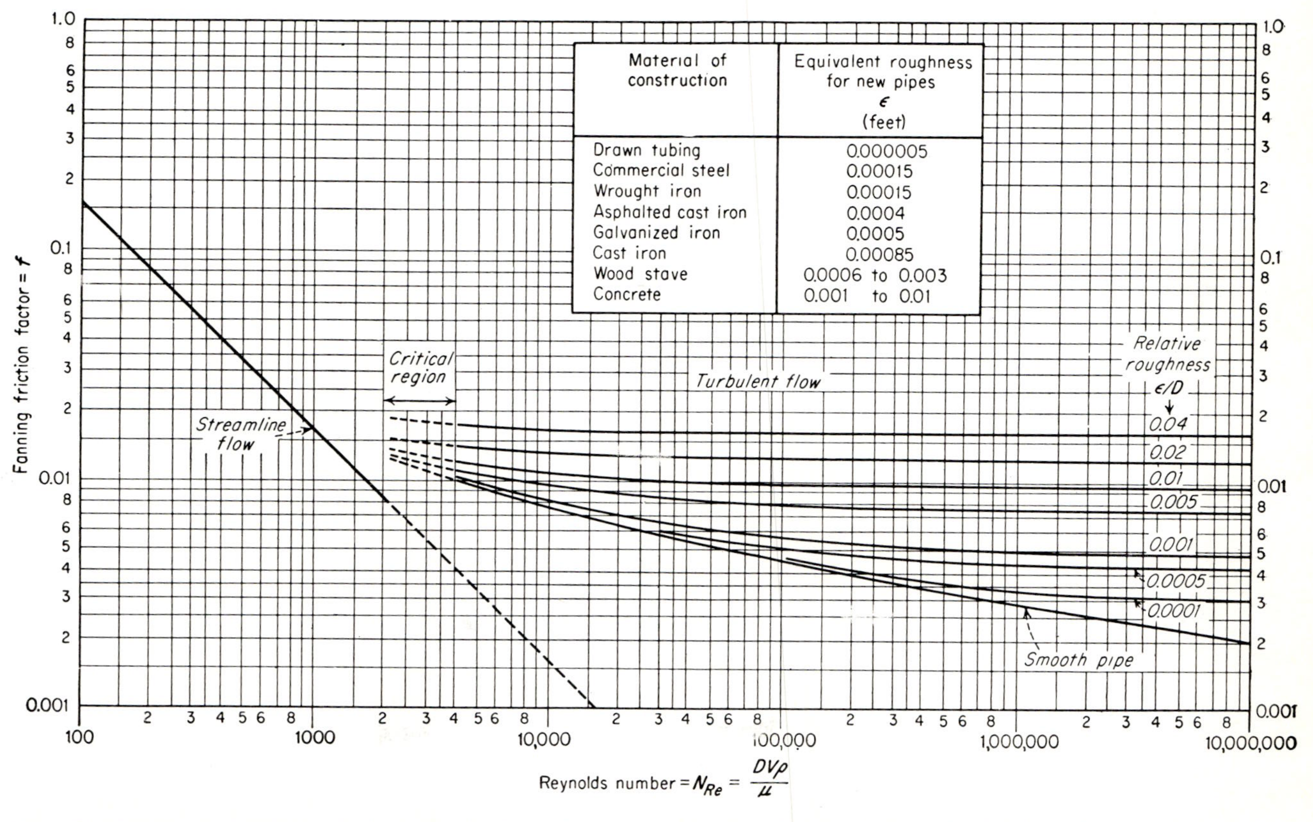

**Figure 5-1** Fanning friction factors for long straight pipes. [*Based on L. F. Moody, Trans ASME*, **66**:671–684 (1944).]

$$\text{Reynolds number} = \frac{DV\rho}{\mu}$$

$$= \frac{2.067\ \text{in}}{} \bigg| \frac{\text{ft}}{12\ \text{in}} \bigg| \frac{15\ \text{ft}}{\text{s}} \bigg| \frac{(\text{ft})(\text{s})}{0.003\ \text{lb}} \bigg| \frac{40\ \text{lb}}{\text{ft}^3}$$

$$= 34{,}500$$

From Fig. 5-1, $f = 0.0063$.

$$\text{Friction} = F = 2f\,\frac{V^2L}{g_cD} = \frac{(2)(0.0063)(15)^2(60)(12)}{(32.17)(2.067)}$$

$$= 31\ \text{ft}\cdot\text{lbf/lbm}$$

(*b*) With SI units:

$$D = 0.0525\ \text{m} \qquad V = 4.57\ \text{m/s} \qquad \rho = 641\ \text{kg/m}^3$$

$$\mu = 0.00446\ \text{kg/(m)(s)} \qquad L = 18.3\ \text{m}$$

$$\epsilon = \frac{0.00015\ \text{ft}}{} \bigg| \frac{0.3048\ \text{m}}{\text{ft}} = 0.000046\ \text{m}$$

$$\frac{\epsilon}{D} = 0.000046/0.0525 = 0.00087$$

$$\text{Reynolds number} = \frac{DV\rho}{\mu} = \frac{0.0525\ \text{m}}{} \bigg| \frac{4.57\ \text{m}}{\text{s}} \bigg| \frac{(\text{m})(\text{s})}{0.00446\ \text{kg}} \bigg| \frac{641\ \text{kg}}{\text{m}^3} = 34{,}500$$

From Fig. 5-1, $f = 0.0063$.

$$\text{Friction} = F = \frac{2fV^2L}{D} = \frac{2}{} \bigg| \frac{0.0063}{} \bigg| \frac{(4.57)^2\ \text{m}^2}{\text{s}^2} \bigg| \frac{18.3\ \text{m}}{} \bigg| \frac{}{0.0525\ \text{m}}$$

$$= 92\,\frac{\text{m}^2}{\text{s}^2} = 92\left[\frac{(\text{m})(\text{kg})}{\text{m}^2}\right]\frac{\text{m}}{\text{kg}} = 92\,\frac{(\text{N})(\text{m})}{\text{kg}} = 92\,\frac{\text{J}}{\text{kg}}$$

Check by comparing answer (*a*) converted to answer (*b*). Since 1 ft · lbf is 1.3588 J and 1 kg is 2.2046 lb, 31 ft · lbf/lbm [answer (*a*)] × 1.3558 J/ft · lbf × 2.2046 lb/kg = 92 J/kg [answer (*b*)], and answer (*b*) is gotten by conversion from answer (*a*).

**Example 5-4: Calculation of Fanning friction factor from experimental flow data** It is desired to determine experimentally the Fanning friction factor for a pipe flow system as part of a laboratory experiment. The pipe used in this experiment is 40 ft long between pressure taps and is horizontal; it has an inside diameter of 2.067 in and no bends, fittings, or diameter change over its 40-ft length. Water at 60°F flows through the pipe at a rate of 7 lb/s. The water density is 62.4 lb/ft$^3$. The pressure drop due to friction over the 40-ft length of pipe is measured by flush taps into the wall at each end of the

40-ft length connected to each side of a mercury manometer, which has mercury in contact with the water. The specific gravity of mercury is 13.6, so its density is $13.6 \times 62.4\ \text{lb/ft}^3$. If the vertical reading on the manometer read as difference in height of mercury in the two legs is 4 in of mercury (water is in the other leg of the manometer), what is the experimental value of the Fanning friction factor ($f$)? Solve the problem using American engineering units and repeat using SI units.

SOLUTION Applying the total mechanical-energy balance, Eq. (5-7), for this system,

$$Z_1 = Z_2 \qquad V_1 = V_2 \qquad v_1 = v_2 = \frac{1}{\rho_{H_2O}}$$

$$W_o = 0 \qquad \int_1^2 p\,dv = 0$$

Therefore, the total mechanical-energy balance becomes

$$\frac{p_1 - p_2}{\rho_{H_2O}} = \Sigma F$$

By the Fanning equation,

$$\Sigma F = \frac{2fV^2L}{g_c D} = \frac{p_1 - p_2}{\rho_{H_2O}}$$

Therefore,

$$f = \frac{(p_1 - p_2)g_c D}{2V^2 L \rho_{H_2O}}$$

(*a*) Solve the preceding equation for $f$ using American engineering units.

The value of $p_1 - p_2$ can be gotten from the reading on the mercury manometer by making a head balance from the base of the manometer, as described in detail in the footnote to Example 5-6. The result is

$$p_1 - p_2 = (\text{manometer height reading})(\rho_{Hg} - \rho_{H_2O})\frac{g}{g_c}$$

Manometer height reading = 4 in

$\rho_{Hg} = (13.6)(62.4)\ \text{lb/ft}^3$

$\rho_{H_2O} = 62.4\ \text{lb/ft}^3$

$g = 32.17\ \text{ft/s}^2$

$g_c = 32.17\ \text{ft} \cdot \text{lbm/(s}^2\text{)(lbf)}$

$$p_1 - p_2 = \frac{4\ \text{in}}{} \bigg| \frac{\text{ft}}{12\ \text{in}} \bigg| \frac{(62.4)(13.6 - 1)\ \text{lbm}}{\text{ft}^3} \bigg| \frac{32.17\ \text{ft}}{\text{s}^2} \bigg| \frac{(\text{s}^2)(\text{lbf})}{32.17\ \text{ft} \cdot \text{lbm}}$$

$$= 262.1\ \text{lbf/ft}^2$$

The average linear velocity of the water in the pipe ($V$) = volumetric flow rate per cross-sectional area of the pipe.

$$\text{Volumetric flow rate} = \frac{7\ \text{lb}}{\text{s}}\left|\frac{\text{ft}^3}{62.4\ \text{lb}}\right. = 0.1122\ \text{ft}^3/\text{s}$$

$D = 2.067$ in

$$\text{Cross-sectional area of the pipe} = \frac{\pi\,(2.067/2)^2\ \text{in}^2}{}\left|\frac{\text{ft}^2}{144\ \text{in}^2}\right. = 0.02329\ \text{ft}^2$$

$$V = \frac{0.1122\ \text{ft}^3}{\text{s}}\left|\frac{}{0.02329\ \text{ft}^2\ \text{cross-sectional area}}\right. = 4.82\ \text{ft/s}$$

$L = 40$ ft

$$f = \frac{(p_1 - p_2)g_c D}{2V^2 L \rho_{H_2O}}$$

$$f = \frac{262.1\ \text{lbf}}{\text{ft}^2}\left|\frac{32.17\ \text{ft}\cdot\text{lbm}}{(\text{s}^2)(\text{lbf})}\right|\frac{2.067\ \text{in}}{}\left|\frac{\text{ft}}{12\ \text{in}}\right|\frac{}{2}\left|\frac{\text{s}^2}{(4.82)^2(\text{ft}^2)}\right|\frac{}{40\ \text{ft}}\left|\frac{\text{ft}^3}{62.4\ \text{lb}}\right.$$

$= 0.013$ (dimensionless)

(*b*) Repeat the preceding solution using SI units.

$$p_1 - p_2 = (\text{manometer height reading})(\rho_{Hg} - \rho_{H_2O})g$$

$$\text{Manometer height reading} = \frac{4\ \text{in}}{}\left|\frac{\text{m}}{39.37\ \text{in}}\right. = 0.1016\ \text{m}$$

By App. A, there are 16.02 kg/m$^3$ per lb/ft$^3$. Therefore,

$$\rho_{Hg} = \frac{(13.6)(62.4)\ \text{lb}}{\text{ft}^3}\left|\frac{16.02\ \text{kg/m}^3}{\text{lb/ft}^3}\right. = 13{,}600\ \text{kg/m}^3$$

$$\rho_{H_2O} = (62.4)(16.02) = 999.6\ \text{kg/m}^3$$

$$g = 9.8\ \text{m/s}^2$$

$$p_1 - p_2 = \frac{0.1016\ \text{m}}{}\left|\frac{(13{,}600 - 999.6)\ \text{kg}}{\text{m}^3}\right|\frac{9.8\ \text{m}}{\text{s}^2} = 12{,}550\ \text{kg/(m)(s}^2)$$

$$= 12{,}550\ \text{N/m}^2 = 12{,}550\ \text{Pa}$$

$V$ = volumetric flow rate per cross-sectional area of pipe

$$\text{Volumetric flow rate} = \frac{7\ \text{lb}}{\text{s}}\left|\frac{0.4536\ \text{kg}}{\text{lb}}\right|\frac{\text{m}^3}{999.6\ \text{kg}} = 0.00318\ \text{m}^3/\text{s}$$

$$D = \frac{2.067\ \text{in}}{}\left|\frac{\text{m}}{39.37\ \text{in}}\right. = 0.0525\ \text{m}$$

Cross-sectional area of pipe $= \pi\,(0.0525/2)^2 = 0.00216\ \text{m}^2$

$$V = \frac{0.00318\ \text{m}^3}{\text{s}} \bigg| \frac{}{0.00216\ \text{m}^2} = 1.47\ \text{m/s}$$

$$L = \frac{40\ \text{ft}}{} \bigg| \frac{0.3048\ \text{m}}{\text{ft}} = 12.2\ \text{m}$$

The expression for $f$ in SI units is

$$f = \frac{(p_1 - p_2)D}{2V^2 L \rho_{H_2O}}$$

or $f =$

$$\frac{12{,}550\ \text{kg}}{(\text{m})(\text{s}^2)} \bigg| \frac{0.0525\ \text{m}}{} \bigg| \frac{}{2} \bigg| \frac{\text{s}^2}{(1.47)^2\ \text{m}^2} \bigg| \frac{}{12.2\ \text{m}} \bigg| \frac{\text{m}^3}{999.6\ \text{kg}} = 0.013\ \text{(dimensionless)}$$

## Poiseuille's Law

By reading the appropriate value of $f$ from a plot of Reynolds number versus $f$, the Fanning equation can be applied to both streamline and turbulent flow. However, in the streamline-flow region an expression has been derived which does not depend on previously determined data.

From the definition of the absolute viscosity of a fluid and by the application of calculus, an expression for mechanical energy lost in a system because of friction can be obtained for streamline flow. When the system is limited to steady flow of a noncompressible fluid in circular pipes of uniform cross section, the following expression is obtained:

$$F = \frac{32\mu VL}{g_c D^2 \rho} \tag{5-10}$$

This is known as *Poiseuille's law*. It can be applied in place of the Fanning equation when the Reynolds number is less than 2100.

## Limitations of the Fanning Equation and Poiseuille's Law

The expressions for the Fanning equation and Poiseuille's law given in the preceding sections are, in a strict sense, limited to point conditions or to conditions involving steady flow of a fluid through long straight pipes when the velocity, density, and viscosity of the fluid are constant. The density and, consequently, the linear velocity of gases and vapors change considerably with changes in pressure. If the pressure drop over the system is large and a compressible fluid is involved, the density and the velocity of the fluid will change appreciably.

When dealing with compressible fluids, such as air, steam, or any gas, it is good engineering practice to use Eqs. (5-8) and (5-10) only if the pressure drop over the system is less than 10 percent of the final pressure. With noncompressible fluids such as water, the density and velocity of the fluids are esentially unaffected by pressure changes and Eqs. (5-8) and (5-10) are applicable.

In nonisothermal flow, the change in temperature tends to cause the viscosity, density, and average linear velocity of the fluid to change. Ordinarily, little error is introduced when the Fanning equation or Poiseuille's law is used for the nonisothermal flow of fluids as long as the temperature change is not more than 20°C. When temperature changes or pressure changes are involved, the best accuracy is obtained by using the linear velocity, density, and viscosity of the fluid as determined at the average temperature and pressure.

Exact results may be obtained from the Fanning equation and Poiseuille's law by integrating the differential forms of these expressions, taking all the changes into consideration.

## End Effects

The preceding discussion has been limited to steady flow in straight pipes of uniform diameter. Frictional losses always occur when there is a sudden enlargement or a sudden contraction of the cross-sectional area of the container in which the fluid is flowing. In the application of the total mechanical-energy balance, these effects must be taken into consideration.

Because of the friction caused by contraction and expansion, it is very important to indicate clearly the points between which the energy balance is to be made. For example, if water were being pumped through a pipe to a large storage tank, the end point for the energy balance could be chosen just outside the exit from the pipe in the storage tank. In this case, the kinetic energy of the fluid at this final point would be zero, since the velocity of the water in the large tank would be negligible. However, there would be a frictional effect due to the sudden expansion of the system from the cross-sectional area of the pipe to the cross-sectional area of the storage tank. If the final reference point were taken just inside the pipe at the water exit, there would be no expansion effect, but the kinetic-energy term at the final point would have a finite value since the velocity of the water in the pipe would be finite.

## Frictional Losses Due to Sudden Enlargement

The loss of mechanical energy as friction due to the sudden enlargement of the cross-sectional area of the duct in which the fluid is flowing can be calculated by the following equation:

$$F_e = \frac{(V_1 - V_2)^2}{2\alpha g_c} \tag{5-11}$$

$V_1$ is the average linear velocity of the fluid in the section preceding the enlargement, and $V_2$ is the average linear velocity of the fluid after the enlargement. The units of $F_e$ are the equivalent of foot-pounds force per pound-mass when $g_c$

is included for American engineering units. With SI units, the $g_c$ conversion factor is 1.0, and the units of $F_e$ are joules per kilogram. When the flow is turbulent, $\alpha$ should be taken as 1.0. For streamline flow, it may be assumed that $\alpha$ is 0.5.

## Frictional Losses Due to Sudden Contraction

The loss of mechanical energy as friction due to a sudden contraction of the cross-sectional area of the flow system under consideration can be calculated by the following equation:

$$F_c = K_c \frac{V_2^2}{2\alpha g_c} \tag{5-12}$$

where $V_2$ is the downstream velocity or the velocity in the pipe after the contraction has occurred and $K_c$ is a constant dependent on the ratio of the two cross-sectional areas involved.

Values of $K_c$ can be estimated from Fig. 5-2, where $S_1$ is the cross-sectional area of the pipe before constriction and $S_2$ is the cross-sectional area of the pipe after constriction. When the sudden contraction occurs from a very large container, as a lake or city reservoir, the ratio of $S_2/S_1$ is zero and $K_c$ is 0.5. For turbulent flow, $\alpha$ is 1.0. For streamline flow, $\alpha$ may be taken as 0.5.

## Frictional Losses Due to Pipe Fittings

Mechanical energy is lost as friction due to the resistance encountered by a fluid flowing through various types of pipe fittings such as elbows, tees, and valves. These losses are accounted for by assigning the various fittings a fictitious length $L_e$. This length is equivalent to the length of pipe, having the same nominal diameter as the fitting, which would cause the same frictional loss as that caused by the fitting itself. The actual length of the pipe plus $L_e$ can be substituted for $L$ in

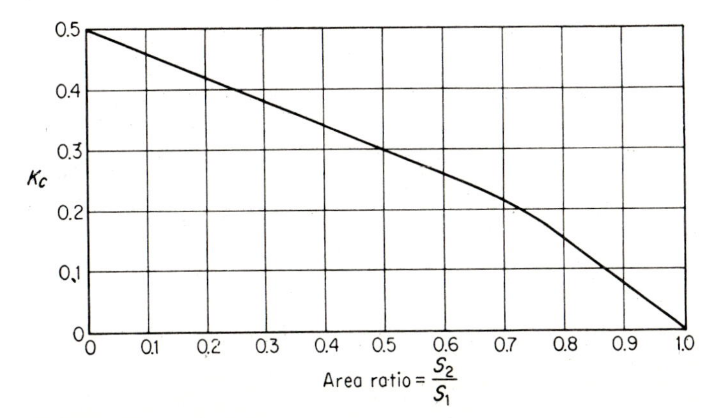


**Figure 5-2** Coefficients for contraction losses.

**Table 5-2 $L_e/D$ Ratios for standard pipe fittings**

| Fitting | $L_e/D$ |
|---|---|
| 90° elbows | 32 |
| 45° elbows | 15 |
| Square elbows (intersection of two cylinders) | 60 |
| Couplings | Negligible |
| Unions | Negligible |
| Gate valve, open | 7 |
| Globe valve, open | 300 |

Eq. (5-8) or (5-10) to give the total frictional loss due to the pipe plus the fittings.

Table 5-2 presents values of $L_e$ for various fittings in the region of turbulent flow. Since $L_e$ varies with the size of the fitting, the information in Table 5-2 is expressed as $L_e/D$. Thus, the value of $L_e$ for one 45° elbow in a 2-in inside-diameter line would be $(15)(2)/12 = 2.5$ ft or $(15)(2)(0.0254) = 0.762$ m, where 0.0254 is meters per inch.

## Some Practical Aspects of Frictional Effects

Frictional effects are extremely important in flow processes. In many cases, friction may be the main cause for resistance to the flow of a fluid through a given system. Consider the common example of water passing through a pipe. If no frictional effects were present, it would be possible to use pipes of very small diameters for all flow rates. Under these conditions, the pumping power costs for forcing 100,000 gal of water per hour through a pipe with a diameter of $\frac{1}{8}$ in would be the same as the power costs for forcing the same amount of water through a pipe of equal length having a diameter of 2 ft. However, frictional effects *are* present, and they must be taken into consideration when dealing with any real flow process.

The linear velocity of a fluid flowing through a pipe is inversely proportional to the cross-sectional area of the pipe. For a given mass rate of flow, the velocity of a fluid passing through a small-diameter pipe will be greater than the velocity of the same fluid passing through a different pipe of larger diameter.

According to the Fanning equation, the mechanical energy lost because of friction is approximately proportional to the square of the linear velocity of the fluid and inversely proportional to the pipe diameter. Consequently, for a given set of operating conditions, the frictional losses are increased if a large-diameter pipe is replaced by one having a smaller diameter. In order to maintain the same flow rate, more pumping power must be supplied to overcome the increased frictional losses. On the other hand, the purchase cost of the small-diameter pipe would be less than that of the larger-diameter pipe.

By balancing power costs against pipe costs, it is possible to determine the optimum pipe diameter where the total of all costs is a minimum for a given rate of fluid flow. This method is presented in detail in Chap. 14 (Chemical Engineering

Economics and Plant Design) and in App. B. In general, liquids of approximately the same density and viscosity as water should not flow in steel pipes at linear velocities higher than 4 or 5 ft/s (1.2 to 1.5 m/s), while vapors such as steam at ordinary pressures (20 to 50 lb/in$^2$ gage, or 240 to 450 kPa) should not flow in pipes at linear velocities higher than 50 to 60 ft/s (15 to 18 m/s).

The presence of valves and fittings in pipelines causes an increase in the amount of frictional flow resistance. An open globe valve causes about 40 times as much frictional loss as an open gate valve. Therefore, in cases where power considerations are important, a gate valve is preferable to a globe valve. If the purpose of the valve is to regulate the rate of fluid flow, it is better to accept the added frictional losses and use a globe valve instead of a gate valve, because a globe valve permits closer control of the flow rate.

Old pipes often become corroded and pitted, or scale may build up on the inside-wall surfaces. As a result, the equivalent effective roughness of the pipe is increased, and the values of $\epsilon$ given in Fig. 5-1 no longer apply. Therefore, frictional effects may increase as the pipe ages. The scale formation may also decrease the available flow area in a pipe, causing even higher frictional losses due to the increase in linear velocity of the fluid if mass flow is maintained at a constant rate.

## CONVERSION OF UNITS IN MECHANICAL-ENERGY BALANCE

The conventional units for the individual energy terms in the total energy balances are foot-pounds force per pound-mass of fluid or joules per kilogram of fluid. It is often desirable to convert these units to pressure-drop units or to standard power units. For example, the pressure drop due to friction in a steady-flow system of known fluid density can be calculated from the friction term $\Sigma F$. Assume the value of $\Sigma F$ has been calculated in the normal manner to be 176.0 ft · lbf/lbm (526.1 J/kg) for water with an average density of 62.3 lbm/ft$^3$ (998.0 kg/m$^3$). The pressure drop due to friction is then equal to the value of $\Sigma F$ multiplied by the average density, as shown in Example 5-4. With American engineering units,

$$\text{Pressure drop due to friction} = \Delta p = \frac{176.0\ \text{ft}\cdot\text{lbf}}{\text{lbm}}\left|\frac{62.3\ \text{lbm}}{\text{ft}^3}\right|\frac{\text{ft}^2}{144\ \text{in}^2}$$

$$= 76.1\ \text{lbf/in}^2$$

With SI units, since 1 J is 1 (N)(m),

$$\Delta p = \frac{526.1\ (\text{N})(\text{m})}{\text{kg}}\left|\frac{998.0\ \text{kg}}{\text{m}^3}\right. = 525{,}000\ \text{N/m}^2 = 525\ \text{kPa}$$

Many practical flow problems require the determination of the size of motor necessary to pump a liquid at a set rate through a given system. The power requirements of the motor can be determined by use of the total mechanical-energy balance. Assume the application of the total mechanical-energy balance over a system has indicated that the magnitude of the term $W_o$ (work supplied from an external source) is 157.0 ft · lbf/lbm (469.3 J/kg) of flowing fluid. This mechanical-energy balance is applied to an oil with a density of 45 lb/ft$^3$ (721 kg/m$^3$) flowing at a rate of 50 gal/min (0.189 m$^3$/min), and the pump and motor can be assumed to have an overall efficiency of 40 percent. The horsepower supplied to the pump motor to handle this flow may be calculated in this way:

The following equivalents can be found in App. C:

$$7.48 \text{ gal} = 1 \text{ ft}^3$$

$$1 \text{ hp} = 550 \text{ ft} \cdot \text{lbf/s}$$

The foot-pound force per second delivered to the fluid as $W_o$ can be determined by multiplying the mass rate of flow times the mechanical work supplied to the fluid in consistent units, or 157 ft · lbf/lbm is equivalent to

$$\frac{157 \text{ ft} \cdot \text{lbf}}{\text{lbm}} \left| \frac{50 \text{ gal}}{\text{min}} \right| \frac{\text{ft}^3}{7.48 \text{ gal}} \left| \frac{45 \text{ lb mass}}{\text{ft}^3} \right| \frac{\text{min}}{60 \text{ s}} = 787 \text{ ft} \cdot \text{lbf/s}$$

The horsepower delivered to the fluid is

$$\frac{787 \text{ ft} \cdot \text{lbf}}{\text{s}} \left| \frac{(\text{s})(\text{hp})}{550 \text{ ft} \cdot \text{lbf}} \right. = 1.43 \text{ hp}$$

Because the equipment is only 40 percent efficient, the total horsepower delivered to the motor, or the power requirement for the motor, is 1.43/0.40 = 3.57 hp delivered to pump motor.

Working the same example with SI units to determine the watts of power delivered to the pump motor, since 1 W is 1 J/s by App. A, the total number of watts as power delivered to the fluid is

$$\frac{469.3 \text{ J}}{\text{kg}} \left| \frac{0.189 \text{ m}^3}{\text{min}} \right| \frac{721 \text{ kg}}{\text{m}^3} \left| \frac{\text{min}}{60 \text{ s}} \right. = 1066 \text{ J/s or } 1066 \text{ W}$$

Because the equipment is only 40 percent efficient, the total number of watts as power requirement for the motor is 1066/0.40 = 2665 W delivered to pump motor. Since 1 hp is 745.7 W,

$$\frac{2665 \text{ W}}{} \left| \frac{\text{hp}}{745.7 \text{ W}} \right. = 3.57 \text{ hp}$$

in agreement with the results obtained with American engineering units.

**Example 5-5: Practical fluid-flow problem involving application of total mechanical-energy balance** Find the cost per hour of operating the pump motor in the system shown in Fig. 5-3. The efficiency of the pump-motor assembly is 45 percent. This includes losses at the entrance and exit of the pump housing. The cost for electrical energy is 1.5 cents/kWh. The water

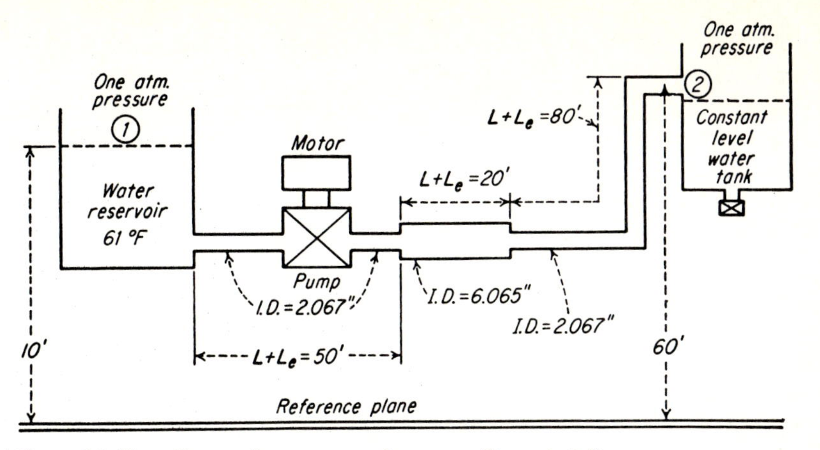


**Figure 5-3** Flow diagram for water-transfer system (Example 5-5).

flow is to be maintained at 150 gal/min, and the water temperature is constant at 61°F. The material of construction for the pipes is steel.

SOLUTION

**Basis**

1 lb of flowing water

Total mechanical-energy balance (between point 1 and point 2):

$$Z_1 \frac{g}{g_c} + p_1 v_1 + \frac{V_1^2}{2\alpha_1 g_c} + \int_1^2 p\, dv + W_o = Z_2 \frac{g}{g_c} + p_2 v_2 + \frac{V_2^2}{2\alpha_2 g_c} + \Sigma F + \Sigma F_e + \Sigma F_c$$

Density of water at 61°F = 62.3 lb/ft$^3$

Viscosity of water at 61°F = 1.12 cP = (1.12)(0.000672) lb/(s)(ft)

Conversion factors:

$$7.48 \text{ gal} = 1 \text{ ft}^3$$

$$60 \text{ s} = 1 \text{ min}$$

$$60 \text{ min} = 1 \text{ h}$$

$$1 \text{ kW} = 738 \text{ ft} \cdot \text{lbf/s}$$

Velocity in 2-in pipe is

| 150 gal | min | ft$^3$ | | 144 in$^2$ |
|---|---|---|---|---|
| min | 60 s | 7.48 gal | $(2.067/2)^2\pi$ in$^2$ cross-sectional area | ft$^2$ |

= 14.34 ft/s

Velocity in 6-in pipe is

$$\frac{(150)(144)}{(60)(7.48)(6.065/2)^2\pi} = 1.67 \text{ ft/s.}$$

Reynolds number in 2-in pipe $= DV\rho/\mu$ or

$$\frac{2.067 \text{ in}}{} \left| \frac{\text{ft}}{12 \text{ in}} \right| \frac{14.34 \text{ ft}}{\text{s}} \left| \frac{62.3 \text{ lb}}{\text{ft}^3} \right| \frac{}{1.12 \text{ cP}} \left| \frac{(\text{ft})(\text{s})(\text{cP})}{0.000672 \text{ lb}} \right. = 204{,}000$$

$$\frac{\epsilon}{D} = \frac{(0.00015)(12)}{2.067} = 0.00087$$

Friction factor for 2-in pipe $= f = 0.0051$
Total equivalent length for 2-in pipe $= 50 + 80 = 130$ ft
Reynolds number in 6-in pipe is

$$\frac{DV\rho}{\mu} = \frac{(6.065)(1.67)(62.3)}{(12)(0.000672)(1.12)} = 70{,}000$$

$$\frac{\epsilon}{D} = \frac{(0.00015)(12)}{6.065} = 0.0003$$

Friction factor for 6-in pipe $= f = 0.0051$
Total equivalent length for 6-in pipe $= 20$ ft

$$\Sigma F = \Sigma 2f \frac{V^2L}{g_cD} = \frac{(2)(0.0051)(14.34)^2(12)(130)}{(32.17)(2.067)} + \frac{(2)(0.0051)(1.67)^2(12)(20)}{(32.17)(6.065)}$$

$$= 49.3 \text{ ft} \cdot \text{lbf/lbm}$$

Flow in the system is turbulent; so $\alpha$ is unity in all cases.
From reservoir to 2-in pipe,

$$F_c = \frac{K_cV_2^2}{2g_c} = \frac{0.5}{2} \left| \frac{(14.34)^2 \text{ ft}^2}{\text{s}^2} \right| \frac{(\text{s}^2)(\text{lbf})}{32.17 \text{ ft} \cdot \text{lbm}} = 1.6 \text{ ft} \cdot \text{lbf/lbm}$$

From 6-in pipe to 2-in pipe,

$$F_c = \frac{K_cV_2^2}{2g_c} = \frac{(0.46)(14.34)^2}{(2)(32.17)} = 1.47 \text{ ft} \cdot \text{lbf/lbm}$$

$$\Sigma F_c = 1.6 + 1.47 = 3.07 \text{ ft} \cdot \text{lbf/lbm}$$

From 2-in pipe to 6-in pipe,

$$F_e = \frac{(V_1 - V_2)^2}{2g_c} = \frac{(14.34 - 1.67)^2}{(2)(32.17)} = 2.49 \text{ ft} \cdot \text{lbf/lbm}$$

From 2-in pipe to constant-level tank,

$$F_e = \frac{(V_1 - V_2)^2}{2g_c} = \frac{(14.34 - 0)^2}{(2)(32.17)} = 3.2 \text{ ft·lbf/lbm}$$

$$\Sigma F_e = 2.49 + 3.20 = 5.69 \text{ ft · lbf/lbm}$$

$$p_1 = p_2 = \text{atmospheric pressure} = (14.7)(144) \text{ lbf/ft}^2$$

$$v_1 = v_2 = \frac{1}{62.3} \text{ ft}^3\text{/lbm}$$

$$Z_1 \frac{g}{g_c} = 10 \text{ ft · lbf/lbm}$$

$$Z_2 \frac{g}{g_c} = 60 \text{ ft · lbf/lbm}$$

$$\int_1^2 p \, dv = 0 \text{ since water is a noncompressible liquid}$$

$$V_1 = V_2 = 0 \text{ since reservoir and constant-level tank are very large}$$

The total mechanical energy balance can now be written as

$$10 + \frac{(14.7)(144)}{(62.3)} + 0 + 0 + W_o = 60 + \frac{(14.7)(144)}{(62.3)} + 49.3 + 3.07 + 5.69$$

$$W_0 = 108.10 \text{ ft · lbf/lbm}, \quad \text{or } \frac{108.1}{0.45} \text{ ft · lbf/lbm}$$

supplied to the pump motor.

Cost per hour to operate pump is

$$\frac{(108.1)}{(0.45)} \left| \frac{(150)}{} \right| \frac{}{(7.48)} \left| \frac{(62.3)}{} \right| \frac{}{(60)} \left| \frac{}{(738)} \right| \frac{(1)}{} \left| \frac{(1.5)}{} \right. = 10 \text{ cents/h}$$

A unit analysis indicates:

$$\frac{(\text{ft · lbf})}{(\text{lbm})} \left| \frac{(\text{gal})}{(\text{min})} \right| \frac{(\text{ft}^3)}{(\text{gal})} \left| \frac{(\text{lbm})}{(\text{ft}^3)} \right| \frac{(\text{min})}{(\text{s})} \left| \frac{(\text{s})(\text{kW})}{(\text{ft · lbf})} \right| \frac{(\text{h})}{(\text{h})} \left| \frac{(\text{cents})}{(\text{kWh})} \right. = \text{cents/h}$$

*Note*: Solution of the preceding problem using SI units would have involved using a basis of 1 kg of flowing water, with all of the $g_c$ values in the equations being 1.0. With the appropriate SI units, the energy terms in the total mechanical-energy balance equation would have had base units of meters squared per second squared, which are the same as

$$\frac{(\text{m})(\text{kg})}{\text{s}^2} \left| \frac{\text{m}}{\text{kg}} \right. \quad \text{or} \quad (\text{N})(\text{m})/\text{kg} \quad \text{or} \quad \text{J/kg}$$

Conversion from joules per kilogram to cents per hour as cost for operating the pump motor would have been accomplished by the same procedure as was used to convert foot-pound force per pound-mass to cents per hour. This final conversion with SI units being used is shown in the following:

$W_o$ supplied to the pump motor (equivalent to 108.1/0.45 ft · lbf/lbm) = 718.0 J/kg

150 gal/min = 0.009464 $m^3/s$

62.3 $lb/ft^3$ = 998.0 $kg/m^3$

1 W = 1 J/s

$$\begin{array}{c|c|c} 718.0\text{ J} & 0.009464\text{ m}^3 & 998.0\text{ kg} \\ \hline \text{kg} & \text{s} & \text{m}^3 \end{array} = 6782\text{ J/s} = 6782\text{ W}$$

$$\begin{array}{c|c|c|c} 6782\text{ W} & \text{kW} & 1\text{ h} & 1.5\text{ cents} \\ \hline & 1000\text{ W} & 1\text{ h} & \text{kWh} \end{array} = 10\text{ cents/h}$$

## MEASUREMENT OF FLOW RATE

Many different types of commercial equipment are available for determining the rate at which fluids are flowing. Several of the more important methods for measuring flow rates will be discussed here, and the theory behind these methods will be presented.

### Orifice Meter

When a fluid, flowing in a system at a uniform mass rate, is forced to pass through a constriction, the average linear velocity at the constriction must increase. This means the kinetic energy of the fluid must increase at the expense of the *pv* mechanical energy. As a result, the static pressure of the fluid must decrease when the fluid passes through a constriction. This may be visualized by reference to Fig. 5-4, which is a schematic diagram of a fluid passing through a constriction.

A constriction, such as the one shown in Fig. 5-4, is known as an *orifice*. The pressure drop across the orifice can be determined by a manometer attached so that the difference in the static pressures is indicated by the height of a liquid of known density in the manometer.

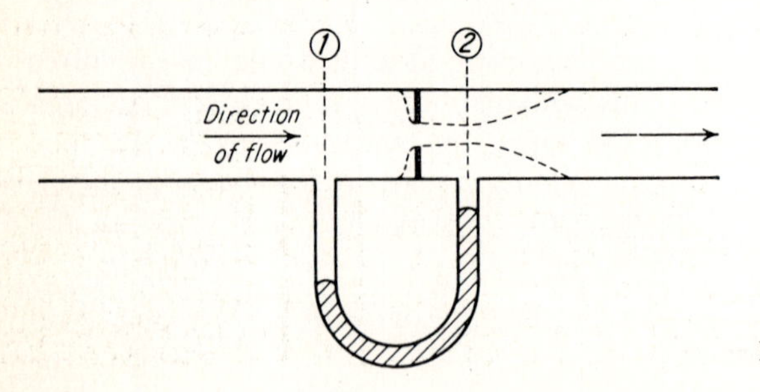


**Figure 5-4** Orifice meter.

A jet issuing from a constriction may contract even farther into a smaller cross section. The point where the cross-sectional flow area is a minimum is known as the *vena contracta*. In Fig. 5-4, point 2 is located at the vena contracta. Theoretically, the downstream tap for the orifice manometer should be located as close as possible to the cross-sectional plane of the vena contracta. This can be accomplished, for all practical purposes, by placing the downstream manometer tap at a distance of three-tenths of the pipe diameter past the orifice. The upstream manometer tap should be located at a distance of approximately one pipe diameter from the upstream side of the orifice plate.

The pressure drop across the orifice increases for a given fluid as the rate of fluid flow increases. Thus, the magnitude of the pressure drop shown by the manometer gives an indication of the rate at which the fluid is flowing in the pipe. An equation expressing the flow rate in terms of the other pertinent variables may be obtained from the total mechanical-energy balance.

By applying the total mechanical-energy balance to fluids of essentially constant density between points 1 and 2, as indicated in Fig. 5-4, the following equation can be derived:

$$V_o^2 - V_1^2 = C^2 2g_c v(p_1 - p_2) \tag{5-13}$$

where $V_o$ = velocity of fluid at orifice, m/s or ft/s
$V_1$ = velocity of fluid in upstream section, m/s or ft/s
$C$ = coefficient of discharge, dimensionless
$p_1$ = upstream static pressure, Pa, $N/m^2$, or $lb/ft^2$
$p_2$ = static pressure at vena contracta, Pa, $N/m^2$, or $lb/ft^2$
$v$ = average volume of fluid per unit mass, $m^3/kg$ or $ft^3/lb$

As in the previous equations presented, $g_c$ is 32.17 ft · lbm/(s²)(lbf) if American engineering units are used and 1.0 if SI units are used. The value of the coefficient of discharge $C$ depends on the type of fluid flow involved and therefore depends on the value of the Reynolds number. It also depends on the ratio of the cross-sectional flow area at the vena contracta to the cross-sectional area of the orifice opening and to the cross-sectional area of the upstream pipe section.

Since most orifice and venturi meters are operated in a horizontal position, the potential-energy term $Z$ does not appear in Eq. (5-13). This equation assumes uniform velocity distribution across the path of fluid flow. Any error introduced by this assumption is absorbed in the coefficient of discharge. Frictional effects are also included in the coefficient of discharge.

The value of $v(p_1 - p_2)$ can be determined from the reading of a manometer attached across the orifice. This value is commonly indicated by $H_v$, the difference in static head between the upstream and vena-contracta sections based on the density of the fluid flowing.

By a material balance, the mass of fluid per second flowing through the orifice must equal the mass of fluid per second flowing through the pipe; therefore

$$V_o S_o = V_1 S_1 \tag{5-14}$$

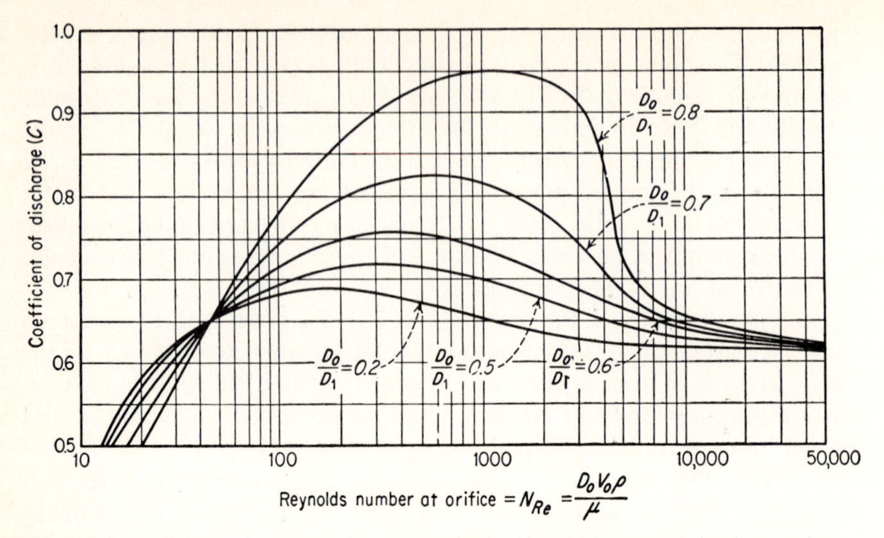


**Figure 5-5** Coefficients of discharge for square-edged orifices with centered circular opening.

where $S_o$ and $S_1$ are the cross-sectional areas of the orifice and the pipe, respectively. Using these definitions, Eq. (5-13) can be rewritten as

$$V_o = \frac{C\sqrt{2g_c H_v}}{\sqrt{1 - (S_o^2/S_1^2)}} \tag{5-15}$$

Figure 5-5 gives values of $C$ for various orifice Reynolds numbers at different $D_o/D_1$ ratios. When the Reynolds number at the orifice exceeds 30,000, $C$ may be considered as constant at 0.61. The values of $C$ given in Fig. 5-5 apply when the downstream manometer tap is located at the vena contracta. In commercial practice, the downstream tap is often located at the orifice plate. The result of this change is a slight increase in the value of the coefficient of discharge. However, within an error of 1.5 percent, the values of $C$ determined from Fig. 5-5 may be used for downstream manometer taps located anywhere between the orifice plate and the vena contracta.

In engineering calculations, Eq. (5-15) is applied to compressible fluids such as air only if the indicated pressure drop over the orifice is less than 5 percent of the upstream pressure. For accurate measurements of flow rates, it is usually advisable to determine the values of $C$ by direct calibration of the orifice meter. In this manner, errors introduced in the Fig. 5-5 values of $C$ as a result of manometer-tap location or velocity distribution are eliminated.

**Permanent pressure loss due to an orifice.** Because of the sudden constriction followed by a rapid expansion, there is usually an appreciable loss of mechanical energy due to friction when an orifice is used. Figure 5-6 indicates the percent of

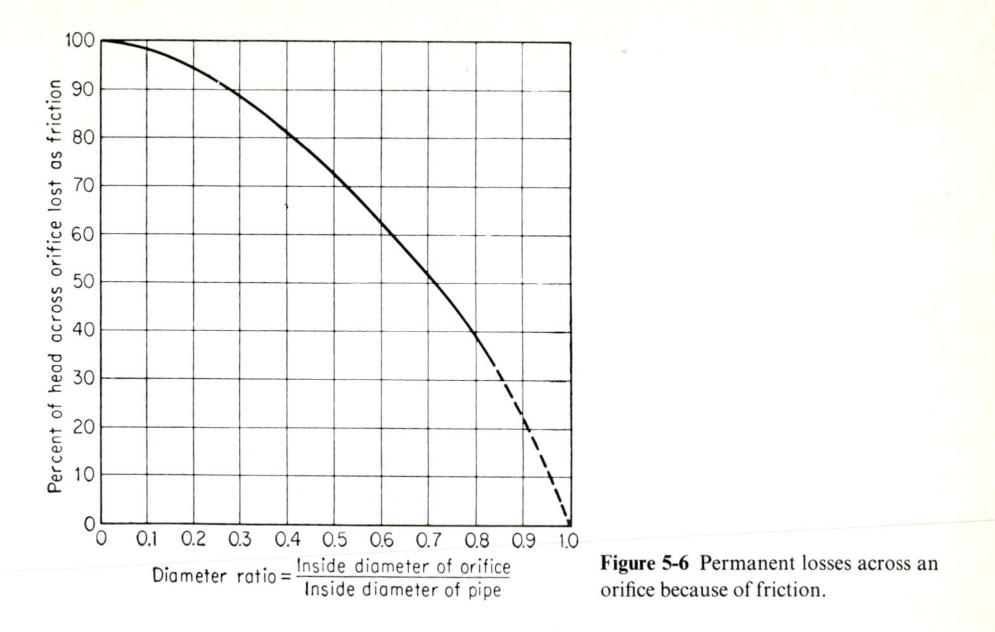


**Figure 5-6** Permanent losses across an orifice because of friction.

permanent head loss across an orifice for different ratios of orifice diameter to pipe diameter.

**Example 5-6: Application of orifice equations for flow-rate determinations** Water with a density of 62.4 lb/ft$^3$ (1000 kg/m$^3$) is flowing through a pipe having an inside diameter of 15 in (0.381 m). An orifice with a 3-in (0.0762-m) diameter opening is placed concentrically in the pipe. A mercury manometer in which the mercury and the flowing water are in contact is attached across the orifice. The manometer indicates a difference in level of 3.0 inHg (0.0762 mHg). The viscosity of the water is 1.12 cP. What is the flow rate of water in cubic feet per minute, and what is the permanent pressure loss across the orifice in pounds per square inch? Solve the problem using both American engineering units and SI units. The specific gravity of mercury is 13.6.

SOLUTION As a first approximation, assume the orifice Reynolds number is such that $C = 0.61$.

$$S_o = (\tfrac{3}{2})^2\,\pi \text{ in}^2 = (0.0762/2)^2\,\pi \text{ m}^2$$
$$S_1 = (\tfrac{15}{2})^2\,\pi \text{ in}^2 = (0.381/2)^2\,\pi \text{ m}^2$$

$$1 - \frac{S_o^2}{S_1^2} = \sqrt{1 - \frac{(3)^4}{(15)^4}} = \sqrt{1 - \frac{(0.0762)^4}{(0.381)^4}} = 1.0$$

Density of mercury = (13.6)(62.4) lb/ft$^3$ = (13.6)(1000) kg/m$^3$
Density of water = 62.4 lb/ft$^3$ = 1000 kg/m$^3$

In American engineering units,[12]

$$H_v = v(p_1 - p_2) = \frac{1}{62.4}\left[\frac{(13.6)(62.4)(3)}{12} - \frac{(62.4)(3)}{12}\right]\frac{g}{g_c} = 3.15\,\frac{\text{ft}\cdot\text{lbf}}{\text{lbm}}$$

The unit analysis is

$$\frac{\text{ft}^3}{\text{lbm}}\left|\frac{\text{lbm}}{\text{ft}^3}\right|\frac{\text{in fluid head}}{}\left|\frac{\text{ft}}{\text{in}}\right|\frac{g = 32.17\ \text{ft/s}^2}{g_c = 32.17\ \text{ft}\cdot\text{lbm/(s}^2)(\text{lbf})}$$

$$= \frac{\text{ft}\cdot\text{lbf}}{\text{lbm}} = \text{ftH}_2\text{O net fluid head}$$

[12] The method for determination of this $p_1 - p_2$ may be visualized by considering that the pressure at the base of the manometer (point $A$) in Fig. 5-7 is the same in all directions. Then, from the concept

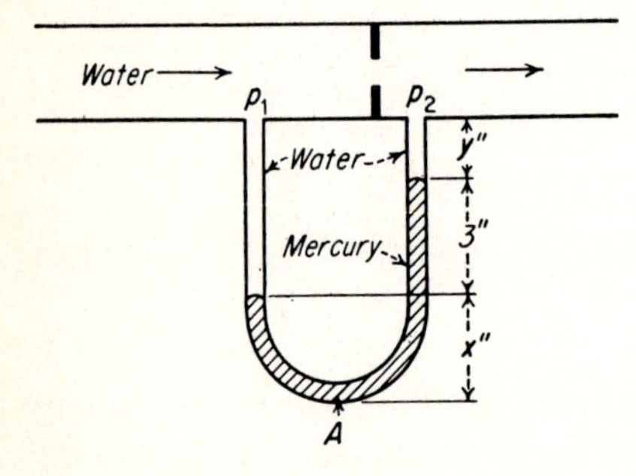


**Figure 5-7** Manometer connected across an orifice (Example 5-6).

of fluid head, the pressure exerted at point $A$ due to the total head plus the top-exerted pressure must be equal for each of the manometer legs, or, using $p$ as pound-force per square foot,

$$p_1 + \frac{(y+3)\ \text{in}}{}\left|\frac{\text{ft}}{12\ \text{in}}\right|\frac{62.4\ \text{lbm}}{\text{ft}^3}\left|\frac{g = 32.17\ \text{ft/s}^2}{g_c = 32.17\ \text{ft}\cdot\text{lbm/(s}^2)(\text{lbf})}\right.$$

$$+ \frac{x}{12}(62.4)(13.6)\frac{g}{g_c} = p_2 + \frac{y}{12}(62.4)\frac{g}{g_c} + \frac{3}{12}(62.4)(13.6)\frac{g}{g_c}$$

$$+ \frac{x}{12}(62.4)(13.6)\frac{g}{g_c}$$

The final result is

$$p_1 - p_2 = \left[\frac{3}{12}(62.4)(13.6) - \frac{3}{12}(62.4)\right]\frac{g}{g_c} = 197\ \text{lbf/ft}^2$$

With SI units, each of the head terms is expressed as

$$\frac{\text{Head (m)}}{}\left|\frac{\rho\ \text{kg}}{\text{m}^3}\right|\frac{g = 9.8\ \text{m/s}^2}{} = \frac{\text{kg}}{(\text{m})(\text{s}^2)} = \frac{(\text{m})(\text{kg})}{(\text{s}^2)}\left(\frac{1}{\text{m}^2}\right) = \frac{N}{\text{m}^2} = \text{Pa}$$

The final result of repeating the full head and pressure balance at point $A$ with SI units is

$$p_1 - p_2 = (0.0762)(1000)(13.6)(9.8) - (0.0762)(1000)(9.8) = 9410\ \text{Pa}$$

In SI units, with $g = 9.80 \text{ m/s}^2$,

$$H_v = v(p_1 - p_2) = \frac{\text{m}^3}{1000 \text{ kg}} [(13.6)(1000)(0.0762) - (1000)(0.0762)]$$

$$\times \frac{(\text{kg})(\text{m})}{\text{m}^3} \left| \frac{9.8 \text{ m}}{\text{s}^2} \right.$$

$$= 9.41 \frac{\text{m}^2}{\text{s}^2} = 9.41 \frac{(\text{m})(\text{kg})}{(\text{s}^2)} \left(\frac{(\text{m})}{(\text{kg})}\right) = 9.41 \frac{(\text{N})(\text{m})}{\text{kg}} = 9.41 \text{ J/kg}$$

In American engineering units,

$$V_o = C\sqrt{2g_c H_v} = 0.61\sqrt{(2)(32.17)(3.15)} = 8.68 \text{ ft/s}$$

In SI units,

$$V_o = C\sqrt{2H_v} = 0.61\sqrt{(2)(9.41)} = 2.65 \text{ m/s}$$

Reynolds number at orifice $= DV\rho/\mu$

$$\frac{3 \text{ in}}{} \left| \frac{\text{ft}}{12 \text{ in}} \right| \frac{8.68 \text{ ft}}{\text{s}} \left| \frac{62.4 \text{ lb}}{\text{ft}^3} \right| \frac{}{1.12 \text{ cP}} \left| \frac{(\text{ft})(\text{s})(\text{cP})}{0.000672 \text{ lb}} \right. = 180{,}000 \text{ (dimensionless)}$$

$$\frac{0.0762 \text{ m}}{} \left| \frac{2.65 \text{ m}}{\text{s}} \right| \frac{1000 \text{ kg}}{\text{m}^3} \left| \frac{}{1.12 \text{ cP}} \right| \frac{(\text{m})(\text{s})(\text{cP})}{0.001 \text{ kg}} = 180{,}000 \text{ (dimensionless)}$$

From Fig. 5-5, $C = 0.61$; so the assumption of $C = 0.61$ was correct.

$$\text{Water-flow rate} = V_o S_o = \frac{8.68 \text{ ft}}{\text{s}} \left| \frac{(\frac{3}{2})^2 \pi \text{ in}^2}{} \right| \frac{\text{ft}^2}{144 \text{ in}^2} \left| \frac{60 \text{ s}}{\text{min}} \right. = 25.6 \text{ ft}^3/\text{min}$$

By SI units,

$$\text{Water-flow rate} = V_o S_o = \frac{2.65 \text{ m}}{\text{s}} \left| \frac{(0.0762/2)^2 \pi \text{ m}^2}{} \right| \frac{60 \text{ s}}{\text{min}} \left| \frac{\text{ft}^3}{0.02832 \text{ m}^3} \right.$$

$$= 25.6 \text{ ft}^3/\text{min}$$

$$\frac{D_o}{D_1} = \frac{3 \text{ in}}{15 \text{ in}} = \frac{0.0762 \text{ m}}{0.381 \text{ m}} = 0.20$$

From Fig. 5-6, the permanent pressure loss over the orifice equals 95 percent of the head, or 95 percent of the actual pressure drop. Permanent pressure loss across the orifice is

$$(p_1 - p_2)(0.95) = \frac{3.15 \text{ ft} \cdot \text{lbf}}{\text{lbm}} \left| \frac{62.4 \text{ lbm}}{\text{ft}^3} \right| \frac{\text{ft}^2}{144 \text{ in}^2} \left| \frac{0.95}{} \right. = 1.3 \text{ lbf/in}^2$$

By SI units, permanent pressure loss across the orifice is

$$(p_1 - p_2)(0.95) = \frac{9.41\ (\text{N})(\text{m}) \mid 1000\ \text{kg} \mid 0.95}{\text{kg} \mid \text{m}^3 \mid} = 8936\ \frac{\text{N}}{\text{m}^2} = 8936\ \text{Pa}$$

$$= 8.936\ \text{kPa}$$

$$= \frac{8.936\ \text{kPa} \mid \text{lbf/in}^2}{\mid 6.895\ \text{kPa}} = 1.3\ \text{lbf/in}^2$$

## Venturi Meter

A *venturi meter* is similar to an orifice meter, but the contraction and expansion sections in a venturi are gradual because of a double-cone-shaped construction. As a result of the gradual velocity changes, the permanent pressure loss is much less with a venturi meter than with an orifice meter.

The contracting portion of a venturi meter is a frustum of a cone with a vertex angle of about 30°. The divergent portion is also cone-shaped with an angle of approximately 7°. The throat diameter is one-half to one-quarter of the upstream pipe diameter, while the actual throat length should be about one-half of the upstream pipe diameter.

The same equations are used for calculations with venturi meters as are used for orifice meters. The coefficient of discharge for a venturi may be taken as 0.98 for all values of Reynolds numbers higher than 5000 (based on conditions in upstream section).

A correctly designed venturi meter should have an overall pressure loss of about 10 percent of the static-head difference reading. A schematic diagram of a venturi meter is shown in Fig. 5-8.

Venturi and orifice meters are widely used in industry for measuring liquid- and gas-flow rates. The equations developed in the preceding sections may be used to estimate flow rates through these installations; however, better accuracy is obtained if the meters are calibrated after installation. In many cases, venturi meters and orifice meters are sold by the manufacturers with accompanying calibration charts. For these charts to be applicable, the meters must be installed carefully according to the manufacturer's instructions. Sometimes corrosion or scaling will change the size of the orifice or venturi openings, and meters of this

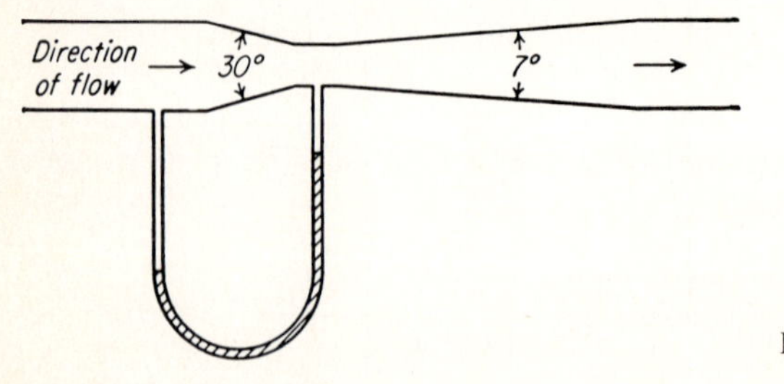


**Figure 5-8** Venturi meter.

type must be checked periodically to make certain the original calibrations are still valid.

## Pitot Tube

A *pitot tube* is a device for measuring local or point velocities. It measures the velocity pressure of a flowing fluid by indicating the difference between the impact pressure and the static pressure of the fluid. Figure 5-9 is a schematic diagram of a pitot tube.

The velocity of the fluid at the point where the tip of the pitot tube is located can be calculated by the equation,

$$V_{\text{point}} = \sqrt{2g_c H_v} \tag{5-16}$$

where $H_v$ is the head equivalent to the specific volume of the flowing fluid times the difference between the impact and static pressures; $g_c$ becomes unity if SI units are used.

While the orifice meter and venturi meter may be used to determine the total rate of flow of a fluid through a pipe, the pitot tube is used to find the rate of flow at just one point in the cross-sectional area of the flowing stream. To give a minimum of flow interference, the size of the pitot tube should be very small relative to the cross-sectional area of the container in which the fluid is flowing. Care must be taken to have the impact opening facing squarely upstream.

Instruments based on the pitot-tube principle are used for measuring airplane speeds. Corrections for variation in air density and prevailing winds must be applied to the pitot tube readings if the true velocity of the airplane is to be indicated.

## Rotameter

The rotameter is one of the most common devices used for the measurement of fluid-flow rates. It is, essentially, an orifice of variable area and constant pressure drop. A typical rotameter, such as the one illustrated in Fig. 5-10, consists of a plummet (or rotor) which is free to rise and fall in a tapered calibrated tube. The plummet is usually grooved so the flow of fluid past it causes it to rotate. Fluid flows into the lower end of the rotameter tube and causes the plummet to rotate and rise

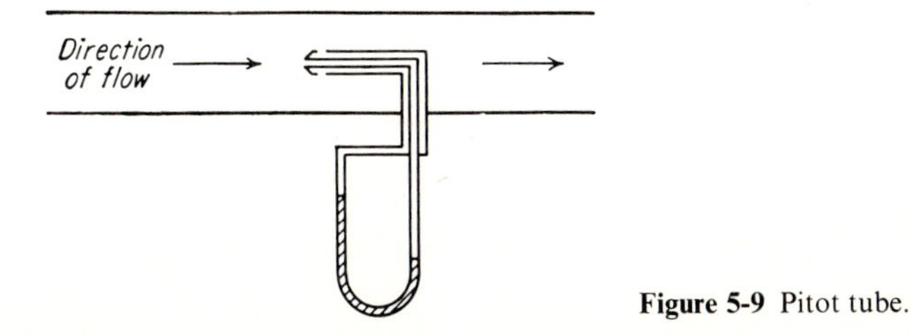


**Figure 5-9** Pitot tube.

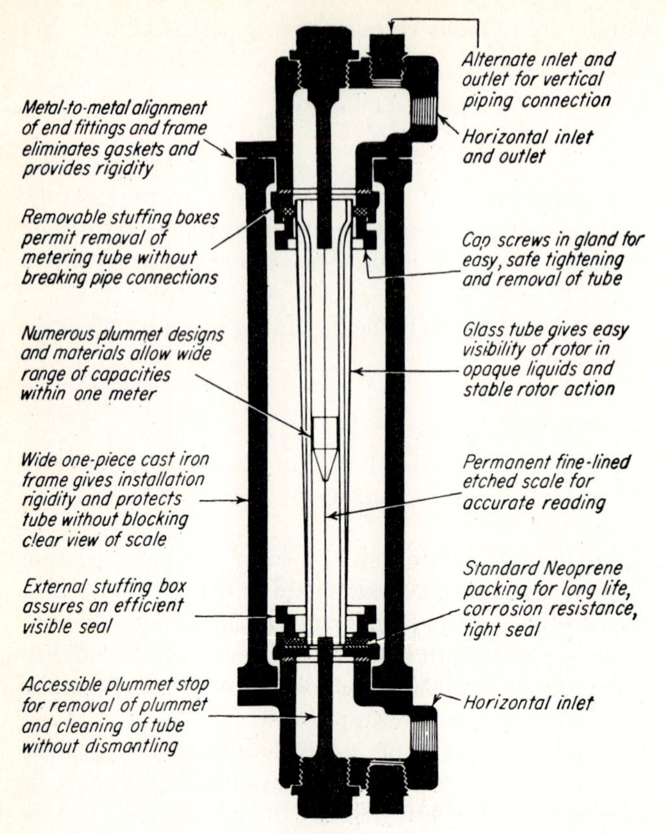


**Figure 5-10** Cutaway view of standard rotameter. (*Courtesy of Schutte & Koerting Co.*)

smoothly until the pressure drop across the space between the tube wall and the plummet is just sufficient to support the effective weight of the plummet.

The downward force of the plummet is due to its weight minus the buoyant effect of the liquid it displaces. This may be expressed in equation form[13] as

$$\text{Downward force of plummet} = V_P(\rho_P - \rho)\frac{g}{g_c} \qquad \text{N or lbf} \tag{5-17}$$

$$\text{Upward force on plummet} = S_P(p_1 - p_2) \qquad \text{N or lbf} \tag{5-18}$$

where $V_P$ = volume of plummet, $m^3$ or $ft^3$
$S_P$ = maximum cross-sectional area of plummet, $m^2$ or $ft^2$
$\rho_P$ = density of plummet, $kg/m^3$ or $lbm/ft^3$
$\rho$ = density of fluid, $kg/m^3$ or $lbm/ft^3$
$p_1 - p_2$ = pressure drop from bottom to top of plummet, Pa or $lbf/ft^2$

[13] When SI units are used, $g_c = 1.0$.

When the plummet is in a stationary position, the downward force must equal the upward force. Therefore, from Eqs. (5-17) and (5-18),

$$p_1 - p_2 = \frac{V_P(\rho_P - \rho)g}{S_P g_c} \tag{5-19}$$

Since the orifice equations are applicable to a rotameter, Eq. (5-13) may be expressed as

$$V_o^2 - V_1^2 = C^2 2g_c v \frac{V_P(\rho_P - \rho)g}{S_P g_c} \tag{5-20}$$

where $V_o$ = velocity of fluid at orifice or (in this case) velocity of fluid past plummet at its maximum area, m/s or ft/s

$V_1$ = velocity of fluid in rotameter tube just below plummet, m/s or ft/s

$v$ = average volume of fluid per unit mass = $1/\rho$, $m^3/kg$ or $ft^3/lbm$

By a material balance,

$$V_o S_o = V_1 S_1 \tag{5-21}$$

$$V_o S_o = q = \text{total volumetric rate of flow, } m^3/s \text{ or } ft^3/s \tag{5-22}$$

where $S_o$ = free cross-sectional area between plummet and rotameter tube wall at maximum area of plummet, $m^2$ or $ft^2$

$S_1$ = cross-sectional inside area of rotameter tube just below plummet, $m^2$ or $ft^2$

Combining Eqs. (5-20), (5-21), and (5-22),

$$V_o S_o = q = S_o C \sqrt{\frac{V_P 2g(\rho_P - \rho)}{S_P[1 - (S_o/S_1)^2]\rho}} \tag{5-23}$$

Equation (5-23) is the general equation for rotameters. Values for $S_o$ and $S_1$ must be determined at each plummet height. The value of $C$ depends on the plummet shape and is a function of the Reynolds number. Curves are available in the literature giving numerical values of $C$ under various conditions.

Industrial rotameters are ordinarily sold with accompanying calibration curves or with special calibrations etched on the indicating tube. If the rotameter is to be used with liquids, the calibration is usually based on water with a density of 62.4 $lb/ft^3$. The calibrations may also be based on the density of air at 70°F and a pressure of 1 atm. The readings from the calibration curves or etched markings may be converted to other fluid densities by applying Eq. (5-23).

As an example, assume a rotameter gives a scale reading of 48 when a mixture of methanol and water weighing 58.0 $lb/ft^3$ passes through it. The calibration curve supplied with the rotameter indicates that 4.8 $ft^3$ of water (density = 62.4 $lb/ft^3$) passes through the rotameter per minute when the scale reading is 48. The plummet is made of stainless steel with a density of 497 $lb/ft^3$. Since the density and viscosity of the methanol-water mixture are approximately the same as for pure water, the value of $C$ is essentially the same for both liquids.

The volumetric flow rate for the methanol-water mixture can now be found by applying Eq. (5-23). For pure-water flow at a scale reading of 48,

$$4.8 = (60)(S_o)(C)\sqrt{\frac{V_P 2g}{S_P[1 - (S_o/S_1)^2]}}\sqrt{\frac{497 - 62.4}{62.4}} \text{ ft}^3\text{/min} \tag{5-24}$$

For the methanol-water flow at a scale reading of 48,

$$q = (60)(S_o)(C)\sqrt{\frac{V_P 2g}{S_P[1 - (S_o/S_1)^2]}}\sqrt{\frac{497 - 58.0}{58.0}} \text{ ft}^3\text{/min} \tag{5-25}$$

Dividing Eq. (5-25) by Eq. (5-24) and canceling the constant values,

$$q = 4.8\sqrt{\frac{(497 - 58.0)(62.4)}{(58.0)(497 - 62.4)}}$$

$$= 5.0 \text{ ft}^3 \text{ of methanol-water mixture per min}$$

## FLUID FLOW IN NONCIRCULAR SECTIONS

Most of the discussion in this chapter has been confined to fluids flowing in pipes of circular cross section. When the cross section is not circular, it is possible to use many of the relationships presented in this chapter by substituting an equivalent diameter for the circular-section diameter.

When a fluid flows through a cross section of any uniform shape, there is a certain perimeter of the container which is in contact with the fluid. This wetted perimeter, taken normal to the direction of flow, is designated by $\psi$. If the cross-sectional area is indicated by $S$, the hydraulic radius $R_H$, defined as the ratio of the cross-sectional area to the wetted perimeter, will be $S/\psi$.

The assumptions made in the development of all the relations presented in this chapter for turbulent flow permit the substitution of four times the hydraulic radius in place of the circular cross-sectional diameter. Thus, when the flow is turbulent, $4R_H$ may be substituted for $D$ with little loss in accuracy as long as the cross section is uniform along the length of conduit considered. When the flow is streamline, the substitution of $4R_H$ for $D$ does not give accurate results, and special methods available in the literature should be used for noncircular cross sections.

## PUMPS

It is necessary to have a driving force to make a fluid flow from one point to another. Sometimes this driving force is supplied by gravity or by the difference in elevation between the entrance and exit of the system. Usually the driving force is supplied by a mechanical device such as a pump.

The pump must be capable of supplying sufficient power to the system to overcome any energy losses incurred through friction, change in head, change in kinetic energy, or changes in pressure-volume energy terms. The energy balances presented in this chapter may be used to determine the amount of outside mechanical energy necessary to carry out a given operation. This mechanical energy added from an outside source represents the amount of energy which must be added to the system, but it is not directly equivalent to the amount of power put into the mechanical-energy source. For example, if calculations indicate that mechanical energy equivalent to 4.0 hp must be supplied to a system to pump a constant amount of water to an overhead tank, this does not mean that a 4.0-hp motor on the pump would be satisfactory. Some energy is lost through shaft friction between the motor and the actual pump, and mechanical energy is lost because of friction

**Table 5-3 Form for pumping-equipment specifications***

- General
  - Service
  - Number of pumps required
    - in regular use
    - as spares
- Process requirements per pump
  - Fluid handled
  - Quality of fluid
    - Corrosive or noncorrosive
    - Corrosive compounds
    - Solids (if any)
  - Quantity handled, gal/min at 60°F and 760 mmHg
  - Specific gravity at 60°F and 760 mmHg
  - Pumping temp, °F
  - Viscosity, cP at pumping temp
- Suction conditions (at pump)
  - Pressure, lb/in$^2$, gage or absolute
  - Vapor pressure at pumping temp, lb/in$^2$, gage or absolute
  - Specific gravity at pumping temp and suction pressure
  - Net positive head above vapor pressure, ft
- Discharge conditions (at pump)
  - Pressure, lb/in$^2$, gage or absolute
  - Specific gravity at pumping temp and suction pressure
  - Capacity at discharge conditions, gal/min
- Design conditions
  - Differential pressure
    - lb/in$^2$, gage or absolute
    - ft
  - Liquid horsepower
  - Maximum temp, °F
  - Maximum suction pressure, lb/in$^2$, gage or absolute
  - Maximum allowable pump $\Delta P$, lb/in$^2$
- Remarks
  - Special requirements of pumps or drivers, etc.
  - Type of pump recommended
  - Type of driver recommended
    - Regular
    - Spare

* American engineering units are shown in this table. SI units may be required in some cases for specifications.

in the pump housing. Therefore, the pump-and-motor efficiency must be taken into consideration. If the overall pump-and-motor efficiency is 80 percent, a 5.0-hp motor would be necessary to supply 4.0 hp to the actual flow system.

There are three types of pumps ordinarily employed in chemical processes. They are (1) rotary pumps, (2) reciprocating pumps, and (3) centrifugal pumps. These three types of pumps have been discussed in Chap. 4 (Industrial Chemical Engineering Equipment).

Engineers have many occasions to list specifications for pumps, and it is important to present all the crucial facts when making these specifications. Table 5-3 presents a standard form showing the essential information which should be supplied when specifying pumping equipment.

## WATER HAMMER

If a liquid is flowing at constant mass rate from a storage tank through a long pipe, the average linear velocity of the fluid is essentially constant throughout the entire length of the pipe. Under these conditions, assume a valve at the pipe outlet is suddenly closed. Since the velocity at the outlet end of the pipe then becomes zero, it would appear that the velocity at all other points throughout the pipe would become zero simultaneously. Actually, this is impossible because the kinetic energy of the flowing fluid cannot be reduced to zero instantaneously. The fluid in the pipe continues to flow, resulting in a compression of the liquid at the valve and a pressure increase in the region just before the valve. When the kinetic energy of the fluid has been dissipated, the pressure at the valve will be higher than the pressure at the entrance to the pipe and the liquid will start to flow back toward the storage tank. However, as the liquid flows toward the storage tank, the liquid will be compressed and the pressure will increase at the entrance to the storage tank. In this manner waves of fluid flow will pulsate back and forth in the pipe until the effect is dampened out by frictional losses.

A phenomenon of the type just described is called *water hammer* from the hammering sound heard when a valve is suddenly closed in a long pipe containing flowing water. While it may be impossible to close a valve instantaneously, it is possible to close it fast enough to cause a water hammer. This results in vibrations and undue stresses on the pipe and may eventually cause the pipe to break. Water hammer may be eliminated by closing the valve slowly or by the use of surge chambers or relief valves.

## PROBLEMS*

**5-1** A liquid with a density of 40.0 $lb/ft^3$ flows through a long straight pipe of circular cross section at a rate of 200 $ft^3/h$. Determine if the flow is turbulent or streamline under the following conditions:

(*a*) When the inside pipe diameter is 2.0 in and the absolute viscosity of the fluid is 1.8 lb/(h)(ft)

(*b*) When the inside pipe diameter is 1.2 ft and the absolute viscosity of the fluid is 3.0 cP.

* All the problems with asterisks should be solved using American engineering units and also SI units.

**5-2*** Determine the height of a vertical column of fluid with a density of 40.0 lb/ft$^3$ (641 kg/m$^3$) that will be supported by a static pressure of 18 lb/in$^2$, absolute (124.1 kPa) if the top of the column is open to the atmosphere. Atmospheric pressure may be taken as 14.7 lb/in$^2$, absolute (101.3 kPa).

**5-3** A fluid with a density of 50.0 lb/ft$^3$ has an absolute viscosity of 0.4 cP. Express this viscosity as (*a*) poises; (*b*) pounds per foot-second; (*c*) pounds per foot-hour; (*d*) centistokes; (*e*) stokes.

**5-4** Air is flowing with a velocity of 300 ft/s at 70°F and an absolute pressure of 150 lb/in$^2$ in the upstream section of a horizontal pipe. The air passes through an orifice. At a point near but definitely beyond the orifice, the linear velocity of the air is 500 ft/s and the absolute pressure is 100 lb/in$^2$. Considering air as a perfect gas, what is the temperature at this point? The mean heat capacity at constant pressure for air over the temperature range involved can be taken as 7.0 Btu/(lb mol)(°F). The flow can be assumed as turbulent throughout the entire system.

**5-5*** The linear velocity of a noncompressible liquid is 20 ft/s (6.1 m/s), and the absolute pressure is 30 lb/in$^2$ (206.8 kPa) at the initial point of a flow system. At the final point of the section under consideration, the linear velocity of the fluid is 30 ft/s (9.1 m/s) and the absolute pressure is 70 lb/in$^2$ (482.5 kPa). The increase in vertical elevation over this section is 25 ft (7.6 m). The total of all frictional losses is 40 ft·lbf per pound of fluid flowing (119.5 J/kg). The flow in this system is turbulent, and the temperature is constant. Calculate the amount of mechanical work put into the system as foot-pounds force per pound of fluid flowing if the density of the liquid is 50 lb/ft$^3$ (801 kg/m$^3$). Answer as joules per kilogram with SI.

**5-6*** A liquid with a density of 70 lb/ft$^3$ (1121 kg/m$^3$) flows through a straight steel pipe having an inside diameter of 2 in (0.0508 m) at the rate of 100 lb/h (45.4 kg/h). The viscosity of the liquid is 0.9 cP. Calculate the pressure drop in pounds per square foot due to friction if the pipe is 1 mi long( 1609 m). Answer as pascals with SI.

**5-7** Calculate the loss of mechanical energy as friction, expressed as foot-pounds force per pound-mass, due to the sudden enlargement from a 2-in-inside-diameter pipe to a 4-in-inside-diameter pipe for water at 65°F flowing at a rate of 1000 lb/min.

**5-8*** Calculate the loss of mechanical energy as friction, expressed as foot-pounds force per pound-mass, on the sudden contraction from a 2-in (0.0508-m)-inside-diameter pipe to a 1-in (0.0254-m)-inside-diameter pipe for a liquid with a density of 40 lb/ft$^3$ (641 kg/m$^3$) flowing at a rate of 50 lb/min (22.7 kg/min). Assume turbulent flow exists. Answer as joules per kilogram with SI.

**5-9** Calculate the horsepower required for a pump assembly with an overall efficiency of 55 percent to deliver 200 ft · lbf/lbm of fluid flowing to a system in which 100 lb of fluid flow per minute.

**5-10*** A liquid with a density of 40.0 lb/ft$^3$ (641 kg/m$^3$) is flowing through a 2-in (0.0508-m)-inside-diameter pipe. A 1-in (0.0254-m)-diameter orifice is located in the line. The pressure drop across the orifice is indicated by a mercury manometer in which the fluid from the pipe and the mercury are in contact. The reading taken from the manometer is 4 in (0.1016 m). Assuming an orifice coefficient of 0.61, calculate the flow in pounds per hour. Answer as kilograms per hour with SI.

**5-11** A venturi meter is used to measure the flow of water from a pumping installation. The venturi has 36-in openings and an 18-in throat. The venturi mercury manometer, in which mercury and water are in contact, has a reading of 3.1 in. What is the flow in gallons per day? The temperature of the water is 65°F.

**5-12** A liquid with a density of 75 lb/ft$^3$ flows through a 4-in sharp-edge orifice in the bottom of a tank at the rate of 75 ft$^3$/min. The diameter of the tank is 10 ft. The orifice coefficient is 0.61. What is the depth of the liquid in the tank? The top of the tank is open to the atmosphere, and the water flows out of the tank to the atmosphere.

**5-13*** A special oil is to be used in an absorption tower. The preliminary design of the unit requires the oil to be pumped from an open tank with a liquid level 10 ft (3.05 m) above the floor and forced through 150 ft (45.7 m) of 3.068-in (0.0779-m)-inside-diameter steel pipe with five 90° elbows into the top of a tower 30 ft (9.14 m) above the floor level. The operating pressure in the tower is to be 64.7 lb/in$^2$, absolute (446.0 kPa), and the oil requirement is estimated to be 50 gal/min (0.1893 m$^3$/min). The viscosity of the oil is 15 cP, and its density is 53.5 lb/ft$^3$ (857 kg/m$^3$). Assuming the pumping outfit operates with an overall efficiency of 40 percent, what horsepower input will be required for the motor? Atmospheric pressure is 14.7 lb/in$^2$, absolute (101.3 kPa). Answer as watts with SI.

**5-14** Oil is flowing through a horizontal pipe with a manometer hooked up between points 1 and 2 in a form similar to that shown in Fig. 5-7, with oil in contact with mercury. The specific gravity of the oil is 0.8, and the specific gravity of mercury is 13.6. If the reading of difference in height of mercury in the two legs of the manometer is 18 in, what is $p_1 - p_2$ in kPa?

**5-15** Water is flowing at a steady-state flow rate of 1.0 kg/s through a pipe that is 20 m long with an inside diameter of 2.1 cm. The pipe is horizontal and straight with a constant diameter; for use of Fig. 5-1, the pipe can be considered as new and smooth. The specific gravity of the water is 1.0, and its viscosity is 1.0 cP. Estimate the total pressure drop over the 20-m length of pipe as pounds force per square foot and as kilopascals. Note that the only cause for pressure drop is skin friction due to the flowing fluid.

**5-16** In an experiment being made to determine the Fanning friction factor, a horizontal rough pipe is being used to conduct a liquid having a specific gravity of 0.8. The pipe is 50 m long between pressure taps and has an inside diameter of 5.25 cm. Taps are connected at each end of the 50-m length flush with the pipe wall so that static pressure is measured, and each of these is connected to one end of a manometer containing an indicating fluid with a specific gravity of 5.0. The liquid of 0.8 specific gravity is in the line leading to the manometer and is in contact with the liquid of 5.0 specific gravity in the manometer. The two liquids are not miscible, so they remain as two separate phases. The manometer reading as height of the 5.0–specific gravity liquid in one leg of the manometer minus the height in the other leg is 200 cm. The 0.8–specific gravity liquid is flowing through the pipe under steady-state conditions at a rate of 3.05 m/s. Its viscosity is 2 cP. Under these conditions, determine the Fanning friction factor from the experimental data.

**5-17** (*This problem is intended for computer-program solution.*) A centrifugal pump is used to transfer kerosene from one tank to another, with both tanks at the same level. The pump raises the pressure of the liquid from $p_1$ (atmospheric) in the first tank to $p_2$ at the pump exit, but this pressure is gradually lost because of friction inside the long pipe, and the pressure at the exit in the second tank ($p_3$) is back down to atmospheric pressure. The pressure rise in kPa across the pump is given approximately by the empirical relation

$$p_2 - p_1 = a - bQ^{1.5}$$

where $a$ and $b$ are constants that depend on the particular pump being used, and $Q$ is the flow rate in $m^3/s$. Also, the pressure drop in a horizontal pipe of length $L$ m and internal diameter $D$ m is given by

$$p_2 - p_3 = 8.104 \times 10^{-4} \frac{f_m \rho L Q^2}{D^5}$$

where $\rho$ is the density ($kg/m^3$) of the liquid being pumped and $f_m$ is a dimensionless friction factor.

Write a program that computes the flow rate $Q$ given the following data:

$$D = 0.06271 \text{ m}$$

$$L = 15.24 \text{ m}$$

$$\rho = 823 \text{ kg/m}^3$$

$$f_m = 0.026 \text{ (dimensionless)}$$

$$a = 265.4 \text{ kPa}$$

$$b = 407{,}325 \text{ kPa/(m}^3\text{/s)}^{1.5}$$

For the computer program algorithm chosen, show the full computer printout to converge on the answer.

CHAPTER

# SIX

# HEAT TRANSFER

Physical processes involving the transfer of heat from one point to another are often encountered in the chemical industry. The heating and cooling of liquids or solids, the condensation of vapors, and the removal of heat liberated by a chemical reaction are common examples of processes which involve heat transfer. Because of the many applications of heat-transfer principles, it is important for the chemical engineer to understand the practical aspects of heat transfer as well as the basic laws governing this operation.

The unit operation of heat transfer is usually only one component part of an overall process, and the interrelations among the different operations involved must be recognized. For example, a sodium hydroxide–water mixture may be concentrated by use of an evaporator. In order for the evaporation to proceed at an appreciable rate, heat must be added to the liquid mixture. This heat may be supplied by steam condensing inside pipes immersed in the liquid. By the use of heat-transfer principles, the steam pressure necessary to carry out the evaporation at a given rate could be determined if the other pertinent variables such as heat-transfer area and the liquid-mixture temperature were known. However, the operation of fluid flow is also involved in that steam must flow through the pipes leading up to the evaporator. In addition, a mass-transfer operation must take place when the water in the sodium hydroxide–water mixture changes from the liquid phase to the vapor phase.

The simple example presented in the preceding paragraph illustrates the interdependence of the different unit operations. The heat-transfer operation deals only with one part of the process in which heat flows from the hot condensing steam through the pipe walls into the colder liquid surrounding the pipes.

In order for heat to flow, there must be a driving force. This driving force is the temperature difference between the points where heat is received and where the heat originates. In the study of the flow of fluids, it has been observed that a fluid tends to flow from a point of high pressure to one of low pressure. The driving force, in this case, is the pressure drop. Similarly, heat tends to flow from a point

## Table 6-1 Nomenclature for heat transfer

$A$ = area of heat-transfer surface, $m^2$ and $ft^2$
$c_p$ = heat capacity (specific heat) at constant pressure, J/(kg)(K) or Btu/(lb)(°F)
$d$ = prefix indicating differential, dimensionless
$D$ = diameter, m or ft
$D_o$ = outside diameter, m or ft
$D'$ = diameter, in
$F_A$ = geometrical-shape factor for black-body radiation, dimensionless
$F_{AE}$ = correction factor dependent on shape and emissivity of radiating surfaces, dimensionless
$g$ = acceleration due to gravity, normally taken as 9.8 $m/s^2$ or 32.17 $ft/s^2$
$G$ = mass velocity (equals mass rate of flow divided by cross-sectional area of flow), kg/(s)($m^2$ of cross section) or lb/(h)($ft^2$ of cross section)
$h$ = individual film coefficient of heat transfer, J/(s)($m^2$)(K) or Btu/(h)($ft^2$)(°F)
$k$ = thermal conductivity, J/(s)($m^2$)(K/m) or Btu/(h)($ft^2$)(°F/ft)
$K$ = empirical constant, dependent on shape of body
$L$ = length of heat-transfer surface, m or ft
$N_{Gr}$ = Grashof number (equals $D^3\rho^2 g\beta\Delta t/\mu^2$), dimensionless
$q$ = rate of heat flow, J/s or Btu/h
$Q$ = quantity of heat, J or Btu
$R$ = individual thermal resistance, $x/kA$, (s)(K)/J or (h)(°F)/Btu
$t$ = temperature, °C or °F
$T$ = absolute temperature, K = °C + 273.16 or °R = °F + 459.7
$U$ = overall coefficient of heat transfer, J/(s)($m^2$)(K) or Btu/(h)($ft^2$)(°F)
$V$ = average linear velocity (equals volumetric flow rate divided by cross-sectional area of flow), m/s or ft/h
$V_s$ = average linear velocity, ft/s
$w$ = mass rate of flow per tube, kg/(s)(tube) or lb/(h)(tube)
$x$ = length of conduction path, m or ft

Greek symbols

$\beta$ = beta, volumetric coefficient of thermal expansion, $m^3/(m^3)$(K) or $ft^3/(ft^3)$(°F)
$\Delta t$ = delta $t$, temperature difference, K or °F
$\theta$ = theta, time, h or s
$\lambda$ = lambda, latent heat (enthalpy) of vaporization, J/kg or Btu/lb
$\mu$ = mu, absolute viscosity, kg/(s)(m) (equals 0.001 times cP) or lb/(h)(ft) (equals 2.42 times cP)
$\pi$ = pi, 3.1416
$\rho$ = rho, density at bulk temperature, $kg/m^3$ or $lb/ft^3$

Subscripts

$d$ refers to dirt or scale
$f$ refers to film
$w$ refers to wall
1 = entering or initial conditions
2 = leaving or final conditions

of high temperature to a point of low temperature because of the temperature-difference driving force.

Many of the relationships used in the treatment of heat-transfer problems involve empirical factors. The empirical constants indicated in the equations of this chapter are the results of numerous experiments by different investigators. These semiempirical equations may be accepted as good representations of the relationships of the variables involved. However, the chemical engineer must realize that any calculation based on predetermined empirical results will not give exactly correct results. If the relationships are applied correctly, the results will be close to the true value.

The first part of this chapter deals with the basic laws of heat transfer and indicates some of the common equations used for design purposes. The last part of the chapter deals with the practical aspects of heat transfer pertaining particularly to industrial operations.

## DIMENSIONLESS GROUPS

Many of the generalized relationships used in heat-transfer calculations have been determined by means of dimensional analysis and empirical considerations. It has been found that certain standard dimensionless groups appear repeatedly in the final equations. The chemical engineer should recognize the more important of these groups. Some of the most common dimensionless groups are given below with their names.

$$\text{Reynolds number} = \frac{DV\rho}{\mu} = \frac{DG}{\mu}$$

$$\text{Prandtl number} = \frac{c_p \mu}{k}$$

$$\text{Nusselt number} = \frac{hD}{k}$$

$$\text{Peclet number} = \frac{DV\rho c_p}{k}$$

$$\text{Grashof number} = \frac{D^3 \rho^2 g \beta \, \Delta t}{\mu^2}$$

$$\text{Stanton number} = \frac{h}{c_p G}$$

In the application of these groups, care must be taken to use the equivalent units so all the dimensions can cancel out. Any system of units may be used in a dimensionless group as long as the final result will permit all units to disappear by cancellation.

## DIFFERENT METHODS OF HEAT TRANSFER

Basically, there are only three means by which heat may be transferred: by conduction, by convection, or by radiation. In many cases, heat transfer occurs by all three of these methods simultaneously.

### Conduction

The molecules of any material are constantly in motion. Even the molecules of a piece of iron are always vibrating back and forth, although they move only a small distance in any one direction. The higher the temperature level of a material, the more violently its molecules vibrate. If one end of a cold iron rod is held in a flame, the rod temperature at that end increases. The molecules at the hot end vibrate more rapidly than the molecules at the cold end of the rod. This vibration is imparted to the neighboring molecules which, in turn, vibrate faster and therefore "heat up." This process is continued on along the entire length of the rod until the end away from the flame has also become hot. This mode of heat transmission is known as *conduction*. There is no actual movement or flow of the material. The molecules vibrate in just one spot, but their energy is transmitted; therefore, it can be said that heat flows through the substance by conduction.

### Convection

Heat may be transferred from one point to another by the actual physical movement of particles. This type of heat transfer is called *convection*. Thus, if a hot and a cold liquid are mixed by a mechanical stirrer, heat can be said to be transferred from the hot liquid to the cold liquid by convection. Heat transfer by convection can occur as a result of any type of physical mixing such as mechanical agitation or flow due to density differences.

### Radiation

A body emits radiant energy in all directions. If this energy strikes a receiver, part of it may be absorbed and part may be reflected. Heat transfer from a hot to a cold body in this manner is known as heat transfer by *radiation*.

## HEAT TRANSFER BY CONDUCTION

### Basic Equation (Fourier's Law)

The instantaneous rate of heat transfer through a homogeneous body by conduction is directly proportional to the temperature-difference driving force across the body and to the cross-sectional area (i.e., the area at right angles to the direction of heat flow) of the body through which the heat flows. The rate of heat transfer is inversely proportional to the thickness of the body (i.e., to the length of the path along which heat flows). These proportionalities can be expressed mathematically as,

$$\frac{dQ}{d\theta} = -kA\frac{dt}{dx} \qquad (6\text{-}1)$$

where $dQ/d\theta$ is the instantaneous rate of heat transfer, $A$ is the cross-sectional area of the body perpendicular to the direction of heat flow, and $-(dt/dx)$ is the rate of change of temperature $t$ with respect to the length of the heat-flow path $x$. The proportionality factor $k$ is defined as the thermal conductivity of the material through which the heat is flowing.

The thermal conductivity of a homogeneous substance depends, essentially, on its temperature. Tables in App. C list values of the thermal conductivities for common materials at various temperatures.

The *thermal conductivity*, as used in this book, means the number of British thermal units conducted in 1 h through 1 $ft^2$ of area measured perpendicular to the direction of heat flow when the distance the heat must flow is 1 ft and the temperature-difference driving force is 1°F. Thus, the units of thermal conductivity are Btu/(h)($ft^2$)(°F/ft). Equivalent units in SI would be J/(s)($m^2$)(K/m), with a multiplication factor of 1.731 needed for conversion from American engineering units to SI units.

## Steady Flow of Heat in Homogeneous Bodies

In most industrial processes, heat is transferred from one point to another under steady conditions of temperature difference, length of heat-flow path, and cross-sectional area. This is known as the *condition of steady flow of heat*. For example, after a furnace has been in operation for several hours, its walls reach a certain temperature and maintain that temperature practically unchanged as long as the furnace is kept in active use. In this case, the difference between the temperature at the inside and outside walls of the furnace remains the same, and a condition of steady flow of heat exists.

For the common cases of steady heat flow, Eq. (6-1) may be expressed as

$$\frac{Q}{\theta} = kA\frac{\Delta t}{x} \tag{6-2}$$

where $Q$ represents the total amount of heat transferred in time $\theta$. The letter $A$ represents the area at right angles to the direction of heat flow, and $x$ is the distance the heat must flow. The temperature-difference driving force, taken as the higher temperature minus the lower temperature, is indicated by $\Delta t$, and $k$ is the thermal conductivity of the material through which heat is being transferred.

The thermal conductivity of most homogeneous substances varies almost linearly with temperature. Therefore, over moderate ranges of temperature, adequate accuracy is obtained if the value of $k$ is determined at the average temperature of the material.

The use of Eq. (6-2) may be illustrated by considering the practical case of heat losses from a furnace. Assume one wall of a furnace is 8 in thick and is made of a brick having an average thermal conductivity of 2.2 Btu/(h)($ft^2$)(°F/ft) over the temperature range involved. The inner side of the furnace wall has a temperature of 1700°F, and the temperature of the outer side of the wall is 700°F. If the area of the wall is known, the amount of heat lost per hour through the wall can easily be calculated by the use of Eq. (6-2). Assuming a wall area of 80 ft$^2$, the heat loss per hour is

$$\frac{Q}{\theta} = q = kA\frac{\Delta t}{x} = \frac{(2.2)(80)(12)(1700 - 700)}{8} = 264{,}000 \text{ Btu/h}$$

## Concept of Resistance to Flow of Heat

When heat flows through a solid by conduction, there is a certain resistance to the flow determined by the thermal conductivity, cross-sectional area, and thickness of the particular substance. To obtain any given rate of heat flow, this resistance must be overcome by setting up a certain temperature-difference driving force.

**Resistance in series.** If one solid is placed in series with another solid so that heat must pass through both of these, the resistance to the flow of heat will be greater than it would be for one of the solids alone.

If three solids of equal cross-sectional area, designated as 1, 2, and 3, are placed in series, when steady flow of heat exists, exactly the same amount of heat per unit time must be transferred through each of the solids. This can be expressed mathematically as

$$\left(\frac{Q}{\theta}\right)_1 = \left(\frac{Q}{\theta}\right)_2 = \left(\frac{Q}{\theta}\right)_3 = \left(\frac{Q}{\theta}\right)_{\text{total}} \tag{6-3}$$

The rate of heat transfer through solid 1, according to Eq. (6-2), must be

$$\left(\frac{Q}{\theta}\right)_1 = \left(kA\frac{\Delta t}{x}\right)_1 = \left(\frac{kA}{x}\right)_1 \Delta t_1 \tag{6-4}$$

Similarly,

$$\left(\frac{Q}{\theta}\right)_2 = \left(kA\frac{\Delta t}{x}\right)_2 = \left(\frac{kA}{x}\right)_2 \Delta t_2 \tag{6-5}$$

$$\left(\frac{Q}{\theta}\right)_3 = \left(kA\frac{\Delta t}{x}\right)_3 = \left(\frac{kA}{x}\right)_3 \Delta t_3 \tag{6-6}$$

The total temperature difference across the three solids must be the sum of the individual temperature differences across each of the solids, or

$$\Delta t_{\text{total}} = \Delta t_1 + \Delta t_2 + \Delta t_3 \tag{6-7}$$

From Eqs. (6-4) to (6-7), it can be seen that

$$\Delta t_{\text{total}} = \frac{(Q/\theta)_1}{(kA/x)_1} + \frac{(Q/\theta)_2}{(kA/x)_2} + \frac{(Q/\theta)_3}{(kA/x)_3} \tag{6-8}$$

The total rate of heat transfer can be expressed in terms of the total temperature difference by combining Eqs. (6-3) and (6-8) as

$$\left(\frac{Q}{\theta}\right)_{\text{total}} = \frac{\Delta t_{\text{total}}}{\dfrac{1}{(kA/x)_1} + \dfrac{1}{(kA/x)_2} + \dfrac{1}{(kA/x)_3}} \tag{6-9}$$

It is also possible to determine the temperature drop across any one of the solids by combining Eqs. (6-3) and (6-9) with Eq. (6-4), (6-5), or (6-6). Thus the temperature drop across solid 1 would be

$$\Delta t_1 = \Delta t_{\text{total}} \frac{\dfrac{1}{(kA/x)_1}}{\dfrac{1}{(kA/x)_1} + \dfrac{1}{(kA/x)_2} + \dfrac{1}{(kA/x)_3}} \tag{6-10}$$

Since the term $x/kA$ is a representation of the resistance encountered by heat flowing through the material, the symbol $R$ is often used for this $x/kA$. Using this notation, Eq. (6-9) becomes

$$\left(\frac{Q}{\theta}\right)_{\text{total}} = \frac{\Delta t_{\text{total}}}{R_1 + R_2 + R_3} \tag{6-11}$$

Figure 6-1 is a diagrammatic representation of resistances in series and resistances in parallel.

**Resistances in parallel.** When several solids are placed side by side with their edges touching in such a manner that the direction of heat flow is perpendicular to the plane of the exposed face surfaces, the solids are said to be placed in parallel. The

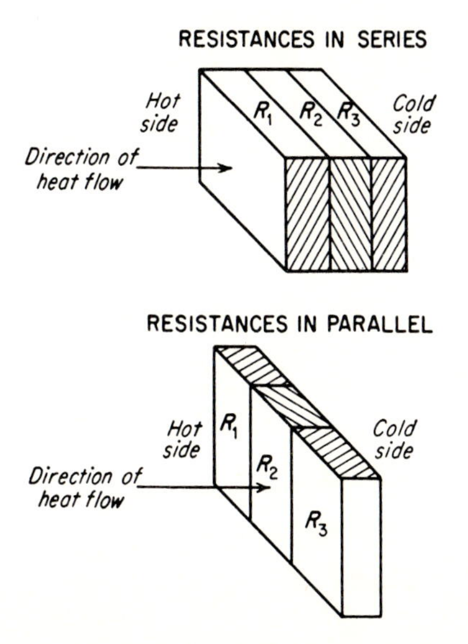


**Figure 6-1** Concept of thermal resistances in series and in parallel.

temperature drop may, ordinarily, be taken as the same across each of the resistances. The total cross-sectional area available for heat transfer with parallel resistances is greater than it would be for any one of the resistances alone. Thus, with a given temperature difference, more heat can be transferred through several solids placed in parallel than could be transferred through any one of them alone.

If three solids, designated as 1, 2, and 3, are placed in parallel, the total heat transferred through these resistances in a given time is the sum of the heat transferred through each one individually; thus, for steady flow of heat through homogeneous bodies,

$$\left(\frac{Q}{\theta}\right)_{\text{total}} = \left(\frac{Q}{\theta}\right)_1 + \left(\frac{Q}{\theta}\right)_2 + \left(\frac{Q}{\theta}\right)_3$$

$$= \left(kA\frac{\Delta t}{x}\right)_1 + \left(kA\frac{\Delta t}{x}\right)_2 + \left(kA\frac{\Delta t}{x}\right)_3 \tag{6-12}$$

Assuming the $\Delta t$ is the same for each of the three solids,

$$\left(\frac{Q}{\theta}\right)_{\text{total}} = \left[\left(\frac{kA}{x}\right)_1 + \left(\frac{kA}{x}\right)_2 + \left(\frac{kA}{x}\right)_3\right]\Delta t \tag{6-13}$$

Substituting $R$ for $x/kA$,

$$\left(\frac{Q}{\theta}\right)_{\text{total}} = \left(\frac{1}{R_1} + \frac{1}{R_2} + \frac{1}{R_3}\right)\Delta t \tag{6-14}$$

**Example 6-1: Determination of amount of heat transferred by conduction through resistances in series** The temperature at the inside surface of an oven is 460°F. The inside wall of the oven is constructed of brick and is 8 in thick. The thermal conductivity of the brick is 2.2 Btu/(h)(ft$^2$)(°F/ft). The outside of the oven is covered with a 3-in layer of asbestos. The thermal conductivity of this asbestos may be taken as 0.11 Btu/(h)(ft$^2$)(°F/ft). If the outside surface of the insulation has a temperature of 100°F, calculate the amount of heat lost through 2 ft$^2$ of wall area in 3 h.

SOLUTION

$$\Delta t_{\text{total}} = 460 - 100 = 360°\text{F}$$

$$\left(\frac{kA}{x}\right)_{\text{brick}} = \frac{(2.2)(2)(12)}{8} = 6.6 \text{ Btu/(h)(°F)}$$

$$\left(\frac{kA}{x}\right)_{\text{asbestos}} = \frac{(0.11)(2)(12)}{3} = 0.88 \text{ Btu/(h)(°F)}$$

$$\theta = 3 \text{ hr}$$

From Eq. (6-9),

$$Q_{\text{total}} = \frac{(360)(3)}{1/6.6 + 1/0.88} = 839 \text{ Btu}$$

This is the amount of heat conducted through 2 $ft^2$ of wall area in 3 h.

**Example 6-2: Determination of amount of heat transferred through resistances in parallel** A glass window with an area of 1 $m^2$ is installed in a wooden wall of a room. The dimensions of the wall are 3 m by 6 m. The wood is 2.54 cm (0.0254 m) thick and has a thermal conductivity of 0.15 J/(s)($m^2$)(K/m). The glass is 0.42 cm (0.0042 m) thick and has a thermal conductivity of 0.69 J/(s)($m^2$)(K/m). If the inside wall-and-glass temperature is 30°C and the outside wall-and-glass temperature is 0°C, calculate the total amount of heat conducted through the wall and glass as joules per hour.

SOLUTION

$$\Delta t = 30 - 0 = 30°\text{C} = 30 \text{ K}$$

$$\left(\frac{kA}{x}\right)_{\text{glass}} = \frac{0.69 \text{ J}}{(\text{s})(\text{m}^2)(\text{K/m})} \left| \frac{1 \text{ m}^2}{} \right| \frac{}{0.0042 \text{ m}} \left| \frac{3600 \text{ s}}{\text{h}} \right. = 591{,}000 \frac{\text{J}}{(\text{h})(\text{K})}$$

$$\left(\frac{kA}{x}\right)_{\text{wood}} = \frac{(0.15)(18 - 1)(3600)}{0.0254} = 361{,}000 \frac{\text{J}}{(\text{h})(\text{K})}$$

$$\theta = 1 \text{ h}$$

From Eq. (6-13),

$$Q_{\text{total}} = (591{,}000 + 361{,}000)(30) = 29 \times 10^6 \text{ J}$$

This is the total amount of heat conducted through the wall in 1 h.

## USE OF MEAN AREA AND MEAN TEMPERATURE DIFFERENCE

In the preceding discussion, cases have been considered where the cross-sectional area through which heat was transferred has been constant along the entire path of the heat flow. In addition, the temperature difference between the inside and outside surfaces has been considered as constant over the complete system.

In many types of heat-transfer equipment, the cross-sectional area varies along the heat-flow path and the temperature difference changes from one point to another. For example, when water flows through straight pipes and is heated by steam condensing on the pipes, the inside area of the pipe wall is smaller than the outside area of the pipe wall. If cold water comes into the heater, the overall temperature difference between the cold water and the steam at the entrance to the heater will be greater than the overall temperature difference between the hot water and the steam at the heater exit. Therefore, to apply the basic heat-transfer equations, some type of mean or average values for the area and temperature difference must be employed.

*Arithmetic-mean values* represent the arithmetic average of the two extremes in area or temperature difference.

$$A_{\text{ar mean}} = \frac{A_1 + A_2}{2} \tag{6-15}$$

$$\Delta t_{\text{ar mean}} = \frac{\Delta t_1 + \Delta t_2}{2} \tag{6-16}$$

*Logarithmic-mean values of heat transfer areas* are applicable to closed cylindrical bodies of circular section. By integration of the basic heat-transfer equation, the following expression may be obtained for the logarithmic-mean area:[1]

$$A_{\text{log mean}} = \frac{A_2 - A_1}{2.3 \log_{10}(A_2/A_1)} \tag{6-17}$$

This mean area may be applied to circular pipes and pipe lagging or to any case in which the cross-sectional area of the heat-transfer path is proportional to the distance along the path. When the value of $A_2/A_1$ does not exceed 2, the arithmetic-mean area differs from the logarithmic-mean area by less than 5 percent. This accuracy is considered sufficient for most heat-transfer calculations.

*The logarithmic-mean temperature difference* may be expressed as

$$\Delta t_{\text{log mean}} = \frac{\Delta t_2 - \Delta t_1}{2.3 \log_{10}(\Delta t_2/\Delta t_1)} \tag{6-18}$$

This mean temperature difference is applicable for the common case of single-pass heat exchangers when the overall heat-transfer coefficient and the heat capacities of the fluids are essentially constant throughout the exchanger. The logarithmic-mean temperature difference can also be used for multipass heat exchangers if the temperature of one of the fluids, the overall heat-transfer coefficient, and the fluid heat capacities are constant. When the value of $\Delta t_2/\Delta t_1$ does not exceed 2, the arithmetic-mean temperature difference differs from the logarithmic-mean temperature difference by less than 5 percent.[2]

## HEAT TRANSFER BY CONVECTION

Heat may be transferred from one material to another by the physical process of mixing a hot substance with a cold substance. This method of heat transfer is

[1] It should be noted that the natural logarithm (ln) is the same as $2.3 \log_{10}$.

[2] For a discussion of other types of mean areas and mean temperature differences, see any standard textbook or handbook on heat transfer.

known as *convection*. A certain amount of heat is transferred by conduction simultaneously with convection heat transfer. Since it is not practical to differentiate between convection and conduction when they are both occurring at the same point, a special concept which takes both of these methods of heat transfer into consideration has been developed.

## Concept of the Film

When a fluid is flowing past a stationary surface, a thin film of the fluid is postulated as existing between the flowing fluid and the stationary surface. It is assumed that all the resistance to transmission of heat between the flowing fluid and the body containing the fluid is due to the film at the stationary surface.

The amount of heat transferred across this film is proportional to the surface area of the film and to the temperature difference across the film. This can be expressed as

$$\frac{Q}{\theta} = hA_f \, \Delta t_f \tag{6-19}$$

where $h$, the proportionality constant, is known as the *individual film coefficient of heat transfer*. In the American engineering system, the units of $h$ are in British thermal units per hour per square foot per degree Fahrenheit. With SI, the units of $h$ are in joules per second per square meter per kelvin, with a multiplication factor of 5.678 needed for conversion from American engineering units to SI units.

## Overall Coefficient

Many of the important cases of heat transfer involve the flow of heat from one fluid through a solid retaining wall into another fluid. This heat must flow through several resistances in series. The resistances to be overcome are the two fluid-film resistances and the solid-wall resistance. The total flow of heat is proportional to the heat-transfer area and to the overall temperature difference, or

$$\frac{Q}{\theta} = UA \, \Delta t_{\text{total}} \tag{6-20}$$

where the proportionality constant $U$ is termed the *overall coefficient of heat transfer*.

Because there are usually several different areas on which the overall coefficient may be based, it is essential to indicate the base area along with the units of $U$. For example, if an overall coefficient of heat transfer is based on the inside area of a pipe, the units of this $U$ should be expressed as joules per second per square meter of inside area per kelvin or Btu per hour per square foot of inside area per degree Fahrenheit.

When heat flows through resistances in series, the same amount of heat per unit area and per unit time must be transferred through each resistance. When heat flows from one fluid to another through a solid retaining wall, the total amount of heat transferred may be expressed as follows:

$$\left(\frac{Q}{\theta}\right)_{\text{total}} = UA\,\Delta t_{\text{total}} = h'A'_f\,\Delta t'_f = h''A''_f\,\Delta t''_f \tag{6-21}$$

$$= kA_w \frac{\Delta t_w}{x_w}$$

where $A$ is the base area chosen for the evaluation of $U$ and the primes refer to the different film resistances involved.

The total temperature difference must equal the sum of the temperature differences across each resistance, or

$$\Delta t_{\text{total}} = \Delta t'_f + \Delta t''_f + \Delta t_w \tag{6-22}$$

From Eqs. (6-21) and (6-22),

$$\Delta t_{\text{total}} = \left(\frac{Q}{\theta}\right)_{\text{total}} \left(\frac{1}{h'A'_f} + \frac{1}{h''A''_f} + \frac{x_w}{kA_w}\right) = \left(\frac{Q}{\theta}\right)_{\text{total}} \left(\frac{1}{UA}\right) \tag{6-23}$$

Therefore

$$\frac{1}{U} = \frac{A}{h'A'_f} + \frac{A}{h''A''_f} + \frac{Ax_w}{kA_w} \tag{6-24}$$

When the wall thickness of the pipe containing the inner fluid is small compared to the pipe diameter, $A$, $A'_f$, $A''_f$, and $A_w$ are all approximately equal, and Eq. (6-24) becomes

$$\frac{1}{U} = \frac{1}{h'} + \frac{1}{h''} + \frac{x_w}{k} \tag{6-25}$$

If there is an additional resistance due to scale or dirt deposits on both sides of the pipe wall, Eq. (6-25) becomes

$$\frac{1}{U} = \frac{1}{h'} + \frac{1}{h''} + \frac{x_w}{k} + \frac{1}{h'_d} + \frac{1}{h''_d} \tag{6-26}$$

**Example 6-3: Application of overall heat-transfer coefficient in determination of rate of heat flow** A standard (Schedule 40) 2-in-diameter steel pipe [$k = 26$ Btu/(h)(ft$^2$)(°F/ft)] has water flowing through it under such conditions that the individual water film coefficient is 500 Btu/(h)(ft$^2$)(°F). Steam, with an individual film coefficient of 2000, is condensing on the outside of the pipe. The overall $\Delta t$ at the point where water enters the heater is 150°F, and the overall $\Delta t$ at the exit from the heater is 50°F. A logarithmic-mean temperature difference should be used. If the pipe is 10 ft long, calculate the amount of heat gained by the water per hour. The inside diameter of standard 2-in steel pipe is 2.067 in, and the wall thickness is 0.154 in. There is no scale on the pipe.

SOLUTION The log-mean

$$\Delta t_{\text{total}} = \frac{150 - 50}{2.3 \log(150/50)} = 91°\text{F}$$

The overall coefficient $U$ will be based on the inside area of the pipe. Using Eq. (6-24),

$$A = \frac{(2.067)(\pi)(10)}{12} \text{ ft}^2$$

$$A'_f = \frac{(2.067)(\pi)(10)}{12} \text{ ft}^2$$

$$h' = 500 \text{ Btu/(h)(ft}^2\text{)(°F)}$$

$$A''_f = \frac{(2.375)(\pi)(10)}{12} \text{ ft}^2$$

$$h'' = 2000 \text{ Btu/(h)(ft}^2\text{)(°F)}$$

$$x_w = \frac{0.154}{12} \text{ ft}$$

$$A_w = \frac{(2.221)(\pi)(10)}{12} \text{ ft}^2$$

$$k = 26.0 \text{ Btu/(h)(ft}^2\text{)(°F/ft)}$$

$$\frac{1}{U} = \frac{2.067}{(500)(2.067)} + \frac{2.067}{(2000)(2.375)} + \frac{(2.067)(0.154)}{(26)(12)(2.221)}$$

$$= 0.002896$$

$$U = \frac{1}{0.002896} = 346 \text{ Btu/(h)(ft}^2 \text{ of inside area)(°F)}$$

$$\theta = 1 \text{ h}$$

$$Q_{\text{total}} = UA\,\Delta t_{\text{total}} = \frac{(346)(2.067)(3.14)(10)(91)}{12}$$

$$= 170{,}000 \text{ Btu/h}$$

Exactly the same final result would have been obtained if the overall coefficient had been based on another area such as the outside area of the pipe.

## EVALUATION OF INDIVIDUAL FILM COEFFICIENTS

### Turbulent Flow of Fluids Inside Circular Pipes

By application of the method of dimensional analysis and empirical considerations, the following equation, where the film coefficient is shown as part of the dimensionless group $hD/k$, has been developed for the determination of the individual coefficient of heat transfer in circular pipes.

$$\left(\frac{hD}{k}\right) = 0.023\left(\frac{DG}{\mu}\right)^{0.8}\left(\frac{c_p\mu}{k}\right)^{1/3} \tag{6-27}$$

This equation gives accurate results for the flow of common fluids when the Reynolds number is greater than 2100.[3] The values of $k$, $\mu$, and $c_p$ should be determined at the average bulk temperature of the fluid. The equation is made up of three dimensionless groups as shown by the parentheses. Thus, any self-consistent set of units can be used in the calculation to determine $h$ as long as the ratio of $hD/k$ is dimensionless.

For the common gases such as air, oxygen, and carbon dioxide, the physical properties of the gases are such that the value of $(c_p\mu/k)^{1/3}\ k/c_p\mu^{0.8}$ is approximately constant, and Eq. (6-27) reduces to the following *dimensional* form:

$$h = 0.0144c_p\frac{G^{0.8}}{D^{0.2}} \tag{6-28}$$

where $h$ = gas-film coefficient of heat transfer, Btu/(h)(ft$^2$)(°F)
$c_p$ = heat capacity of gas at constant pressure, Btu/(lb)(°F)
$G$ = mass flow rate of gas, lb/(h)(ft$^2$)
$D$ = inside diameter of pipe, ft

For water at ordinary temperatures, Eq. (6-27) becomes

$$h = 150(1 + 0.011t)\frac{V_s^{0.8}}{(D')^{0.2}} \tag{6-29}$$

where $t$ = average bulk temperature of the water, °F
$V_s$ = average linear velocity of the water, f/s
$D'$ = inside diameter of the pipe, in

Both Eqs. (6-28) and (6-29) are *dimensional equations*. This means they can only be used with the specific units for the variables, as indicated immediately after the equations.

**Example 6-4: Direct calculation of individual film coefficients of heat transfer**
Water flows through a 2-in standard (Schedule 40) steel pipe with an average linear velocity of 5 ft/s. The average bulk temperature of the water is 100°F.

[3] When the viscosity of the fluid at the average bulk temperature ($\mu$) divided by the viscosity of the fluid at the inside-wall temperature ($\mu_w$) is greater than $\frac{3}{2}$ or less than $\frac{2}{3}$ a correction factor of approximately $(\mu/\mu_w)^{0.14}$ should be applied to the right-hand side of Eq. (6-27).

(*a*) Determine whether the flow is turbulent or streamline.
(*b*) Calculate the water-film coefficient by means of Eq. (6-27).
(*c*) Calculate the water-film coefficient by means of Eq. (6-29).
Solve the problem first using American engineering units and then repeat using SI units.

SOLUTION First with American engineering units:
$\mu$ = viscosity of water at 100°F = 0.684 cP
$\rho$ = density of water at 100°F = 62.0 lb/ft$^3$
$V$ = 5.0 ft/s
$D$ = 2.067/12 ft
$k$ = thermal conductivity of water at 100°F = 0.363 Btu/(h)(ft$^2$)(°F/ft)
$c_p$ = heat capacity of water at 100°F = 1.0 Btu/(lb)(°F)

$$(a)\ \text{Reynolds number} = \frac{DV\rho}{\mu} = \frac{(2.067)(5.0)(62.0)}{(12)(0.684)(0.000672)}$$
$$= 116{,}200 \text{ (dimensionless)}$$

Since the Reynolds number is greater than 2100, the flow is turbulent.
(*b*) From Eq. (6-27),

$$h = 0.023\frac{k}{D}\left(\frac{DG}{\mu}\right)^{0.8}\left(\frac{c_p\mu}{k}\right)^{1/3}$$
$$= (0.023)\left[\frac{(0.363)(12)}{2.067}\right](116{,}200)^{0.8}\left[\frac{(1)(0.684)(2.42)}{0.363}\right]^{1/3}$$
$$= 907 \text{ Btu/(h)(ft}^2\text{)(°F)}$$

(*c*) From Eq. (6-29),

$$h = 150(1 + 0.011t)\frac{(V_s)^{0.8}}{(D')^{0.2}}$$
$$= 150[1 + (0.011)(100)]\frac{(5)^{0.8}}{(2.067)^{0.2}}$$
$$= 987 \text{ Btu/(h)(ft}^2\text{)(°F)}$$

Repeat of solution using SI units:
100°F = 37.8°C
$\mu$ = viscosity of water at 37.8°C = 0.684 cP = 0.000684 kg/(m)(s)
$\rho$ = density of water at 37.8°C = 62.4 lb/ft$^3$ = 993.1 kg/m$^3$
$V$ = 5.0 ft/s = 1.524 m/s
$D$ = 2.067/12 ft = (2.067)(0.0254) = 0.0525 m
$k$ = thermal conductivity of water at 37.8°C
= 0.363 Btu/(h)(ft$^2$)(°F/ft) = 0.628 J/(s)(m$^2$)(K/m)
$c_p$ = heat capacity of water at 37.8°C = 1 Btu/(lb)(°F)
= 4187 J/(kg)(K)

$$(a)\ \text{Reynolds number} = \frac{DV\rho}{\mu} = \frac{(0.0525)(1.524)(993.1)}{0.000684}$$

$$= 116{,}200 \text{ (dimensionless)}$$

Since the Reynolds number is greater than 2100, the flow is turbulent.
($b$) From Eq. (6-27),

$$h = 0.023\frac{k}{D}\left(\frac{DG}{\mu}\right)^{0.8}\left(\frac{c_p\mu}{k}\right)^{1/3}$$

$$= 0.023\left[\frac{0.628\ \text{J}}{(\text{s})(\text{m}^2)(\text{K/m})} \left| \frac{}{0.0525\ \text{m}}\right.\right](116{,}200)^{0.8}\left[\frac{(4187)(0.000684)}{0.628}\right]^{1/3}$$

$$= 5148\ \text{J/(s)(m}^2\text{)(K)}$$

Since there is 5.678 J/(s)(m$^2$)(K) for each Btu per hour per square foot per degree Fahrenheit, 5148/5.678 = 907 Btu/(h)(ft$^2$)(°F), and the same answer is gotten by American engineering units or SI units.
($c$) From Eq. (6-29),

$$h = 150(1 + 0.011t)\frac{(V_s)^{0.8}}{(D')^{0.2}}$$

Because this is a dimensional equation,

$t = 100°\text{F}$
$V_s = 5\ \text{ft/s}$
$D' = 2.067\ \text{in}$
$h = 987\ \text{Btu/(h)(ft}^2\text{)(°F)} = (5.678)(987) = 5604\ \text{J/(s)(m}^2\text{)(K)}$

## Turbulent Flow of Fluids in Annular Spaces

The individual coefficient of heat transfer in annular spaces may be obtained by using Eq. (6-27) with an equivalent diameter in place of $D$. This equivalent diameter should be taken as four times the cross-sectional flow area divided by the heated perimeter.

## Streamline Flow of Fluids Inside Circular Pipes

The following empirical equation, involving only dimensionless groups, has been developed for streamline flow inside circular pipes.

$$\left(\frac{hD}{k}\right) = 1.75\left(\frac{wc_p}{kL}\right)^{1/3}\left[\left(\frac{\mu}{\mu_f}\right)^{1/3}(1 + 0.015\sqrt[3]{N_{Gr}})\right] \qquad (6\text{-}30)$$

where $\mu_f$ represents the viscosity of the fluid at the average temperature of the film and $\mu$ is the viscosity of the fluid at the bulk temperature. This equation should be used for flowing fluids when the Reynolds number is less than 2100.[4]

### Forced Convection of Gases at Right Angles to a Single Cylinder

When mechanical means are used to force a fluid to flow past a surface which is hotter or colder than the fluid, the method of heat transfer between the fluid and the surface is designated as *forced convection.* For forced convection of gases flowing at right angles to a single cylinder, the following equation is recommended for Reynolds numbers ($D_o G/\mu_f$) from 0.1 to 1000.

$$\left(\frac{hD_o}{k_f}\right) = \left[0.35 + 0.47\left(\frac{D_o G}{\mu_f}\right)^{0.52}\right]\left(\frac{c_p \mu_f}{k_f}\right)^{0.3} \tag{6-31}$$

For Reynolds numbers higher than 1000, the following equation can be used to get an approximate value of the individual heat-transfer coefficient for gases.

$$\left(\frac{hD_o}{k_f}\right) = 0.26\left(\frac{D_o G}{\mu_f}\right)^{0.6}\left(\frac{c_p \mu_f}{k_f}\right)^{0.3} \tag{6-32}$$

### Natural Convection to Gases

When a cool gas is exposed to a warmer surface, the temperature of the gas close to the surface increases. This warm gas has a smaller density than the surrounding gases. As a result, the warmer gas flows upward from the hot surface, setting up a natural flow of the surrounding gases past the hot surface. Heat transfer from one material to another when the flow is of this type is termed *natural convection.*

For the common gases, the individual coefficient of heat transfer due to natural convection may be calculated by the following dimensional equation.

$$h_c = K(\Delta t)^{0.25} \tag{6-33}$$

where $h_c$ has the units of Btu/(h)(ft$^2$)(°F) and $\Delta t$ is in °F. Values of $K$ for natural convection to air are given in Table 6-2.

### Condensing Vapors

When a saturated vapor condenses on a cool surface, the condensation may be either dropwise or as a liquid film. When the condensation is in drops, a much higher film coefficient is obtained than when the condensation takes place to give a liquid film over the cool surface. Theoretical methods have been developed for estimating individual coefficients of heat transfer for film-type condensation.

[4] Another equation that is often used for streamline flow of fluids inside circular pipes is

$$\left(\frac{hD}{k}\right) = 1.86\left(\frac{DG}{\mu}\frac{c_p \mu}{k}\frac{D}{L}\right)^{1/3}\left(\frac{\mu}{\mu_w}\right)^{0.14}$$

**Table 6-2 Natural convection to air at room temperature and pressure**

| Condition | Value of $K$ |
| --- | --- |
| Horizontal plates, facing upward | 0.38 |
| Horizontal plates, facing downward | 0.20 |
| Vertical plates more than 2 ft high | 0.27 |
| Vertical plates less than 2 ft high | $\dfrac{0.28}{(\text{Vertical height in ft})^{0.25}}$ |
| Vertical pipes more than 1 ft high | $\dfrac{0.27}{(D_o \text{ in ft})^{0.25}}$ |
| Horizontal pipes | $\dfrac{0.27}{(D_o \text{ in ft})^{0.25}}$ |

For the case of film-type condensation of a saturated vapor on horizontal tubes, the following equation, with consistent use of American engineering units or SI units, may be used to estimate the film coefficient.

$$h = 0.73 \sqrt[4]{\frac{k_f^3 \lambda g \rho_f^2}{D \mu_f \, \Delta t}} \qquad (6\text{-}34)$$

The individual heat-transfer coefficient for dropwise condensation may be ten or more times larger than the coefficient for film-type condensation when all other variables are held constant. There are no general equations available for the direct determination of film coefficients for dropwise condensation.

## Boiling Liquids

A liquid boils when heat is supplied so rapidly that a portion of the liquid becomes vapor at the heating surface. Bubbles of vapor rise through the liquid, resulting in the phenomenon designated as "boiling."

As the temperature difference between the heat source and the liquid is increased, the boiling becomes more violent. When the temperature difference becomes sufficiently high, the surface of the heat source becomes covered with a film of vapor. This vapor film has a higher resistance to heat transfer than the liquid film. Thus, it is quite possible for the flow of heat to be reduced by the use of too high a temperature difference. This can be illustrated by consideration of the rate of evaporation when a drop of water is placed on an electrically heated hot plate. If the temperature of the plate is high enough, say 400°F, a drop of water will dance on the plate on an insulating film of steam. If the plate temperature is lowered to 240°F, a drop of water will wet the surface, and it can be observed that the water will evaporate more rapidly at this lower temperature than at the higher plate temperature.

**Table 6-3 Order of magnitude of various individual film coefficients**

| Condition | $h$, Btu/(h)(ft)(°F) | J/(s)($m^2$)(K) |
|---|---|---|
| Dropwise condensation of steam | 10,000–20,000 | 57,000–114,000 |
| Film-type condensation of steam | 1000–3000 | 5700–17,000 |
| Boiling water | 300–9000 | 1700–51,000 |
| Film-type condensation of organic vapors | 200–400 | 1100–2200 |
| Heating or cooling water | 50–3000 | 300–17,000 |
| Superheated steam | 5–20 | 30–110 |
| Heating or cooling air | 0.2–8 | 1.1–45 |

The temperature difference above which the individual heat-transfer coefficient starts to decrease markedly through formation of an insulating vapor film is known as the *critical temperature difference*, For boiling water, the critical temperature difference is approximately 45°F.

## RADIATION

Radiant energy is emitted in all directions by a body. If this energy can contact a receiver, some of the energy will be absorbed by this receiver. The flow of heat from a hot body to a cold body in this manner is known as *heat transfer by thermal radiation.*

### Black Body

When radiant energy strikes a surface, part of the energy is absorbed, part is reflected, and some may pass through the receiver unaffected if the body is not completely opaque. We may postulate an ideal case where all the radiant energy impinging on a surface is absorbed and none is reflected or transmitted. Such a receiver is known as a *black body*. A piece of rough black cloth is a common example of a material closely approximating a black body.

### Stefan-Boltzmann Law of Radiation for Black Bodies

It has been observed experimentally that the amount of thermal radiation emitted from a body increases rapidly with increase in temperature. Stefan and Boltzmann found that the amount of radiant energy emitted is proportional to the fourth power of the absolute temperature of the heat-source body. The proportionality constant was found to be $0.173 \times 10^{-8}$ when the energy emitted had the units of Btu per hour per square foot and the temperature was expressed in degrees Rankine.[5] This can be expressed in equation form as

$$q = \text{Btu/h} = 0.173A\left(\frac{T}{100}\right)^4 \tag{6-35}$$

[5] For units of joule per second per square meter, with $T$ in K, the proportionality constant of 0.173 in Eqs. (6-35) through (6-38) becomes 5.73.

If a cold black body at $T_1$ is surrounded by a hot black body at $T_2$, each of the bodies will emit radiant energy according to Eq. (6-35). The net transfer of radiant energy from the hot to the cold body will be the difference between the radiant energy absorbed by body 1 and the energy radiated away from body 1. Thus the net gain of radiant energy by body 1 will be

$$q_{1,\text{net}} = 0.173A_1\left[\left(\frac{T_2}{100}\right)^4 - \left(\frac{T_1}{100}\right)^4\right] \tag{6-36}$$

The net loss of heat from the hot black body 2 cannot be calculated by direct application of Eq. (6-36) using $A_2$ in place of $A_1$. Since not all the heat radiated from body 2 is necessarily received by body 1, a certain shape factor must be applied to Eq. (6-36) when it is used for determining the net radiant heat loss of the surrounding black body; thus

$$q_{2,\text{net}} = 0.173A_2\left[\left(\frac{T_2}{100}\right)^4 - \left(\frac{T_1}{100}\right)^4\right]F_A \tag{6-37}$$

## Absorptivity and Emissivity

The fraction of incident radiation which is absorbed by a material is called its *absorptivity*.

The total radiant energy emitted from a unit area of a body in unit time, divided by the total radiant energy emitted per unit time from a unit area of a black body at the same temperature, is known as its *emissivity*. At thermal equilibrium, the emissivity and the absorptivity of a body are equal.

The radiant heat exchange between two nonblack bodies can be calculated by the following equation:

$$q = 0.173A\left[\left(\frac{T_2}{100}\right)^4 - \left(\frac{T_1}{100}\right)^4\right]F_{AE} \tag{6-38}$$

where $F_{AE}$ is a correction factor which allows for the angle at which the surfaces view each other and for the emissivities and absorptivities of the surfaces. The factor $F_{AE}$ is also dependent on which surface is chosen to evaluate the area term $A$.[6]

**Example 6-5: Heat transfer by radiation between black bodies** A sphere 1 ft in diameter is placed in a completely closed box. The temperature of the sphere is maintained at 1000°F, and the inside surface of the box is kept at 400°F. Assuming the sphere and the inside surface of the box are black bodies, calculate the radiation heat loss from the sphere to the box in Btu per hour.

[6] For methods of determining these correction factors, see any textbook or handbook on radiation heat transfer.

SOLUTION

$$\text{Surface area of the sphere} = \pi D^2 = (3.14)(1)^2 = 3.14 \text{ ft}^2$$

$$\text{Heat loss from sphere to box} = (0.173)(3.14)\left[\left(\frac{1460}{100}\right)^4 - \left(\frac{860}{100}\right)^4\right]$$

$$= 21{,}700 \text{ Btu/h}$$

## PRACTICAL APPLICATIONS OF HEAT-TRANSFER PRINCIPLES

There are many practical items which must be considered in the design, installation, and operation of heat-transfer equipment. For example, if a heat exchanger is not properly installed, it may be possible for inert gases to be trapped in the unit. These gases cause a reduction in the overall heat-transfer coefficient or a reduction in the effective area for transfer of heat. Consequently, it is necessary to install the equipment so that inert gases can be swept out by the entering fluid or to include a bleed valve to be used for bleeding off any trapped gases.

The rest of this chapter deals with heat-transfer equipment and the practical applications of heat-transfer principles.

### Fouling

In most heat exchangers, heat is transferred from a hot fluid through a confining wall into a colder fluid. Each of the fluids and the confining wall offer resistance to heat flow. However, there is another resistance which may be a major factor in determining the rate of heat transfer. This is the resistance due to the formation of scale or dirt deposits on either side of the solid wall separating the two fluids.

As the amount of scale or dirt deposited on the surfaces increases, the rate of heat transfer must decrease unless a greater temperature-difference driving force is used. Therefore, it is necessary to shut down the equipment from time to time and clean the heat-transfer surfaces to keep the resistance due to scale or dirt deposits from becoming excessive.

**Dirt deposits.** Liquid mixtures containing suspended solids tend to foul heat exchangers very rapidly. At low velocities, the solids settle out and bake into a cake on the hot walls. Water containing dirt or slime is a common example of this type of fouling mixture. It is standard practice to maintain a fluid velocity of at least 3 ft/s (0.9 m/s) if the liquid involved contains any suspended solids. Such liquids should be passed through the inside of the tubes in a tube-and-shell heat exchanger, since it is very difficult to eliminate low-velocity pockets on the shell side of an exchanger.

As far as possible, suspended solids should be removed from the liquid before it enters a heat exchanger. This can be accomplished by the use of settling tanks or filters. Screen filters are commonly used on water lines to remove debris such as sticks, pebbles, or pieces of algae, but a screen filter will not remove finely dispersed solids.

**Scale.** Hard water is the source of most scale. As the water is heated, the scale formation increases. At temperatures higher than 120°F (322 K), the formation of scale from hard water becomes excessive. Therefore, when hard water is used as a heat-transfer medium, the temperature of the water is generally not permitted to exceed 120°F. In many cases, it is advisable to soften the water by a chemical treatment before using it in a heat exchanger. Hot water leaving a heat exchanger may be reused by employing a cooling tower to lower the temperature of the water. Cooling towers are widely used by industrial concerns to reduce the amount of water purchased and to decrease the softening costs.

**Cleaning of heat exchangers.** Chemical methods and mechanical methods are used for removing scale and dirt deposits from heat exchangers. Inhibited acids and other types of chemicals may be employed for cleaning purposes, but they require expert handling to prevent damage to the metal parts of the exchanger.

Mechanical cleaning methods involve a large amount of manual labor. If the deposit is on the inside of straight tubes, the cleaning can be accomplished by merely forcing a long worm or wire brush through each tube. Much more labor is entailed if the deposit is on the shell side of the tubes, since the tube bundle must be removed from the shell and special cleaning methods such as sandblasting must be employed.

If possible, fluids which have a high fouling effect should be used on the inside of the tubes rather than on the shell side of a tube-and-shell exchanger, in order to permit easier cleaning. The final decision as to which fluid should pass through the inside of the tubes also depends on the operating pressures involved, the corrosive properties of the fluids, the allowable pressure drop across the system, and the arrangement which will give the greatest overall heat-transfer coefficient or allow the maximum rate of heat transfer.

The necessary cleaning and maintenance of heat exchangers should be considered when the equipment is installed. Sufficient space must be available to permit removal of the tube bundle and to allow inside cleaning of the individual tubes. The need for this extra space is illustrated by Fig. 6-2, which shows the tube bundle being lowered into the shell of a tube-and-shell heat exchanger.

**Allowances for fouling.** In the design of heat exchangers, it is customary to make an allowance for fouling effects by adding extra heating-surface area. This is accomplished by including a fouling-resistance term when the overall heat-transfer coefficient is determined, as is indicated in Eq. (6-26).

Table 6-4 gives fouling coefficients for different fluids. These coefficients may be used in the design of heat exchangers and should permit the unit to maintain design capacity for about one year. The capacity of a clean heat exchanger normally exceeds the process requirements, but the capacity will steadily decrease while the

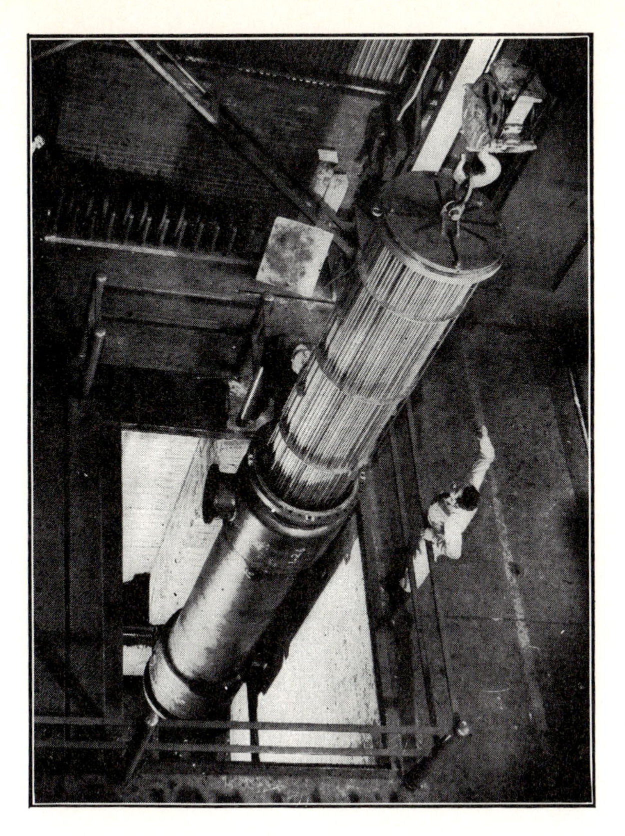

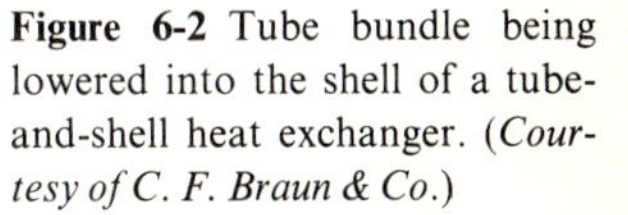

**Figure 6-2** Tube bundle being lowered into the shell of a tube-and-shell heat exchanger. (*Courtesy of C. F. Braun & Co.*)

unit is in service because of the formation of scale and dirt deposits. The use of the correct fouling coefficients permits a reasonable time interval between cleanings.

## Insulation

When insulation is used for the purpose of reducing heat losses, the optimum thickness of the insulation may be determined by means of an economic balance. If the amount of money saved through reduction in heat losses by the use of insulation is balanced against the fixed costs for the installed insulation, it is possible to determine the optimum insulation thickness. This method is presented in Chap. 14 (Chemical Engineering Economics and Plant Design).

Insulation does not always reduce heat losses. In fact, there are many cases where the addition of insulation actually increases heat losses. For example, consider a bare wire of small diameter in contact with air. If an electric current is passed through the wire, the temperature of the wire will increase and heat will be transferred from the wire surface to the surrounding air. If the wire is insulated, the temperature at the surface of the insulation will be less than the wire-surface temperature, but the increased area available for heat transfer to the air may more than offset the reduced temperature-difference driving force. As a result, the rate

**Table 6-4 Fouling heat-transfer coefficients**

| Medium | $h_d$, Btu/(h)(ft$^2$)(°F) | $h_d$, J/(s)(m$^2$)(K) |
|---|---|---|
| Sea-water | 1000–2000 | 5700–11,400 |
| Brackish water | 300–500 | 1700–2800 |
| Cooling-tower water: | | |
| Treated | 500–1000 | 2800–5700 |
| Untreated | 200–400 | 1100–2300 |
| River water: | | |
| Clean | 500–1000 | 2800–5700 |
| Silty | 200–500 | 1100–2800 |
| Hard water (over 15 grains/gal) | 200–350 | 1100–2000 |
| Distilled water | 1000–2000 | 5700–11,400 |
| Steam condensate | 1000–2000 | 5700–11,400 |
| Steam: | | |
| Clean | 2000 | 11,400 |
| Oily | 1000 | 5700 |
| Oils (liquid and vapor): | | |
| Propane, butane, pentanes | 1000–2000 | 5700–11,400 |
| Gasoline—Sweet | 1000–2000 | 5700–11,400 |
| Corrosive | 300–1000 | 1700–5700 |
| Lean oil—Sweet | 500–1000 | 2800–5700 |
| Corrosive | 200–500 | 1100–2800 |
| Fuel oils—Clean | 300–500 | 1700–2800 |
| Dirty | 200–300 | 1100–1700 |
| Crudes—Clean | 300–500 | 1700–2800 |
| Dirty | 200–300 | 1100–1700 |
| Salty | 100–200 | 550–1100 |
| Air | 500 | 2800 |

of heat transfer may be greater from the insulated wire than from the bare wire. Consequently, there should be an insulation thickness at which the heat losses are a maximum. The actual value of this maximum-heat-loss thickness depends on the thermal conductivity of the insulation and the heat-transfer coefficients (including radiation effects) at the insulation surface. The electrical engineer applies this principle by covering bare wires with a thin insulating material which increases the cooling of the wires and simultaneously provides protection against shock or grounding if a person or metallic object should touch the insulated wire.

## OPERATION OF HEAT-TRANSFER EQUIPMENT

The volume of metals increases as their temperature increases, and the extent of the volume change depends on the type of metal involved as well as on the total change in temperature. When a heat exchanger is put into service after a shutdown period, it is important to warm the equipment slowly. Otherwise, undue stresses will be exerted on the metallic parts because of unequal expansion. The same rule applies when the equipment is shut down, and the cooling-down period should be just as slow as the initial warming period.

Most heat exchangers are equipped with floating heads or expansion joints to handle the volume changes of the metallic parts with temperature variations. When dealing with the flow of hot fluids through pipes, it is essential to recognize the dangers due to linear expansion with temperature changes. If the temperature of a steel pipe is increased from 60 to 530°F, the length of the pipe will increase by 4 in for every 100 ft of pipe length. This expansion could easily cause a straight pipe to buckle if it were fastened firmly at both ends. The dangers due to expansion can be minimized by the use of loops in the lines, by bellows joints, or even by including several 90° bends in the piping system.

Most of the operating difficulties with heat exchangers arise during the start-up or shutdown periods. Some general rules regarding the standard methods for operating heat exchangers follow:

1. *Start-up.* Always introduce the colder fluid first. Add the hotter fluid slowly, until the unit is up to the operating conditions. Be sure the entire unit is filled

**Table 6-5 Design data for heat exchangers***

| Service: | | |
|---|---|---|
| Item: | In shell | In tubes |
| Fluid | | |
| Total influent, lb/h: | | |
| Liquid | | |
| Vapor | | |
| Noncondensables | | |
| Steam | | |
| Fluid condensed or vaporized, lb/h | | |
| Specific gravity of liquid at 60°F | | |
| Viscosity, cP | _at_°F | _at_°F |
| Molecular weight of vapor | | |
| Heat capacity, Btu/(lb)(°F) | _at_°F | _at_°F |
| Latent heat, Btu/lb | _at_°F | _at_°F |
| Temperature in, °F | | |
| Temperature out, °F | | |
| Design temperature, °F | | |
| Pressure, lb/in², gage, at inlet | | |
| Design pressure, lb/in² gage | | |
| Allowable pressure drop, lb/in² | | |

Tube layout: Fouling factor $= \frac{1}{h_d}$

Tube material:

Tube size: inch od: Gage: Length:

Space limitations, if any:

Position of exchanger: Horizontal Vertical

* This table uses American engineering units. SI units may also be required in specification tables of this type.

with fluid and there are no pockets of trapped inert gases. Use the bleed valve to bleed off trapped gases.

2. *Shut-down.* Shut off the hot fluid first but do not allow the unit to cool too rapidly. Drain any materials which may cause trouble if left in the exchanger. Examples of this would be water (if the temperature may drop below freezing) or any material which would become solid when cold.
3. *Steam Condensate.* Always drain any steam condensate from heat exchangers when starting up or shutting down. This reduces the possibility of water hammer caused by steam forcing the trapped water through the lines at high velocities.

## Heat Exchangers

Table 6-5 indicates the basic information necessary to carry out the design of a heat exchanger. Many industrial concerns prefer to let the fabricators work out the details of a heat-exchanger design from overall specifications, and Table 6-5 shows the type of overall specifications that should be presented to the fabricator.

## PROBLEMS*

**6-1** A flat hard-rubber wall [$k_{ave} = 0.092$ Btu/(h)(ft$^2$)(°F)/ft)] is 3 in thick. If the difference in surface temperatures from one side of the wall to the other is 100°F, calculate the rate of heat flow in Btu per hour per square foot of wall surface.

**6-2** A copper rod is 3 ft long and 2 in in diameter. The rod is insulated so that radial heat losses may be neglected. If one end of the rod has a temperature of 70°F and the other end is at 220°F, calculate the rate of flow axially along the rod as Btu per hour. For copper, $k_{ave}$ may be taken as 220 Btu/(h)(ft$^2$)(°F/ft).

**6-3** The walls of a large furnace are made up of three layers of different-type bricks. The inside layer is made of 6 in of a special refractory brick [$k = 0.07$ Btu/(h)(ft$^2$)(°F/ft)]. The second layer is 3 in of a Sil-O-Cel brick ($k = 0.04$). The outside layer is 10 in of common brick ($k = 0.70$). The inside surface temperature is 1400°F and the outside surface temperature is 100°F. Calculate (*a*) the loss of heat as Btu per hour per square foot of wall surface; (*b*) the temperature at the interface between the refractory and Sil-O-Cel bricks.

**6-4*** A steam pipe with an outside diameter of 3.3 in (0.0838 m) is lagged with a cork insulation. The insulation is 3 in (0.0762 m) thick, and $k_{ave}$ for the cork is 0.025 Btu/(h)(ft$^2$)(°F/ft)[0.0433 J/(s)(m$^2$)(K/m)]. Thermocouples located at the inside and outside surfaces of the cork indicate temperatures of 230°F (383 K) and 100°F (311 K), respectively. Calculate the heat loss as Btu per hour per linear foot of pipe. Answer as joules per hour per meter with SI.

**6-5** Water flows through a 2-in standard (Schedule 40) steel pipe. The pipe is jacketed with steam at 230°F. The conditions are such that the overall heat-transfer coefficient is 250 Btu/(h)(ft$^2$ of outside pipe area) (°F). If the temperature of the water is 80°F at the pipe entrance and 180°F at the pipe exit, calculate the average amount of heat gained by the water per hour for each foot of pipe length.

**6-6*** An oil with a heat capacity at constant pressure of 0.85 Btu/(lb)(°F)[3559 J/(kg)(K)] flows through a standard (Schedule 40) 3-in (0.0762-m)-diameter pipe at the rate of 4000 lb/h (1814 kg/h). The pipe is 20 ft (6.10 m) long and is jacketed with saturated steam at 230°F (383 K). The oil enters the pipe at 60°F (289 K) and leaves at 190°F (361 K). Assuming a constant overall heat-transfer coefficient for

* All the problems with asterisks should be solved using American engineering units and also SI units.

this system, calculate the value of the overall heat-transfer coefficient based on the outside area of the pipe.

**6-7** Water flows through a copper pipe with an outside diameter of 5.0 in and an inside diameter of 4.5 in. The pipe is jacketed with steam, and the overall coefficient of heat transfer is found to be 425 Btu/(h)(ft$^2$ of inside pipe area)(°F). The water-film coefficient is 600 Btu/(h)(ft$^2$)(°F). The average thermal conductivity of copper is 220 Btu/(h)(ft$^2$)(°F/ft). Calculate the value of the steam-film heat-transfer coefficient.

**6-8** A standard (Schedule 40) 2-in-diameter steel pipe carries water under such conditions that the water-film coefficient is 450 Btu/(h)(ft$^2$)(°F). The pipe is jacketed with steam having a film coefficient of 2000 Btu/(h)(ft$^2$)(°F). What is the rate of heat transfer in Btu per hour per foot of pipe length if the mean overall temperature difference is 100°F? The average thermal conductivity of steel may be taken as 26 Btu/(h)(ft$^2$)(°F/ft).

**6-9*** Water is heated from 80 to 150°F (300 to 339 K) in a 3-in (0.0762-m) standard (Schedule 40) steel pipe by saturated steam at 218°F (376 K) condensing outside the pipe. The water-film coefficient is 500 Btu/(h)(ft$^2$)(°F) [2839 J/(s)(m$^2$)(K)], and the steam-film coefficient is 2000 Btu/(h)(ft$^2$)(°F) [11,356 J/(s)(m$^2$)(K)]. What length of pipe is necessary if the water flows at a rate of 20,000 lb/h (9072 kg/h)?

**6-10** Calculate the overall heat-transfer coefficient based on the inside pipe area for the following system: Alcohol flows through a standard (Schedule 40) 2-in-diameter steel pipe. The alcohol-film coefficient is 400 Btu/(h)(ft$^2$)(°F). A scale deposit with a thickness of 0.003 in and a value of $k$ of 0.5 Btu/(h)(ft$^2$)(°F/ft) is formed on the outside of the pipe. Steam with a film coefficient of 1900 Btu/(h)(ft$^2$)(°F) is condensing on the pipe. The mean difference between the alcohol bulk temperature and the steam tempeature is 90°F. There is no scale on the inside of the pipe.

**6-11*** Anhydrous ethyl alcohol enters a section of standard (Schedule 40) 1-in (0.0254-m) steel pipe at 70°F (294 K) and leaves at 125°F (325 K). The average linear velocity of the alcohol in the pipe is 3 ft/s (0.914 m/s). The pipe is jacketed with saturated steam at 212°F (373 K). The steam film coefficient is 1800 Btu/(h)(ft$^2$)(°F) [10,221 J/(s)(m$^2$)(K)]. The average heat capacity at constant pressure at the bulk temperature of the alcohol may be taken as 0.685 Btu/(lb)(°F) [2868 J/(kg)(K)], average density as 0.78 g/cc, average thermal conductivity as 0.104 Btu/(h)(ft$^2$)(°F/ft) [0.180 J/(s)(m$^2$)(K/m)], and average viscosity as 0.92 cP. Calculate (*a*) the individual film coefficient for the alcohol; (*b*) the length of pipe required.

**6-12** A horizontal circular plate with a diameter of 8 in is located in a room where the air temperature is 70°F. The upper surface of the plate has a temperature of 400°F. Calculate the amount of heat lost from the upper surface of the plate to the air by natural convection in Btu per hour.

**6-13*** A heated flat plate is placed in a large room under such conditions that the region surrounding the plate can be considered as representing a black body. The heated plate is 10 in (0.254 m) in diameter and 2 in (0.0508 m) thick. All surfaces of the plate are maintained at 800°F (700 K). The temperature of the surroundings is 70°F (294 K). If heat is lost by radiation from the top, bottom, and sides of the plate, calculate the total radiation heat loss from the plate in Btu per hour. Answer as joules per hour with SI. The plate may be assumed as a black body.

**6-14** A hot plate with an exposed area of 3 ft$^2$ is placed in a room. The emissivity of the metal making up the plate is 0.85. The plate surface is maintained at 500°F, and the room temperature is kept at 90°F. Calculate the amount of heat lost from the plate by radiation in Btu over a period of 10 h. The nonblack-body correction factor ($F_{AE}$) for this case is 0.85.

**6-15** Water is to flow through the inside of tubes in a one-pass heat exchanger with steam condensing on the outside of the tubes at a temperature of 394 K. The water is to be heated from 283 K to 339 K with a water flow rate of 454 kg/min. The overall heat transfer coefficient ($U$) based on the outside area of the tubes is 852 J/(s)(m$^2$)(K). What will be the cost for the exchanger if the purchase cost for the unit is \$323/m$^2$ of outside tube area? Heat capacity for $H_2O$ = 4187 J/(kg)(K).

**6-16** A pipe containing steam is covered with a cork insulation with thermal conductivity ($k$) of 0.069 J/(s)(m$^2$)(K/m). The outside diameter of the pipe is 0.0838 m, and the insulation thickness is 0.1016 m, making the outside diameter of the insulation 0.287 m. The steady-state temperatures at the inside and outside surfaces of the cork are 394 K and 311 K, respectively. What is the heat loss through the insulation as joules per hour per linear meter of pipe length?

**6-17*** A standard (Schedule 40) 3-in (0.0762-m)-diameter steel pipe carries a liquid under such conditions that the liquid-film coefficient on the inside of the pipe is 400 Btu/(h)(ft$^2$)(°F) [2272 J/(s)(m$^2$)(K)]. The pipe is jacketed with steam having a film coefficient of 2000 Btu/(h)(ft$^2$)(°F) [11,357 J/(s)(m$^2$)(K)] at a constant condensing temperature of 250°F (394 K). The heated section of the pipe is 20 ft (6.096 m) long, and the fluid enters the section at 50°F (283 K) and leaves at 200°F (366 K). There is no scale or dirt on the pipe. If each pound of steam condensed gives up 945 Btu (997,000 J) of heat to the fluid, how many pounds of steam are needed per hour? Answer as kilograms per hour with SI.

CHAPTER
# SEVEN

# EVAPORATION

Evaporation may be defined as an operation whereby a fluid changes from the liquid state into the vapor state. In chemical engineering practice, the term *evaporation* is used to denote the removal of a valueless component from a mixture by a process involving vaporization. The mixture ordinarily consists of a nonvolatile solid or liquid and a volatile liquid.

Because water is the only volatile material available in such large quantities that it can be considered valueless, the operation of evaporation almost always refers to the separation of water from a mixture by vaporization. Common examples of evaporation processes are found in the concentration of aqueous solutions of sodium hydroxide, sodium chloride, glycerol, sugar, or glue.

The transfer of heat and the transfer of mass are the two basic processes involved in evaporation. Heat must be supplied to the mixture to furnish the energy necessary for the vaporization. The volatile liquid changes to a vapor, and this vapor must be removed.

The heat may be supplied by exposing the liquid directly to the source of heat, as, for example, in evaporation by the use of the sun's rays. Heat may also be supplied indirectly by transferring it through a suitable retaining medium. An example of this is evaporation with steam as the heat source, where the steam flows through the inside of tubes immersed in the evaporating mixture.

The vapor evolved in the course of an evaporation operation is generally removed in an undiluted form. Vapor may also be removed by passing an inert gas over the surface of the evaporating liquid.

Many different types of equipment are used for carrying out evaporation operations. Open pans exposing a large liquid area to the sun's rays are commonly used for obtaining salt by the evaporation of seawater. Horizontal-tube evaporators contain steam tubes arranged horizontally on the inside of the evaporator body. The steam passes through these tubes, and the steam condensation supplies heat indirectly through the tube walls to the evaporating mixture.

Forced-circulation evaporators usually contain vertical tubes surrounded by steam. The mixture to be evaporated is forced through the inside of the tubes, where heat is gained indirectly from the condensing steam. Good heat-transfer coefficients can be obtained as a result of the high velocity of the liquid through the tubes.

Long-tube vertical evaporators are often used with natural circulation. Their construction is similar to the forced-circulation evaporators. The liquid rises up through the tubes because of the decrease in liquid density with increase in temperature. The heat-transfer coefficients generally are not so high with natural circulation as with forced circulation. However, the expenses of a circulation pump and the accompanying power costs are not incurred when natural circulation is used.

The first part of this chapter presents the basic principles involved in the unit operation of evaporation where heat is transferred indirectly from condensing steam to the evaporating liquid mixture. Some practical applications of these

**Table 7-1 Nomenclature for evaporation**

$a$ = constant
$A$ = area of heat-transfer surface, $m^2$ or $ft^2$
$b$ = constant
B.P.R. = boiling-point rise, K or °F
$h_{shv}$ = total heat content of a superheated vapor, J/kg or Btu/lb
$h_{sv}$ = total heat content of a saturated vapor, J/kg or Btu/lb
$K$ = constant in Table 7-2 (represents the mass of water evaporated per hour when there is no boiling-point rise), kg/h or lb/h
$n$ = number of effects in a multiple-effect evaporator
$n_a$, $n_b$, $n_c$ = number of moles of component $a$, $b$, or $c$ in a solution
$p$ = equilibrium vapor pressure of a component in a solution, Pa or $lb/in^2$
$P°$ = equilibrium vapor pressure of a component in a solution if the component were pure and at the same temperature as the solution, Pa or $lb/in^2$
$q$ = rate of heat transfer through the heating surface, J/s or Btu/h
$t_{cond}$ = condensing temperature of vapors from last effect of a multiple-effect evaporator (equals saturation steam-table temperature corresponding to pressure in vapor space of final effect), K or °F
$t_{sh}$ = amount of superheat in a vapor (equals actual vapor temperature minus temperature of the vapor if it were saturated at the same pressure), K or °F
$t_{steam}$ = temperature of condensing original steam, K or °F
$U$ = standard overall heat-transfer coefficient, $J/(s)(m^2)(K)$ or $Btu/(h)(ft^2)(°F)$
$x$ = mole fraction of a component in a solution

Greek symbols

$\Delta t$ = delta $t$, standard temperature difference (equals temperature of steam condensing in steam chest minus temperature of boiling liquid at liquid-vapor interface in evaporator body), K or °F
$\Delta t_1$ = delta $t$, standard temperature difference in 1st effect, K or °F
$\Delta t_2$ = delta $t$, standard temperature difference in 2d effect, K or °F
$\Delta t_3$ = delta $t$, standard temperature difference in 3d effect, K or °F
$\theta$ = theta, time, h, min, or s
$\Sigma$ = sigma, symbol indicating a summation

principles are presented in the last part of the chapter with special emphasis on operating the equipment.

## HEAT TRANSFER IN EVAPORATORS

The concept of an overall heat-transfer coefficient is employed in the treatment of heat transfer in evaporators. The general equation as presented in Chap. 6 (Heat Transfer) can be used; that is,

$$q = UA\,\Delta t \tag{7-1}$$

In Eq. (7-1), $q$ represents the rate at which heat is transferred through the heating surface, $U$ is the overall heat-transfer coefficient, and $A$ is the heat-transfer area.

## THE TEMPERATURE-DIFFERENCE DRIVING FORCE ($\Delta t$)

The temperature-difference driving force ($\Delta t$) in Eq. (7-1) represents the difference between the temperature of the condensing steam in the steam chest and the temperature of the boiling liquid in the evaporator body. The pressure of the condensing steam in the steam chest can be determined very easily. From this pressure, by the use of steam tables (see Table C-11 in App. C), it is a simple matter to determine the temperature of the saturated steam.

Because of the effect of liquid head, the temperature of the boiling liquid is not constant throughout the entire body of the liquid. For example, the boiling temperature 2 m below the exposed surface of a liquid must be higher than the boiling temperature at the exposed surface because of the pressure increase caused by the 2 m of liquid head. Therefore, it is necessary to choose a standard basis for defining the temperature of the boiling liquid in an evaporator body.

The accepted definition for the temperature of the boiling liquid in an evaporator body is the temperature of the boiling solution at the pressure of the vapor space, or, in other words, the temperature of the boiling solution at the exposed surface of the liquid. With this definition, the standard (or net) temperature difference ($\Delta t$) becomes the difference between the temperature of the condensing steam in the steam chest and the temperature of the boiling solution at the liquid-vapor interface in the evaporator body.

### Apparent Temperature Difference

The pressure of the vapor over the liquid in an evaporator body can be determined quickly and accurately. By the use of steam tables, it is possible to find the temperature of the vapor, assuming it is saturated at the existing pressure. This temperature is defined as the apparent saturation temperature of the vapor leaving the evaporating mixture.

The difference between the steam temperature and the apparent saturation temperature of the vapor leaving the evaporating mixture, as determined from the pressure of the vapor, is designated as the *apparent temperature difference.*

In the past, the apparent temperature difference has been used in Eq. (7-1) to calulate a so-called "apparent overall coefficient." Actually, this apparent value is only an approximation, as it is based on the assumption that the liquid is pure water. Since the liquid mixture in an evaporator ordinarily is not pure water, the temperature of the liquid is not exactly the same as the steam-table saturation temperature based on the pressure in the vapor space. This effect will be discussed in more detail in the next section.

## Boiling-Point Rise due to Material in Solution

A pure liquid (or solvent) exerts a certain vapor pressure at any given temperature. If another substance is dissolved in this pure liquid, the vapor pressure of the solvent over the mixture will be less than that of the pure solvent at the same temperature. This can be visualized by considering that the molecules of the dissolved substance are distributed evenly throughout the solvent. Therefore, not all the surface area of the mixture is available for transfer of the solvent molecules from the liquid to the vapor state. In its simplest form, then, this means the vapor pressure of the solvent is less than it would be if the solvent were pure.

A solution boils when its vapor pressure equals the pressure of the surroundings. Thus, a solution of a nonvolatile material in water must be heated above the boiling point of pure water before boiling can occur. The *boiling-point rise due to material in solution* is defined as the actual surface temperature of the mixture minus the temperature of the pure solvent if it exerted the same vapor pressure as the mixture.

As an example of boiling-point rise due to material in solution, consider the case of a 30% by weight sodium hydroxide–in–water solution at a surface temperature of 175°F. If this solution were pure water at 175°F, the water-vapor pressure above the mixture would be 6.715 lb/in$^2$, absolute. The actual water-vapor pressure above a solution containing 30% by weight sodium hydroxide at 175°F is 3.718 lb/in$^2$, absolute. The steam-table temperature of water corresponding to a pressure of 3.718 lb/in$^2$ is 150°F. Therefore, if the temperature of the pure solvent were 150°F, it would exert exactly the same vapor pressure as the 30 weight percent mixture does at 175°F. From the basic B.P.R. definition, the boiling-point rise due to material in solution, for this case, is 175°F − 150°F, or 25°F.

**Dühring's rule.** A valuable empirical law known as Dühring's rule has been developed. It is very useful for determining boiling-point rise due to material in solution. This rule states that a plot of the temperature of a constant-concentration solution versus the temperature of a reference substance, where the reference substance and the solution exert the same pressure, results in a straight line. Pure water is often used as the reference material. Only two experimental values of vapor pressure and temperature are necessary for a solution of fixed concentration

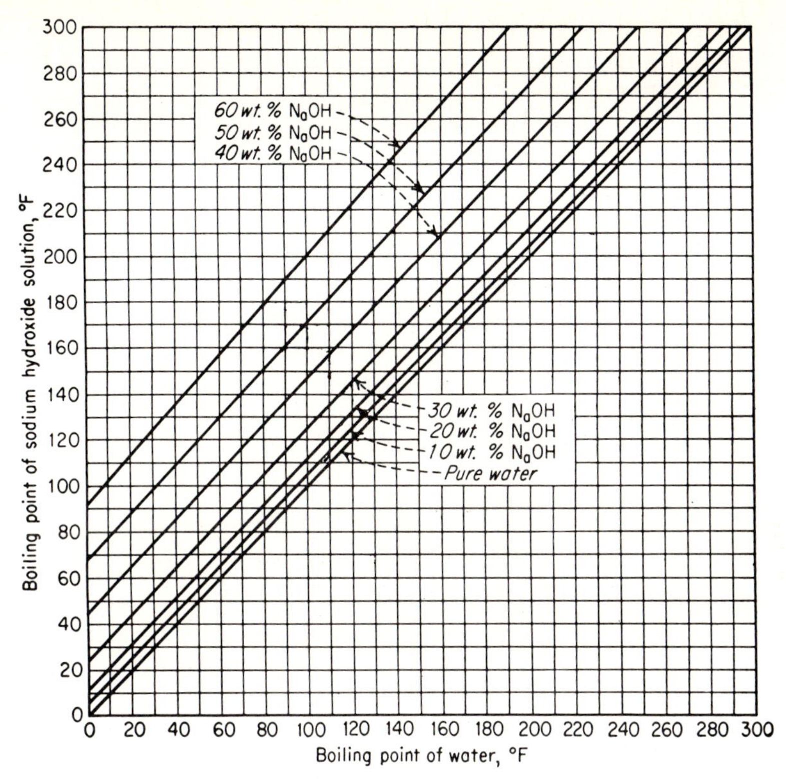


**Figure 7-1** Dühring lines for aqueous solutions of sodium hydroxide.

to determine the entire Dühring line for the given mixture. While it is better to determine more than two experimental values in order to eliminate small errors, theoretically the entire line can be set by two points. Figure 7-1 presents a Dühring plot for solutions of sodium hydroxide in water.

**Raoult's law.** For certain types of liquid solutions, a generalization known as Raoult's law may be used to determine the boiling-point rise due to material in solution. Raoult's law states that the *equilibrium* vapor pressure which is exerted by a component in a liquid solution equals the mole fraction of that component in the solution times the *equilibrium* vapor pressure the component would exert if it were pure and at the same temperature as the solution. Thus

$$p_a = P_a^\circ\left(\frac{n_a}{n_a + n_b + n_c + \cdots}\right) = x_a P_a^\circ \tag{7-2}$$

where $p_a$ = *equilibrium* vapor pressure of component $a$ in solution with components $a, b, c, \ldots$

$P_a^\circ$ = *equilibrium* vapor pressure of component $a$ if it were pure and at same temperature as the solution

$n_a, n_b, n_c, \ldots$ = moles of components $a, b, c, \ldots$ in the solution

$x_a$ = mole fraction of component $a$ in the liquid solution

Raoult's law applies only to mixtures in which the components are similar chemically and in which the molecules of the components do not interact in any way. A solution of sodium hydroxide in water would not follow Raoult's law, since the sodium hydroxide dissociates to form sodium and hydroxyl ions as soon as it is dissolved in water. On the other hand, a solution of glycerol in water would follow Raoult's law fairly closely, because there is no appreciable ionization or interaction when glycerol is dissolved in water and because the two materials are not widely different in their chemical properties.

The boiling-point rise due to material in solution can be determined by the use of Raoult's law if this law is applicable to the components of the solution. For example, consider the case of a water solution containing 30% by weight glycerol ($C_3H_5(OH)_3$) at a temperature of 200°F. The vapor pressure of glycerol may be assumed as negligible, and the solution can be considered as following Raoult's law. The mole fraction of water in this mixture is

$$\frac{\frac{70}{18}}{\frac{70}{18} + \frac{30}{92}} = 0.923$$

The vapor pressure of pure water at 200°F is 11.526 lb/in$^2$, absolute. From Raoult's law, the equilibrium vapor pressure of water over the 30% mixture must be (11.526)(0.923) = 10.64 lb/in$^2$, absolute. If the mixture boils at 200°F, the pressure in the vapor space due to water vapor will be 10.64 lb/in$^2$, absolute. The steam-table temperature corresponding to saturated water vapor at 10.64 lb/in$^2$, absolute, is 196.1°F. Therefore, the boiling-point rise due to material in solution for a 30 wt% mixture of glycerol in water boiling at 200°F is 200 − 196.1 = 3.9°F.

**Example 7-1: Determination of boiling-point rise for Raoult's law mixture using SI units** A solution of sugar ($C_{12}H_{22}O_{11}$, MW = 342) in water as a 67.5 wt% sugar mixture follows Raoult's law. In an evaporation operation where the mixture is at its boiling point, the temperature of the solution at the surface is 366.5 K. What is the boiling-point rise for the mixture in K?

SOLUTION Raoult's law is

$$p_{H_2O} = \frac{n_{H_2O}}{n_{H_2O} + n_{sugar}} P^{\circ\text{ at }366.5\text{ K}}_{H_2O}$$

$P^{\circ\text{ at }366.5\text{ K}}$ = 79.5 kPa by the steam tables in App. C (Table C-11) where 366.5 K = 200°F and absolute pressure in pounds per square inch can be read off directly as 11.526 lb/in$^2$, absolute = (11.526 lb/in$^2$, absolute)(6.895 kPa/lb/in$^2$, absolute) = 79.5 kPa.

**Basis**

100 kg of mixture

$$\text{kg mol of water} = n_{H_2O} = \frac{(0.325)(100)}{18} = 1.806$$

$$\text{kg mol of sugar} = n_{sugar} = \frac{(0.675)(100)}{342} = 0.197$$

By Raoult's law,

$$p_{H_2O} = \frac{1.806}{1.806 + 0.197}(79.5) = 71.7\text{ kPa} = \frac{71.7}{6.895} = 10.399\text{ lb/in}^2\text{, absolute}$$

By the steam table in App. C (Table C-11), with linear interpolation between temperatures of 195°F and 200°F, the temperature of pure water to produce the same vapor pressure as is exerted by the mixture (or to exert a pressure of

$$10.399\text{ lb/in}^2\text{, absolute}) = 195 + \frac{10.399 - 10.385}{11.526 - 10.385}(5) = 195.1°\text{F} = 363.8\text{ K}$$

Boiling-point rise = actual surface temperature of the mixture − temperature of pure water to produce the same vapor pressure as is exerted by the mixture = 366.5 − 363.8 = 2.7 K

## Boiling-Point Rise due to Hydrostatic Head

In many types of evaporators, the heat source (e.g., steam tubes) is submerged below the surface of the evaporating liquid. The solution at the outside of the steam tubes is at a higher pressure than the solution at the top surface of the liquid because of the presence of the existing head of liquid. If boiling occurs at the steam tubes, the liquid temperature at this point of higher pressure must be greater than at the top liquid surface.

*Boiling-point rise due to hydrostatic head* may be defined as the difference in temperature of the liquid at the heat source and the liquid at the top evaporating surface. If the average density and concentration of the liquid solution are known, it is possible to determine the B.P.R. due to hydrostatic head for any given liquid head and vapor-space pressure.

It is impossible to make an accurate determination of the average head for most types of evaporators. For example, with forced-circulation vertical-tube evaporators, the total head varies over the entire length of the evaporator. In addition, the liquid entering the bottom of the evaporator is not at its boiling point. Thus, boiling does not occur throughout the entire evaporator.

Ordinarily, the presence of B.P.R. due to hydrostatic head has little effect on overall heat-transfer calculations. However, its presence should be recognized. Because of the difficulties encountered in determining the head and the corresponding boiling-point rise, it has become standard practice in evaporator calculations to neglect the effect of hydrostatic head. Overall coefficients and heat-transfer rates are calculated by assuming the temperature of the boiling liquid to be that at the interface between the liquid and the vapor.

In the past, a so-called "corrected overall coefficient" was often used. This coefficient was obtained by assuming the temperature of the boiling liquid to be the temperature at the point halfway between the top and bottom of the boiling material. The use of the "corrected overall coefficient" and the "apparent overall coefficient" has resulted in considerable confusion. The former requires extra calculation labor without appreciable gain in accuracy, while the latter is usually only a rough approximation. The use of the standard overall coefficient, based on the difference in temperature between the steam in the steam chest and the boiling liquid at the liquid-vapor interface in the evaporator body, is now generally accepted.

## Standard Overall Coefficients

The overall heat-transfer equation, as indicated by Eq. (7-1), is used to calculate the standard overall coefficient. In this case, the $\Delta t$ is called the *standard (or net) temperature difference* and equals the apparent temperature difference minus the B.P.R. due to material in solution. As an example, consider the case of an evaporator producing a 30% by weight solution of sodium hydroxide in water when the pressure in the evaporator vapor space is 3.718 lb/in$^2$, absolute, and the steam-chest pressure is 25 lb/in$^2$, absolute. The steam-table temperature for saturated water vapor exerting a pressure of 3.718 lb/in$^2$, absolute, is 150°F, and the temperature of saturated steam at 25 lb/in$^2$, absolute, is 240°F. In this example, the apparent temperature difference is $240 - 150 = 90$°F. This value of $\Delta t$ would be used to calculate the apparent overall coefficient. However, as indicated in the section on boiling-point rise due to material in solution, the actual B.P.R. due to material in solution is 25°F for this case. Therefore, the temperature of the liquid at the liquid-vapor interface is $150 + 25 = 175$°F, and the standard temperature difference is $240 - 150 - 25 = 240 - 175 = 65$°F. This value of $\Delta t$ would be used to calculate the standard overall coefficient.

## Direct Determination of Amount of Heat Transferred

In many evaporator calculations, it is necessary to estimate the amount of heat-transfer area required to carry out a given evaporation. In calculations of this type, the standard overall coefficient is usually given or can be assumed from past experience with evaporators. The standard temperature difference can be calculated from the steam pressure, the vapor-space operating pressure, and the concentration of the mixture being evaporated. The equation $q = UA\,\Delta t$ can then be used to calculate the area if a value of $q$ is available.

The rate of heat transfer in a continuous evaporator operating under steady-state conditions is determined essentially by two factors. They are (1) the amount of sensible heat required to heat the feed from its entering temperature to the boiling temperature and (2) the amount of latent heat required to vaporize the water. The amount of sensible heat may be calculated from the heat capacity of the entering solution and the difference between the boiling temperature in the evaporator and the entering feed temperature.

For feeds of inorganic salts in water up to about 25 wt% salt concentration, the heat capacity may be assumed as equal to that of the water alone. Thus, the heat capacity of a sodium hydroxide–water solution containing 10% by weight sodium hydroxide would be 0.9 Btu/(lb)(°F) or 3768 J/(kg)(K), taking the heat capacity for pure water as 1 Btu/(lb)(°F) or 4186.8 J/(kg)(K).

The latent heat of evaporation may be taken as the heat necessary to vaporize 1 lb (or 1 kg) of water at the exposed-surface temperature of the solution. For the case of a 30 wt % sodium hydroxide–in–water solution with a vapor pressure of 3.718 lb/in$^2$, absolute, the surface temperature of the liquid is 175°F. Therefore, the heat of vaporization can be taken as the value for water at 175°F, or 993.3 Btu/lb. Heats of vaporization for water at various temperatures can be found in the steam tables. (See Table C-11 in App. C.)

It should be realized that the methods of calculation as indicated in the preceding discussion have neglected any unaccounted-for heat losses such as losses from outside the evaporator. Heats of dilution have also been neglected.

## Assumptions in Evaporator Calculations

In making calculations dealing with evaporators, it is generally acceptable to make certain assumptions in order to simplify the work. Some of the more important of these are summarized here.

1. The standard (or net) temperature difference ($\Delta t$) is based on the temperature of the boiling liquid at the liquid-vapor interface in the evaporator body.
2. The heat required to vaporize 1 lb (or 1 kg) of the solvent is taken as the latent heat of vaporization at the exposed-surface temperature of the solution.
3. For feeds of inorganic salts in water, the heat capacity may be assumed as equal to that of the water alone.
4. The effect of boiling-point rise due to hydrostatic head is commonly neglected.
5. Boiling-point-rise calculations for continuous evaporators are based on a mixture having the same concentration as the liquid leaving the evaporator.
6. The sensible heat necessary to heat the feed to the boiling point may be approximated by assuming the heat capacity of the original feed is constant until the feed has reached the boiling temperature of the mixture.

**Example 7-2: Calculation of heat-transfer area required in an evaporator** A continuous single-effect evaporator is to concentrate 25,000 lb/h of a 5 percent by weight sodium hydroxide–in–water solution to a final concentration of 20 wt % sodium hydroxide. The vapor space of the evaporator will be maintained at atmospheric pressure (14.7 lb/in$^2$, absolute), and the steam pressure will be kept at 35 lb/in$^2$, gage. The standard overall heat-transfer coefficient can be assumed as 150 Btu/(h)(ft$^2$)(°F). The feed enters the evaporator at 70°F. Determine the area of heat-transfer surface theoretically necessary to carry out the evaporation.

SOLUTION The $\Delta t$ for use in the equation $q = UA\,\Delta t$ will be calculated first. A pressure of 35 lb/in$^2$, gage, is equivalent to $35 + 14.7 = 49.7$ lb/in$^2$ absolute.

From the steam tables, the temperature of saturated steam at 49.7 lb/in$^2$, absolute is 280.6°F.

The vapor-space pressure is 14.7 $lb/in^2$, absolute. This corresponds to a pure-water temperature of 212°F. From Fig. 7-1, the exposed-surface temperature of a 20 wt % sodium hydroxide–in–water solution must be 224°F to give a vapor of the same pressure as water at 212°F. Thus, the boiling-point rise due to dissolved sodium hydroxide is 12°F.

The standard temperature difference = $\Delta t = 280.6 - 224 = 56.6°F$.

The value of $q$, as Btu per hour, can be calculated as follows: Since the feed has a concentration of 95 wt % water, its heat capacity can be taken as 0.95 Btu/(lb)(°F). It is assumed that the original feed must be heated to the boiling-liquid temperature; therefore, the sensible heat added to the feed is

$$(25{,}000)(0.95)(224 - 70) = 3{,}660{,}000 \text{ Btu/h}$$

By applying the principles of stoichiometry, the total mass of water evaporated per hour is mass of water in minus mass of water out, or

$$(0.95)(25{,}000) - \frac{(25{,}000)(0.05)(0.80)}{(0.20)} = 18{,}750 \text{ lb/h}$$

From the steam tables, the latent heat of vaporization of water at 224°F is 962.6 Btu/lb. Therefore, the heat required for actual vaporization of the water is

$$(18{,}750)(962.6) = 18{,}050{,}000 \text{ Btu/h}$$

$$\text{Total value of } q = 3{,}660{,}000 + 18{,}050{,}000$$
$$= 21{,}710{,}000 \text{ Btu/h}$$

$$U = 150 \text{ Btu/(h)(ft}^2\text{)(°F)}$$

$$\Delta t = 56.6°\text{F}$$

$$\text{Area} = A = \frac{q}{U\,\Delta t} = \frac{21{,}710{,}000}{(150)(56.6)} = 2560 \text{ ft}^2$$

The total heat-transfer area theoretically necessary is 2560 $ft^2$.

## MULTIPLE-EFFECT EVAPORATORS

When a series of evaporator bodies is connected so that the vapors from one body act as the heat source for the next body, the evaporator system is called *multiple effect*. By an arrangement of this type, it is possible for 1 lb (or 1 kg) of initial steam to evaporate more than 1 lb (or 1 kg) of water.

Figure 7-2 presents a drawing of a typical three-effect evaporator. The vapors evolved in effect No. 1 are condensed in the steam chest of the second effect. The heat given up during this condensation serves to evaporate some of the volatile component of the liquid in effect No. 2. These evolved vapors are, in turn, condensed in the steam chest of the third effect, releasing heat which is used for evaporation of the liquid in effect No. 3.

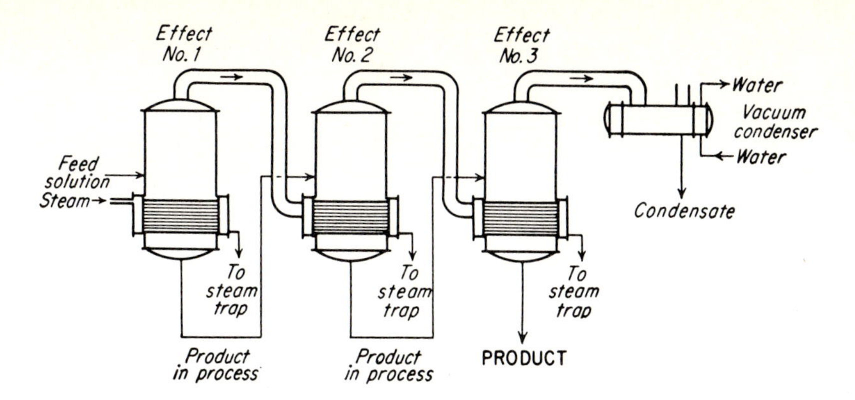


**Figure 7-2** Triple-effect evaporator.

From the discussion of heat transfer in evaporators, it is clear that there must be a temperature-difference driving force between the temperature in the steam chest and the temperature of the boiling liquid. Therefore, the boiling temperature of the liquid in effect No. 3 must be lower than the liquid boiling temperature in effect No. 2. The same reasoning indicates that the liquid boiling temperature in the second effect must be lower than the temperature of the liquid in the first effect.

One way to obtain the decrease in boiling temperature from one effect to another is to put a vacuum pump or vacuum condenser on the vapor space of the final effect. As an example, consider the case of pure water being evaporated in a three-effect evaporator. A steam ejector or vacuum pump is used to maintain a pressure of 4 lb/in$^2$, absolute, in the vapor space of the third effect, and steam at a saturation temperature of 230°F is supplied to the first effect. The steam-table temperature corresponding to 4 lb/in$^2$, absolute, is 153°F. If the effect of hydrostatic head is neglected, the temperature of the boiling water in the third effect must be 153°F.

Designating the standard temperature differences in effects 1, 2, and 3 by $\Delta t_1$, $\Delta t_2$, and $\Delta t_3$, the temperature of the vapors condensing in the steam chest of effect No. 3 must be $153 + \Delta t_3$ °F. Since the vapor temperature and, therefore, the liquid temperature in the second effect are $153 + \Delta t_3$ °F, the temperature of the vapors condensing in the steam chest of the second effect must be $153 + \Delta t_3 + \Delta t_2$. Similarly, the temperature of the original steam condensing in the steam chest of the first effect must be $153 + \Delta t_3 + \Delta t_2 + \Delta t_1 = 230$°F. Thus, the total available temperature drop across the three effects is 230°F − 153°F, or the original steam temperature minus the steam-table temperature corresponding to the vapor-space pressure in the final evaporator.

If, in the preceding example, 1 lb of steam condenses in the steam chest of the first effect, enough heat will be released to evaporate approximately 1 lb of water into the vapor space of effect No. 1. This pound of water vapor condenses in the steam chest of the second effect, where the heat of condensation is sufficient to evaporate approximately 1 lb of water from the liquid in effect No. 2. Thus, in a two-effect evaporator, 1 lb of original steam can evaporate approximately 2 lb of water. In a three-effect evaporator, 1 lb of original steam can evaporate approximately 3 lb of water.

The heat of vaporization of water increases as the equilibrium pressure is decreased, and the original feed to an evaporator is seldom at the boiling temperature. Consequently, under ordinary conditions it is not possible to obtain exactly $n$ lb of water evaporated per pound of original steam supplied to an $n$-effect evaporator. This is the reason why the word "approximately" was used in the preceding paragraph.

## Economy

The total number of pounds (or kilograms) of water vaporized in an evaporator per pound (or kilogram) of original steam supplied is known as the *economy*. The actual economy of a multiple-effect evaporator may approach the number of effects, but it will never reach it unless the temperature of the feed is higher than the boiling point of the liquid in the evaporator body which the feed enters. In a three-effect evaporator with feed entering at its boiling point, the economy may be above 2.9.

## Capacity

The *capacity* of an evaporator may be defined as the total amount of evaporation it is capable of producing in a unit time. The capacity may be determined by measurement of the total heat transferred to the evaporating mixture. This amount of heat can be found by use of the equation $q = UA\,\Delta t$.

In a three-effect evaporator, the total rate of heat transfer to the evaporating mixture is

$$q = q_1 + q_2 + q_3 = U_1A_1\,\Delta t_1 + U_2A_2\,\Delta t_2 + U_3A_3\,\Delta t_3 \tag{7-3}$$

If the areas and overall coefficients are approximately equal in all three evaporators, the total rate of heat transfer may be expressed as follows:

$$q = U_1A_1(\Delta t_1 + \Delta t_2 + \Delta t_3) \tag{7-4}$$

Thus, as long as the total of all the temperature-difference driving forces is kept constant, the capacity of a single-effect evaporator with a heat-transfer area of $A$ ft$^2$ is practically the same as the capacity of an $n$-effect evaporator in which each of the effects has a heat-transfer area of $A$ ft$^2$.

The difference between economy and capacity should be clearly understood. The number of evaporator effects largely determines the economy but has little effect on the capacity.

In designing an evaporator system for a given capacity, the best number of effects is governed by an economic balance between the savings on the amount of steam required for operation with multiple effects and the extra cost of investment for the additional effects.

## Boiling-Point Rise in Multiple-Effect Evaporators

The presence of dissolved material in an evaporating mixture causes a boiling-point rise. This effect must be taken into consideration in determining the temperature-difference driving force to use in the equation $q = UA\,\Delta t$. There is also a boiling-point rise due to the hydrostatic head in the individual evaporators; however, this may be neglected in ordinary calculations where the standard overall heat-transfer coefficient is used.

Consider the example of a two-effect evaporator in which the steam entering the steam chest has a saturation temperature of 235°F and the pressure in the vapor space of the second effect is 6 lb/in$^2$, absolute. Since the steam-table temperature corresponding to 6 lb/in$^2$, absolute, is 170°F, the apparent overall temperature difference over both effects is $235 - 170 = 65$°F. Assume the vapor-space pressure in the first effect is 13 lb/in$^2$, absolute, and the boiling-point rise in this effect due to material in solution has been calculated to be 10°F. The steam-table temperature corresponding to 13 lb/in$^2$, absolute, is 205.9°F. Therefore, the actual temperature of the evaporating liquid is $205.9 + 10 = 215.9$°F at the exposed liquid surface, and the standard temperature-difference driving force for this effect is $235 - 205.9 - 10 = 235 - 215.9 = 19.1$°F.

The actual temperature of the vapor leaving the first effect is 215.9°F; i.e., it is superheated by 10°F. This vapor goes to the steam chest of the second evaporator, where it gives up heat by condensation. In cases of this type, where the vapor entering the steam chest is superheated, the superheat is neglected, and the temperature of the vapor is assumed as that at its saturation or condensing point. Thus, in the present example, the temperature of the condensing vapor may be taken as 205.9°F for use in determining the temperature-difference driving force in the second effect.

Assume that in the second effect the B.P.R. due to material in solution has been calculated to be 14°F. The standard temperature difference in this effect must then be $205.9 - 170 - 14 = 205.9 - 184 = 21.9$°F.

From the preceding example, it can bc seen that the overall apparent temperature difference must equal the overall standard temperature difference plus the total boiling-point rise due to material in solution. This statement applies to multiple-effect or single-effect evaporators.

**Results of boiling-point rise.** If minor influences (such as feed temperatures and change in latent heats of vaporization with change in pressure) are neglected, the economy of multiple-effect evaporators is not affected by boiling-point rise. One pound of steam condensing in the steam chest of the first effect must evaporate approximately 1 lb of water, and, in turn, this pound of water vapor must evaporate approximately 1 lb of water in the next effect. Therefore, the economy is independent of the boiling-point rise.

Considering the product of the overall heat-transfer coefficient and heat-transfer area in each effect as constant, the following equations are applicable for the cases of evaporation with and without boiling-point rise: With B.P.R.:

$$q = UA(\Delta t_{\text{total (apparent)}} - \Sigma \text{B.P.R.}) \tag{7-5}$$

**Table 7-2 Approximate effect of boiling-point rise in multiple-effect evaporators**

$n$ = number of evaporator effects

| | Without B.P.R. | With B.P.R. |
|---|---|---|
| Economy | $n$ | $n$ |
| Mass of water evaporated per hour (capacity) | $K$ | $K\left[1 - \frac{\Sigma \text{B.P.R.}}{(t_{\text{steam}} - t_{\text{cond}})}\right]$ |
| Mass of steam necessary per hour | $\frac{K}{n}$ | $\frac{K}{n}\left[1 - \frac{\Sigma \text{B.P.R.}}{(t_{\text{steam}} - t_{\text{cond}})}\right]$ |
| Mass of vapor to final condenser per hour | $\frac{K}{n}$ | $\frac{K}{n}\left[1 - \frac{\Sigma \text{B.P.R.}}{(t_{\text{steam}} - t_{\text{cond}})}\right]$ |

Without B.P.R.

$$q = UA\,\Delta t_{\text{total (apparent)}} \tag{7-6}$$

$$\Delta t_{\text{total (apparent)}} = t_{\text{steam}} - t_{\text{cond}} \tag{7-7}$$

From these equations, it is obvious that the capacity of an evaporator is decreased by the presence of a boiling-point rise. The amount of heat transferred per hour and, therefore, the total amount of water evaporated per hour decrease with increasing B.P.R. as long as the initial steam temperature and final condensing temperature are held constant.

Table 7-2 presents mathematical relationships showing the approximate effect of B.P.R. on economy, mass of water evaporated per hour, mass of steam necessary per hour, and mass of vapor to the final condenser per hour.

## Direction of Feed

The individual effects of a multiple-effect evaporator are numbered in the order of decreasing temperature and pressure of the solution in the effect. If the initial feed to the system enters effect No. 1, then passes to No. 2, No. 3, No. 4, etc., the unit is said to be operating with *forward feed*.

When forward feed is used, the most concentrated solution is in the last effect where the temperature of the solution is the lowest. If the final product has a high concentration of dissolved material, it may become very viscous at low temperatures. In such cases, it is often advantageous to reverse the direction of feed in order to have the most concentrated material in the effect with the highest solution temperature. This may be accomplished by operating the equipment with *backward feed*; i.e., with the feed entering the last effect and being pumped progressively to effects at higher temperature and pressure.

Combinations of forward and backward feed in evaporators containing three or more effects are designated as *mixed feed*. If feed enters each effect and product is withdrawn from each effect, the equipment is said to be operating with *parallel feed*.

## Calculations for Multiple-Effect Evaporators

In making practical calculations dealing with multiple-effect evaporators, the results which are usually desired are these: (1) the area of heating surface necessary in each effect, (2) the pounds of steam per hour which must be supplied, and (3) the amount of vapor leaving the last effect and going to the condenser. These results can be obtained from a knowledge of the steam pressure to be used in the steam chest of the first effect, the final condenser pressure to be maintained, the amount and concentration of material to be fed to the evaporator per unit time, the final concentration desired, physical properties of the mixture being evaporated, and the standard overall heat-transfer coefficient in each effect.

To simplify the construction of the units comprising a multiple-effect evaporator, all the individual effects are ordinarily identical. This means the heat-transfer area in all the effects may be assumed as equal.

The basic method of making calculations for multiple-effect evaporators is as follows:

1. A convenient basis (such as 1 h) is chosen.
2. From the known concentration leaving the last effect and the pressure in the vapor space of this effect, the boiling-point rise due to material in solution and the boiling-liquid temperature in the last effect are calculated.
3. Values are assumed for the standard temperature-difference driving force in each effect.
4. Using the temperature differences assumed in step 3, heat balances are made around each effect. These heat balances yield the amount of evaporation in each effect and the amount of heat transferred in each effect.
5. By use of the equation $q = UA\,\Delta t$, the heat-transfer area in each effect is calculated.
6. If the areas calculated in step 5 are not all essentially equal, new standard temperature differences are assumed. By noting the way the areas vary, it is possible to make a good assumption for the new standard temperature differences by using the following relationship:

$$\Delta t_{\text{revised}} = \Delta t_{\text{original}}\,(A_{\text{calc}}/A_{\text{ave of calc areas}})$$

This method of solution is not so tedious as it appears. With a little experience, the necessary assumptions can be made quite accurately on the first trial. The heat balances around the effects involve no new principles. A convenient level is chosen as the zero heat (or enthalpy) point. All the heat content above this point entering the effect must equal all the heat content above this zero point leaving the effect. To simplify the development of the heat balances, it is recommended that the zero heat level be chosen as liquid water at 32°F. This corresponds to the base energy level chosen for the common steam tables and permits direct use of the values read from the steam tables.

For ordinary evaporator problems, the following equation may be used to calculate the total heat content of superheated steam:

$$h_{shv} = h_{sv} + 0.45t_{sh} \qquad \text{Btu/lb} \tag{7-8}$$

where $h_{shv}$ is the total heat content of 1 lb of the superheated vapor, $h_{sv}$ is the total heat content of 1 lb of saturated vapor at the pressure of the superheated vapor, and $t_{sh}$ is the amount of superheat in degrees Fahrenheit. All heat-content values should be expressed as Btu per pound. At pressures below 40 lb/in$^2$, absolute, and superheats of less than 100°F, Eq. (7-8) predicts the heat content of superheated steam to within $\frac{1}{2}$ percent of the correct value.

The application of the overall method of calculation for evaporator problems is best shown by an example.

**Example 7-3: Calculations for multiple-effect evaporator** A two-effect evaporator is to be used to concentrate 20,000 lb/h of a solution containing 5 percent by weight sodium hydroxide in water to a product containing 20 wt % sodium hydroxide. Forward feed will be used. The feed enters the first effect at 80°F. Saturated steam at 220°F enters the steam chest of the first effect. A pressure of 1.275 lb/in$^2$, absolute, is maintained in the vapor space of the last effect. The standard overall coefficient of heat transfer may be taken as 265 Btu/(h)(ft$^2$)(°F) in the first effect and 200 Btu/(h)(ft$^2$)(°F) in the second effect. The heat-transfer area is to be the same in each effect.

Estimate the values of the following:

(*a*) The heat-transfer area in each effect
(*b*) The pounds of original steam necessary per hour
(*c*) The pounds of vapor leaving the last effect per hour

SOLUTION

(1) Choose a basis of 1 h.
(2) The steam-table temperature corresponding to 1.275 lb/in$^2$, absolute, is 110°F. From Fig. 7-1, for a 20% by weight solution of sodium hydroxide in water, the temperature of the solution in the final effect must be 120°F to give a vapor having the same pressure as water at 110°F. The boiling-point rise in this effect is then 10°F.
(3) The apparent temperature difference over both evaporators is 220 − 110 = 110°F. The boiling-point rise in the first effect will be about 3°F, since the concentration will be much less than 20 percent in this effect. Therefore

$$\Delta t_1 + \Delta t_2 + 3 + 10 = 110°\text{F}$$

Assume $\Delta t_1 = 47$°F and $\Delta t_2 = 50$°F. Using these assumptions:
Liquid and vapor temperature in second effect = 120°F
Degrees superheat of vapor in second effect = 10°F

Steam-table total heat content of vapor at 120°F containing 10°F superheat $= 1109.5 + 0.45(10) = 1114.0$ Btu/lb

Saturated vapor temperature in first effect $= 120 + 50 = 170$°F

Degrees superheat of vapor in first effect = 3°F

Liquid and vapor temperature in first effect = 173°F

Steam-table total heat content of vapor at 173°F containing 3°F superheat = 1135.5 Btu/lb

Steam-table heat content of liquid water at 170°F = 137.9 Btu/lb

Saturated steam temperature in first-effect steam chest $= 173 + 47 = 220$°F

Steam-table total heat content of saturated steam at 220°F = 1153.4 Btu/lb

Steam-table heat content of liquid water at 220°F = 188.1 Btu/lb

(4) Heat balances:

$x$ = pounds water evaporated in first effect per hour

$y$ = pounds water evaporated in second effect per hour

$w$ = pounds of steam supplied to first effect per hour

Base the energy balances on a zero heat-content level of liquid water at 32°F.

Heat balance around first effect:

Heat input:

Heat in steam $= w(1135.5)$ Btu/h

Heat in feed $= (20{,}000)(80 - 32)(0.95)$ Btu/h

Heat output:

Heat in evolved vapors $= x(0135.5)$ Btu/h

Heat in liquid mixture leaving the effect $= (20{,}000 - x)(173 - 32)(0.92)$ Btu/h. (This assumes the mixture is concentrated to 8 wt % sodium hydroxide in the first effect or a B.P.R. of 3°F.)

Heat in steam condensate $= w(188.1)$ Btu/h

$$w(1153.4) + (20{,}000)(80 - 32)(0.95) = x(1135.5) + (20{,}000 - x)(173 - 32)(0.92) + w(188.1) \qquad \text{(A)}$$

Heat balance around second effect:

Heat input:

Heat in vapors to steam chest $= x(1135.5)$ Btu/h

Heat in feed $= (20{,}000 - x)(173 - 32)(0.92)$ Btu/h

Heat output:

Heat in evolved vapors $= y(1114.0)$ Btu/h

Heat in liquid product $= (20{,}000 - x - y)(120 - 32)(0.80)$ Btu/h

Heat in steam condensate $= x(137.9)$ Btu/h

$$x(1135.5) + (20{,}000 - x)(173 - 32)(0.92) = y(1114.0) + (20{,}000 - x - y)(120 - 32)(0.80) + x(137.9) \qquad \text{(B)}$$

By the application of stoichiometry:

$$\text{Total pounds of water evaporated} = 20{,}000 - \frac{(20{,}000)(0.05)}{(0.20)}$$

$$= x + y = 15{,}000 \qquad \text{(C)}$$

Equations (A), (B), and (C) represent three separate equations involving three unknowns. Solving simultaneously,

$$x = 7300 \text{ lb/h}$$
$$y = 7700 \text{ lb/h}$$
$$w = 9350 \text{ lb/h}$$

$q_1$ = Heat released from original steam to effect No. 1 steam chest
$= 9350(1153.4 - 188.1) = 9{,}030{,}000$ Btu/h

$q_2$ = Heat released from vapors into effect No. 2 steam chest
$= 7300(1135.5 - 137.9) = 7{,}280{,}000$ Btu/h

(5) Using the equation $q = UA\,\Delta t$,

$$A_1 = \frac{9{,}030{,}000}{(265)(47)} = 724 \text{ ft}^2$$

$$A_2 = \frac{7{,}280{,}000}{(200)(50)} = 728 \text{ ft}^2$$

Since $x = 7300$ lb, the concentration of sodium hydroxide leaving the first effect is

$$\frac{(20{,}000)(0.05)}{(20{,}000 - 7300)}(100) = 7.88\%$$

therefore, the 8 percent concentration assumption made in obtaining Eqs. (A) and (B) is satisfactory.

(6) The two areas, 724 and 728 ft$^2$, are practically equal. Therefore, the temperature-difference assumptions are adequate. The final area can be taken as the average of the $A_1$ and $A_2$ obtained.

ANSWERS

(*a*) Heat-transfer area in each effect = 726 ft$^2$
(*b*) Pounds of steam necessary per hour = 9350 lb/h
(*c*) Pounds of vapor leaving last effect = 7700 lb/h

## IMPORTANT FACTORS IN OPERATION OF EVAPORATORS

The efficient operation of an evaporator requires a high degree of skill and a complete understanding of the unit. An experienced operator can quickly locate the cause for a disruption in the steady conditions of an evaporator operation if he or she understands the principles of evaporation.

For example, consider a continuous three-effect evaporator which is being operated with fixed feed and product compositions. The operator notices that the capacity of the unit has suddenly decreased (i.e., the rate of concentrated solution leaving the final effect suddenly decreases). A check of the unit reveals that the only other changes are an increase in the temperature and pressure of the vapors leaving effect No. 1 and a decrease in the temperature and pressure of the vapors leaving effect No. 2. This immediately indicates that the temperature-difference driving force has increased in the second effect. Since the concentration in the various effects and the overall temperature difference have remained constant, the only possible conclusion is that the rate of heat transfer in the second effect has decreased, causing a corresponding decrease in rate of heat flow to the other effects.

The decrease in heat-transfer rate in the second effect must have been caused by a reduction in the $UA$ product of the equation $q = UA\,\Delta t_2$. There are four possible reasons why the $UA$ product would decrease. They are as follows:

1. Inadequate removal of the condensate from the second-effect steam chest
2. Collection of noncondensable gases in the second-effect steam chest
3. Drop in liquid level in the second effect, exposing the steam tubes to the vapor space
4. Fouling of the heating surface in the second effect

The operator has several possible choices to get the unit back up to its original capacity. The operator may bypass the steam trap on the second-effect steam chest to make certain the trap has not become plugged or may open the bleed line on the steam chest to remove trapped noncondensables. The operator can also check the liquid level in the body of the second effect by means of the sight glass always attached to each effect. In order to make certain the lines leading to the sight glass are not plugged, the operator should drain the sight glass and note whether the level returns to its original value when the sight glass is put back into use. If none of these simple procedures reveals the source of the trouble, it may be assumed that the heat-transfer area in the second effect has become fouled, and it will probably be impossible to regain the original operating conditions until the unit is shut down and cleaned.

Reasoning processes such as those presented in the preceding example are very valuable for understanding the basic principles involved in evaporation. Table 7-3 presents similar conditions which might be encountered in the actual operation of a three-effect evaporator. The results of the disrupting conditions are indicated in the table. It would be beneficial to the reader's understanding of evaporation principles to reason out the results presented in Table 7-3.

## Scale Formation

During the course of an evaporation, solids often deposit on the heat-transfer surfaces, forming a scale. The presence of the scale causes an increase in the resistance to the flow of heat and, consequently, a decrease in the capacity of the unit if the same temperature-difference driving forces are maintained.

Scale formation occurs to some extent in all types of evaporators, but it is of particular importance when the solution being evaporated contains a dissolved material which has an inverted solubility. The expression "inverted solubility"

**Table 7-3 Results of disrupting occurrences in continuous evaporator operation**

Three-effect evaporator operating under steady conditions and controlled to maintain a constant concentration of product leaving the third effect; $L$ = less, $C$ = constant, $H$ = higher.

$n$ = number of evaporation effects

| | Comparative result if following occurs: | | | | | | | | | | | |
|---|---|---|---|---|---|---|---|---|---|---|---|---|
| Evaporator variable | Condenser temperature rises | | | Steam temperature drops | | | Steam chest in 2d effect one-half filled with water | | | Air leak into vapor line leading to effect No. 2 steam chest | | |
| In effect No. | 1 | 2 | 3 | 1 | 2 | 3 | 1 | 2 | 3 | 1 | 2 | 3 |
| $\Delta t$ | $L$ | $L$ | $L$ | $L$ | $L$ | $L$ | $L$ | $H$ | $L$ | $L$ | $H$ | $L$ |
| $U$* | $C$ | $C$ | $C$ | $C$ | $C$ | $C$ | $C$ | $C$ | $C$ | $C$ | $L$ | $C$ |
| Temperature and pressure of vapor leaving | $H$ | $H$ | $H$ | $L$ | $L$ | $C$ | $H$ | $L$ | $C$ | $H$ | $L$ | $C$ |
| Concentration leaving | $C$ | $C$ | $C$ | $C$ | $C$ | $C$ | $C$ | $C$ | $C$ | $C$ | $C$ | $C$ |
| Capacity | $L$ | $L$ | $L$ | $L$ | $L$ | $L$ | $L$ | $L$ | $L$ | $L$ | $L$ | $L$ |
| Economy | $C$ | $C$ | $C$ | $C$ | $C$ | $C$ | $C$ | $C$ | $C$ | $C$ | $C$ | $C$ |

* It is assumed here that $U$ is not affected by changes in $\Delta t$.

means the solubility decreases as the temperature of the solution is increased. For a material of this type, the solubility is least near the heat-transfer surface where the temperature is the greatest. Thus, any solid crystallizing out of the solution does so near the heat-transfer surface and is quite likely to form a scale on this surface.

The most common scale-forming substances are calcium sulfate, calcium hydroxide, sodium carbonate, sodium sulfate, and calcium salts of certain organic acids. While it is impossible to prevent the formation of scale if scale-forming materials are present, the rate of formation may be decreased by using high liquid velocities past the heat-transfer surfaces.

Where true scale formation occurs, it can be shown[1] that the overall coefficient may be related to the time the evaporator has been in operation by the straight-line equation

$$\frac{1}{U^2} = a\theta + b \tag{7-9}$$

where $a$ and $b$ are constants for any given operation and $U$ is the overall heat-transfer coefficient at any time $\theta$ since the beginning of the operation. A typical scale-formation plot of $1/U^2$ versus $\theta$ is presented in Fig. 7-3. Equation (7-8) may be used to determine the optimum length of time an evaporator should be operated between cleanings in order to give the maximum overall capacity.

[1] W. McCabe and C. Robinson, *Ind Eng Chem*, **16**:478 (1924).

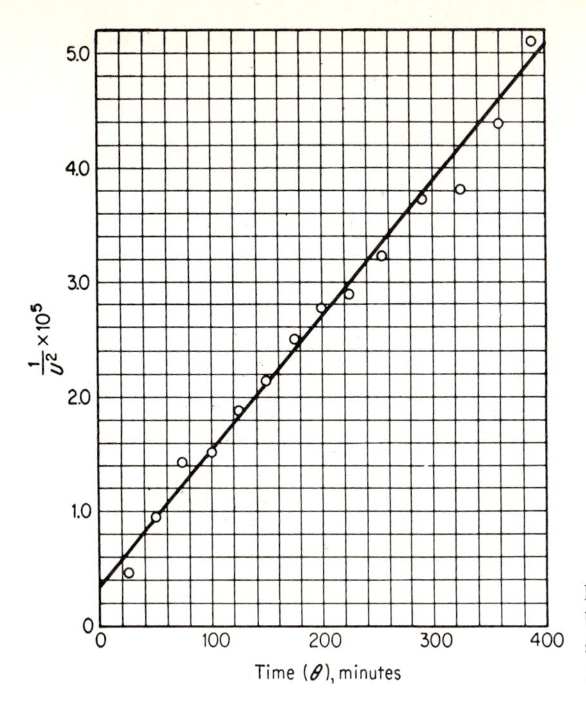


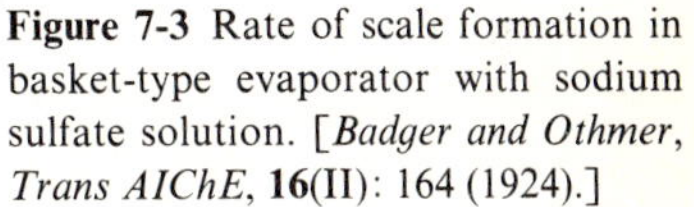
**Figure 7-3** Rate of scale formation in basket-type evaporator with sodium sulfate solution. [*Badger and Othmer, Trans AIChE*, **16**(II): 164 (1924).]

## Entrainment and Frothing

*Entrainment* may be defined as the carrying over of liquid droplets into the vapor space of an evaporator—a carrying over that occurs because the vapor velocity is high enough to overcome the tendency for the liquid drops to settle. *Frothing* means that a mass of bubbles forms on the surface of the liquid.

Entrainment may be controlled by lowering the velocity or rate of vapor formation, by increasing the available vapor space above the liquid, or by causing the vapor carrying the entrained droplets to impinge upon a baffle plate. Many different methods have been used to control frothing. A baffle plate located above the liquid surface will tend to break the bubbles and reduce frothing. The foam may sometimes be broken by the use of steam jets directed against the surface of the foam. Certain fatty oils and fatty acids, when injected into the boiling solution, will decrease the amount of frothing.

## Effect of Operating Methods on Overall Coefficients

One of the important advantages of forced-circulation evaporators is that the overall coefficient may be increased by increasing the liquid velocity through the evaporator tubes. However, the gain in the possible rate of heat transfer must be balanced against the cost of the extra power required for forcing the liquid through the tubes at the higher velocities.

The rate of heat transfer in basket-type evaporators is affected by the depth of liquid in the unit. If the liquid depth is so low that a large portion of the heating surface is exposed to the vapors, the heat-transfer rate will be decreased because of the high heat-transfer resistance offered by the vapors. On the other hand, if the liquid level is too high, the rate of heat transfer will be reduced by the combined effects of decreased temperature-difference driving force and changes in the liquid-film coefficients. Consequently, there is an optimum liquid level at which the heat-transfer rate is a maximum.

## TREND IN MODERN EVAPORATORS

The use of forced-circulation evaporators and natural-circulation long-tube evaporators has become very widespread. Because of the excellent results obtainable with these two types of evaporators, they may eventually replace the older types of basket evaporators almost completely.

The chemical engineer should understand the basic principles involved in the operation of all types of evaporators. However, because of the wide range of factors affecting heat-transfer coefficients and the many changes being made in modern evaporators, the final design of an evaporator should be carried out only by a person with wide experience in this field.

## PROBLEMS

**7-1** From the steam tables in App. C, determine the following:

(*a*) The temperature of saturated steam corresponding to a pressure of 5.2 $lb/in^2$, absolute

(*b*) The temperature of superheated steam if the pressure is 12.0 $lb/in^2$, absolute, and the steam has 10°F of superheat

(*c*) The temperature of the vapor being evolved in an evaporator if the pressure of the vapor is 55.14 kPa and the boiling-point rise due to material in solution is 11.7 K

(*d*) The latent heat of vaporization of water at 348 K

**7-2** Using liquid water at 32°F as the zero heat (or enthalpy) level, determine the total heat content of the following in Btu per pound: (*a*) saturated steam at 8.5 $lb/in^2$, absolute; (*b*) liquid water at 150°F; (*c*) water vapor at 200°F containing 10°F of superheat.

**7-3** The boiling temperature of a liquid at the liquid-vapor interface in an evaporator may be taken as 170°F. Calculate the standard overall heat-transfer coefficient in the evaporator if 10,000,000 Btu/h is transferred to the liquid through an area of 430 $ft^2$. Saturated steam at a pressure of 20 $lb/in^2$, gage, is used as the heat source.

**7-4** A water solution containing 25 wt % sodium chloride boils at 200°F when the vapor pressure of water over the solution is 8.85 $lb/in^2$, absolute. The same solution boils at 140°F when the vapor pressure of water over the solution is 2.22 $lb/in^2$, absolute. Construct a Dühring line for a 25 wt % sodium chloride solution relating boiling points of water to boiling points of the solution.

**7-5** A mixture containing 30 wt % sodium hydroxide and 70 wt % water is in equilibrium with a water-vapor pressure of 51.7 kPa. Determine the boiling temperature of the liquid at the liquid-vapor interface.

**7-6** What are the boiling point in degrees Fahrenheit at the liquid-vapor interface and the boiling-point rise due to material in solution of a mixture containing 35 percent by weight glycerol in water when the pressure in the vapor space over the solution is 12.0 lb/in$^2$, absolute? It may be assumed that Raoult's law is followed by the water and that the vapor pressure of the glycerol is negligible.

**7-7** The specific gravity of a liquid solution is 1.75. If the pressure at the upper surface of the liquid is 100 kPa, calculate the pressure at the region where the hydrostatic head is 0.914 m.

**7-8** A solution of sodium hydroxide in water with a concentration of 10 wt % sodium hydroxide and a temperature of 100°F enters an evaporator. The solution is concentrated to 15 wt % sodium hydroxide. If the pressure in the vapor space is 4.0 lb/in$^2$, absolute, estimate the total number of Btu necessary to evaporate 1 lb of water from the entering solution.

**7-9** A single-effect evaporator is to be used for the continuous concentration of a sodium hydroxide–water solution. The mixture enters the evaporator at a rate of 10,000 lb/h and a temperature of 70°F. The entering concentration is 8 wt % sodium hydroxide, and the leaving concentration is 18 wt % sodium hydroxide. If the available heat-transfer area is 400 ft$^2$, calculate the standard overall heat-transfer coefficient and the pounds of steam theoretically necessary per hour. Saturated steam is available at 230°F, and the vapor space in the evaporator is maintained at a pressure of 8.0 lb/in$^2$, absolute.

**7-10** A three-effect evaporator uses original steam at a saturation pressure of 174.4 kPa gage. The total boiling-point rise is 8.3 K in the first effect, 13.9 K in the second effect, and 22.2 K in the third effect. What is the limiting pressure which must be maintained in the vapor space of the third effect in order for the evaporator to operate?

**7-11** In a double-effect evaporator system having bodies of equal heating-surface area, the saturated steam entering the chest of the first effect is at 250°F. The effective condenser temperature for the vapors leaving the last effect is 130°F. The feed enters the first effect at 200°F containing 5 percent by weight of a nonvolatile material in a water solution. The total amount of feed is 12,000 lb/h. The concentrated liquor leaving the second effect contains 30 wt % nonvolatile material.

The standard overall heat-transfer coefficient in the first effect is 350 Btu/(h)(ft$^2$)(°F), and that in the second effect is 205 Btu/(h)(ft$^2$)(°F). In the first evaporator body, the boiling-point rise is 10°F. In the second body, the boiling-point rise is 22°F. The heat content of any solution involved may be taken as that of the water present in it. Tabulate the temperature and pressure in the vapor space of each effect, along with the area of heating surface and the evaporation in pounds per hour. What is the economy of the system?

**7-12** A continuous three-effect evaporator is operating under steady conditions. The evaporator is controlled to keep the concentration of the final product constant. The unit is constructed so that a line leads from the steam chest of each effect to the vapor space of that effect. The control valve on this line ordinarily is kept closed, but while the unit is operating smoothly, this valve on the second effect suddenly starts to leak. This permits steam to leak from the steam chest of the second effect to the vapor space of the second effect. The original steam temperature and the final condenser temperature remain constant. Will the value of the following stay constant, increase, or decrease?

(*a*) Overall heat-transfer coefficient in each effect if the coefficient is not affected by changes in $\Delta t$
(*b*) Temperature and pressure of vapors leaving each effect
(*c*) Overall capacity (Neglect the effect of B.P.R.)

**7-13** (*This problem is intended for computer program solution.*) In the production of sugar (sucrose) from sugar beets, the sugar is dissolved from sliced beets by warm water, the solution is treated to remove impurities, concentrated by evaporating off water, and then the sugar is crystallized and removed by centrifuging. In the evaporation step, the solution is concentrated from about 6 % sugar by weight to about 12 % sugar by weight.

How much water must be evaporated per 100 kg of feed?

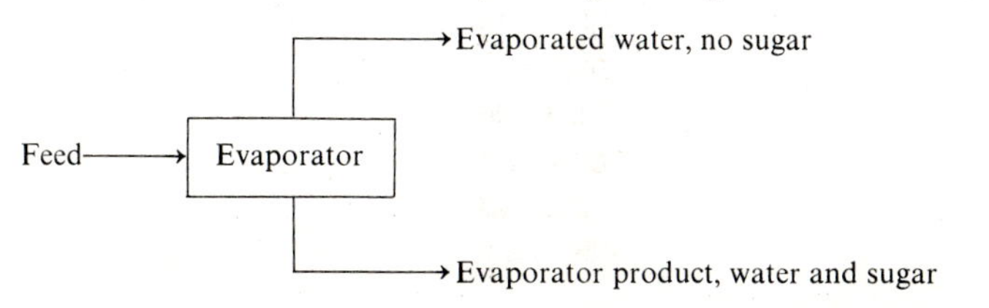

Program this problem to determine amount of water evaporated for feeds of 6, 7, 8, 9, and 10% by weight, and products of 11, 12, and 13 percent.

Print the results in a table laid out in this form and with the labels shown:

| Product concentration, % | 11 | 12 | 13 |
|---|---|---|---|
| Feed concentration | | | |
| 6 | | | |
| 7 | | | |
| 8 | | (Results) | |
| 9 | | | |
| 10 | | | |

CHAPTER

# EIGHT

# DISTILLATION

The chemical industry employs many different methods for separating mixtures into the various components. One of the most important and widely used physical separational operations is distillation. Common examples of other separational operations are evaporation, extraction, absorption, and filtration.

*Distillation* may be defined as the separation of the components of a liquid mixture by a process involving partial vaporization. In general, the vapor evolved is recovered by condensation. The more volatile constituents of the liquid mixture are obtained in the vapor at a higher concentration than in the original liquid, and the less volatile components are recovered in the liquid in increased concentration. The extent of the separation is governed by the properties of the components involved and by the physical arrangement used for the distillation.

To obtain a clear concept of the factors affecting distillation operations, it is necessary to understand how the different variables involved are interrelated. The composite relationships among the variables may be developed by the application of material and energy balances, and most of the equations presented in this chapter are based on the law of conservation of matter or the law of conservation of energy.

Many of the final equations pertaining to distillation operations are directly applicable for design purposes. As a result, the study of distillation often degenerates into a study of design methods. However, the application of the final equations is not of major importance. Anyone can plug numbers into a formula. The important thing is to understand the methods for developing the various relationships and to realize the true physical significance of the final equations.

The person who understands distillation can go out in the plant and operate a distillation unit. That person knows why a unit may appear to give a better separation in winter than in summer and perhaps could suggest to the operator that it would be better to obtain an increased separation by increasing the amount of liquid returned to the top of the column instead of reducing the amount of heat

## Table 8-1 Nomenclature for distillation

A and B = amount of components $a$ and $b$ in a mixture, mol
$B$ = rate of bottoms removal, mol/h
$D$ = rate of distillate removal, mol/h
$F$ = rate of feed, mol/h
H.E.T.P. = height equivalent to a theoretical plate, m or in
H.T.U. = height equivalent to a transfer unit, m or in
$K$ = equilibrium constant, dimensionless
$L$ = liquid reflux rate, mol/h
$L_n$ = liquid reflux rate above feed plate, mol/h
$L_m$ = liquid reflux rate below feed plate, mol/h
$M$ = molecular weight, kg or lb
$n_a, n_b$ = moles of components $a$ and $b$ in a mixture, mol
$N_p$ = number of theoretical plates in a distillation column (not including still pot)
$N.T.U.$ = number of transfer units
$p$ = partial pressure, Pa or lb/in$^2$, absolute, mmHg
$P$ = total vapor pressure, Pa or lb/in$^2$, absolute, mmHg
$P^o$ = vapor pressure of a pure material, Pa or lb/in$^2$, absolute, mmHg
$q$ = total heat required to vaporize 1 mol of feed divided by molal latent heat of vaporization of feed, dimensionless
$R$ = reflux ratio = $L_n/D$, dimensionless
$R_{min}$ = minimum reflux ratio, dimensionless
$v$ = volatility = equilibrium partial pressure divided by mole fraction
$V$ = vapor rate ascending distillation column, mol/h
$V_n$ = vapor rate ascending distillation column above feed plate, mol/h
$V_m$ = vapor rate ascending distillation column below feed plate, mol/h
$W$ = weight of a material, kg or lb
$x$ = mole fraction in liquid state
$y$ = mole fraction in the vapor state
$y^*$ = mole fraction in the vapor state in equilibrium with a liquid

Greek symbols

$\alpha$ = alpha, relative volatility, dimensionless

Subscripts

1, 2, or 3 refers to a stage or period in a process
$a$ refers to one component of a mixture
$ab$ relates the property of component $a$ to that of component $b$ in a mixture
$b$ refers to one component of a mixture
$B$ refers to bottoms in a distillation unit
$D$ refers to distillate in a distillation unit
$F$ refers to feed in a continuous-distillation unit
$f'$ refers to point where feed is introduced
$m$ refers to a theoretical plate or variable in stripping section of a distillation column
$m + 1$ refers to next plate below plate $m$
$n$ refers to a theoretical plate or variable in enriching section of a distillation column
$n + 1$ refers to next plate below $n$
$s$ refers to still pot
$T$ refers to top of the distillation column

put into the still pot. In short, the good chemical engineer knows *why*, as well as how, a distillation column works.

The last part of this chapter deals with some of the practical considerations involved in the actual operation of distillation equipment, while the first part of the chapter presents the basic information necessary for an understanding of distillation.

## PHYSICAL CONCEPT OF DISTILLATION

### Distillation of a Mixture Following Raoult's Law

As indicated in Chap. 7 (Evaporation), Raoult's law states that

$$p_a = x_a P_a^o \tag{8-1}$$

where $p_a$ = *equilibrium* vapor pressure of component $a$ in a liquid solution with components $a, b, c, \ldots$

$P_a^o$ = *equilibrium* vapor pressure of component $a$ if it were pure and at the same temperature as the solution

$x_a$ = mole fraction of component $a$ in the liquid solution

Although Raoult's law applies only to so-called "perfect" or "ideal" liquid solutions, it can be used to give a clear understanding of what takes place in a distillation operation. As an example, consider the case of a liquid mixture of benzene and toluene containing 0.40 mol fraction benzene and 0.60 mol fraction toluene. This mixture is in equilibrium with a vapor containing only benzene and toluene at a temperature of 95.25°C. At this temperature, the equilibrium vapor pressure of pure benzene is 1180 mmHg, and the equilibrium vapor pressure of pure toluene is 481 mmHg. Since a mixture of benzene and toluene follows Raoult's law, the actual vapor pressure of the benzene is $(0.4)(1180) = 472$ mmHg, and the actual vapor pressure of the toluene is $(0.6)(481) = 288.6$ mmHg.

According to Dalton's law, the mole fraction of a component in a *gaseous* mixture equals the vapor pressure of the component divided by the total vapor pressure. Thus the mole fraction of benzene in the vapor of this example is $472/(472 + 288.6) = 0.62$. If a portion of this vapor were suddenly condensed, a liquid would be obtained containing a benzene mole fraction of 0.62 and a toluene mole fraction of 0.38. Therefore, by a process of partial vaporization followed by condensation, a product can be obtained which is richer in benzene than the original mixture. This simple example illustrates a process of physical separation by distillation.

It is important to understand the difference between Raoult's law and Dalton's law. Raoult's law is a liquid law and must be expressed as applying only to the liquid solution and to vapors in equilibrium with the liquid solution. Dalton's law is a gas law and is only applicable to gases.

## VAPOR-LIQUID EQUILIBRIUM RELATIONSHIPS

In order to determine the possible extent of separation between the components of a mixture by the use of distillation, it is necessary to know the relative ease of vaporization of the individual components. If a liquid mixture contains two components and one of these vaporizes easily while the other does not, it should be a simple matter to obtain a high degree of separation by means of distillation.

For mixtures which follow Raoult's law, the composition of the equilibrium vapors evolved at any temperature from a liquid of known concentration can be calculated by use of the vapor pressures of the pure components at the temperature involved. However, since most mixtures do not follow Raoult's law, it is usually necessary to determine the equilibrium liquid and vapor compositions experimentally.

Many distillation columns are operated at atmospheric pressure, and the pressure drop over the length of the column is usually small when compared to the total pressure. Therefore, it is very convenient to have data giving the concentration of liquid mixtures and the corresponding equilibrium concentration of the vapors when the total pressure of the evolved vapors is 1 atm.

Consider the example of a binary (two-component) mixture consisting of methyl alcohol and water. Experimental data taken under equilibrium conditions, where the total vapor pressure of methyl alcohol and water over the liquid mixture is 760 mmHg, are presented in Table 8-2.

The data presented in Table 8-2 indicate that methyl alcohol is more volatile than water; therefore a partial separation of this mixture can be effected by the use of distillation.

Figure 8-1 presents the equilibrium data for methyl alcohol–water mixtures in graphical form. In plots of this type, the abscissa represents the mole fraction of the

**Table 8-2 Vapor-liquid equilibrium data for methyl alcohol–water mixtures at a total pressure of 760 mmHg**

| Mol % methyl alcohol | | Mol % water | | | Total pressure, |
|---|---|---|---|---|---|
| Liquid | Vapor | Liquid | Vapor | Temperature °C | mmHg |
| 0.0 | 0.0 | 100.0 | 100.0 | 100.0 | 760 |
| 4.0 | 23.0 | 96.0 | 77.0 | 93.5 | 760 |
| 10.0 | 41.8 | 90.0 | 58.2 | 87.7 | 760 |
| 20.0 | 57.9 | 80.0 | 42.1 | 81.7 | 760 |
| 40.0 | 72.9 | 60.0 | 27.1 | 75.3 | 760 |
| 60.0 | 82.5 | 40.0 | 17.5 | 71.2 | 760 |
| 80.0 | 91.5 | 20.0 | 8.5 | 67.6 | 760 |
| 90.0 | 95.8 | 10.0 | 4.2 | 66.0 | 760 |
| 95.0 | 97.9 | 5.0 | 2.1 | 65.0 | 760 |
| 100.0 | 100.0 | 0.0 | 0.0 | 64.5 | 760 |

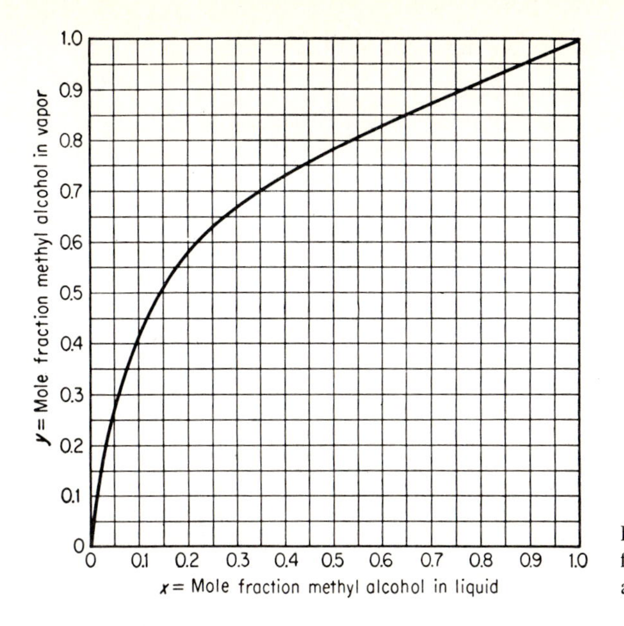


**Figure 8-1** Equilibrium $y$ vs. $x$ plot for methyl alcohol–water system at a total pressure of 760 mmHg.

more volatile component in the liquid, and the ordinate represents the mole fraction of the more volatile component in the vapor. Such a plot is commonly referred to as an *equilibrium y vs. x plot at constant pressure*, where $y$ represents the mole fraction of one component in the vapor and $x$ represents the mole fraction of the same component in the liquid. Any point on the curve gives the equilibrium relationship between the liquid composition and the vapor composition. Thus, by reference to Fig. 8-1, it can be seen that a methyl alcohol–water vapor containing 0.78 mol fraction methyl alcohol is in equilibrium with a liquid methyl alcohol–water mixture containing 0.5 mol fraction methyl alcohol when the total pressure of the vapors is 760 mmHg.

## Volatility

The *volatility* of any substance in a liquid solution may be defined as the equilibrium partial pressure of the substance in the vapor phase divided by the mole fraction of the substance in the liquid solution, or

$$v_a = \text{volatility of component } a \text{ in a liquid solution} = \frac{p_a}{x_a} \tag{8-2}$$

The volatility of a material in the pure state is equal to the vapor pressure of the material in the pure state. Similarly, the volatility of a component in a liquid mixture which follows Raoult's law must be equal to the vapor pressure of that component in the pure state.

## Relative Volatility

The ratio of the volatilities of two components of a liquid mixture gives an indication of the ease with which the two components can be separated by distillation. *Relative volatility* may be defined as the volatility of one component of a liquid mixture divided by the volatility of another component of the liquid mixture. Relative volatilities are commonly expressed with the higher of the two volatilities in the numerator. This means the relative volatility should never have a numerical value less than 1.0.

The symbol used for designating relative volatility is $\alpha$ (alpha).

$$\alpha_{ab} = \frac{v_a}{v_b} = \frac{p_a}{x_a}\frac{x_b}{p_b} \tag{8-3}$$

If the vapors follow Dalton's law,

$$p_a = y_a P \tag{8-4}$$

and

$$p_b = y_b P \tag{8-5}$$

where $P$ equals the total pressure of the vapors.

Combination of Eqs. (8-3), (8-4), and (8-5) gives

$$\alpha_{ab} = \frac{y_a}{y_b}\frac{x_b}{x_a} \tag{8-6}$$

Equation (8-6) is often given as the definition of relative volatility.

By the use of Eq. (8-6), the value of the relative volatility can be calculated directly from vapor-liquid equilibrium data. For example, with a binary mixture of methyl alcohol and water having a total vapor pressure of 760 mmHg, when the liquid contains 0.40 mol fraction methyl alcohol the equilibrium vapor contains 0.729 mol fraction methyl alcohol. In this case, methyl alcohol is the more volatile, and $y_a = 0.729$, $y_b = 1.000 - 0.729$ (since it is a binary mixture), $x_a = 0.40$, and $x_b = 1.00 - 0.40$. Therefore, from Eq. (8-6),

$$\alpha = \frac{0.729}{0.271}\frac{0.60}{0.40} = 4.035$$

It must be noted that the relative volatility, as obtained in the preceding example, is correct only at the one concentration for which it was calculated. The value of the relative volatility may change as the concentration changes.

Some mixtures have relative volatilities which remain essentially constant as the concentrations of the components in the mixture are varied. Binary solutions which follow Raoult's law often show only slight changes in the relative volatility with concentration variations when the total vapor pressure is held constant. However, this is not true for all Raoult's law binary solutions.

For binary mixtures,

$$y_b = 1 - y_a \tag{8-7}$$

and

$$x_b = 1 - x_a \tag{8-8}$$

The following equations, where subscript $a$ refers to the more volatile component, may be obtained by substituting Eqs. (8-7) and (8-8) in Eq. (8-6) and rearranging:

$$y_a = \frac{\alpha x_a}{1 + (\alpha - 1)x_a} \tag{8-9}$$

$$x_a = \frac{y_a}{\alpha + (1 - \alpha)y_a} \tag{8-10}$$

## BATCH DISTILLATION

Distillation operations are often carried out batchwise. In such operations, a charge of liquid is placed in the distillation unit, and material is distilled off until the original charge is all gone or until the operation is no longer doing any beneficial separating. In batch-distillation operations, the liquid in the original charge is constantly being depleted of its more volatile constituents. Since no new material is charged to replace the losses, the concentration of the liquid and vapor must change continuously throughout the entire operation.

### Simple-Batch (Differential) Distillation

A *simple batch distillation* refers to a batch distillation in which only one vaporization stage (or one exposed liquid surface) is involved. This means that a partial separation is obtained by removing the equilibrium vapors as fast as they are formed.

A mathematical relationship between the amounts of any two components present in the liquid at two separate times may be obtained as follows:

According to Eq. (8-6), the relative volatility between components $a$ and $b$ in a mixture may be expressed as

$$\alpha_{ab} = \frac{y_a}{y_b}\frac{x_b}{x_a}$$

At any given time during the distillation, the still contains A mol of component $a$ and B mol of component $b$. Over a differentially small unit of time, the amounts of each component distilled off must be $-dA$ and $-dB$. Consider that the vapor is composed of $dA$ mol of component $a$ and $d$B mol of component $b$ during the small time interval. Therefore, by Dalton's law,

$$y_a = \frac{d\text{A}}{d\text{A} + d\text{B}}$$

and

$$y_b = \frac{dB}{dA + dB}$$

Since the small amount of liquid distilled off during the unit of time under consideration is negligible in comparison to the total liquid present,

$$x_a = \frac{A}{A + B}$$

and

$$x_b = \frac{B}{A + B}$$

Substituting in the relative-volatility equation,

$$\alpha_{ab} = \frac{(dA)(dA + dB)(B)(A + B)}{(dA + dB)(dB)(A + B)(A)} = \frac{dA}{dB}\frac{B}{A} \tag{8-11}$$

and

$$\frac{dA}{A} = \alpha_{ab}\frac{dB}{B} \tag{8-12}$$

If $\alpha_{ab}$ remains constant between time 1 and time 2, Eq. (8-12) may be integrated between the limits of 1 and 2 to give

$$\log\frac{A_1}{A_2} = \alpha_{ab}\log\frac{B_1}{B_2} \tag{8-13}$$

where subscripts 1 and 2 refer to the amount of the particular component in the liquid at times 1 and 2.

Equation (8-13) is one form of the *Rayleigh equation.* Equation (8-13) is applicable only to cases involving simple batch distillations where $\alpha_{ab}$ is constant.

The general form of the Rayleigh equation, which is applicable to all cases of simple batch distillation, is

$$\ln\frac{L_1}{L_2} = \int_{x_2}^{x_1}\frac{dx}{y - x} \tag{8-13a}$$

where $L_1$ and $L_2$ represent the total moles of liquid in the distillation unit at times 1 and 2 [see App. B for derivation of Eq. (8-13*a*)].

**Example 8-1: Application of Rayleigh equation to a simple batch distillation** One hundred moles of a liquid mixture is charged to a distillation unit. This liquid mixture contains 0.20 mol fraction of component *a* and 0.80 mol fraction of component *b*. If the mixture is subjected to a simple batch distillation, what will the liquid composition be after 8 mol of component *a* has been removed with the vapors? Assume $\alpha_{ab}$ as constant at 1.414.

SOLUTION

**Basis**

100 mol of original liquid mixture
Since the relative volatility is constant, Eq. (8-13) may be used.
$\log A_1/A_2 = \alpha_{ab} \log B_1/B_2$
$A_1 = (100)(0.20) \doteq 20$ mol
$B_1 = (100)(0.80) = 80$ mol
$A_2 = 20 - 8 = 12$ mol
$\alpha_{ab} = 1.414$
$\log \frac{20}{12} = 1.414 \log 80/B_2$
$B_2 = 55.8$ mol
The liquid composition after 8 mol of component $a$ has been removed is

$$\frac{55.8}{55.8 + 12} = 0.823 \text{ mol fraction of component } b$$

$$\frac{12}{55.8 + 12} = 0.177 \text{ mol fraction of component } a$$

If the relative volatility of the mixture had not been constant, Eq. (8-13$a$) could have been used to solve the problem. It would have been necessary to have $y$ vs. $x$ equilibrium data for the mixture, and a graphical integration would have been required.

## CONTINUOUS DISTILLATION

By continuously feeding a distillation unit with the material to be separated, it is possible to obtain a so-called "continuous distillation." In an operation of this type, the unit can be brought back to steady operating conditions where the amount of feed exactly equals the amount of material removed, and the vapor and liquid concentrations at any point in the unit remain constant.

### Simple Equilibrium Distillation

Figure 8-2 represents a continuous process in which a single vaporization occurs under conditions where the vapors are in equilibrium with the liquid. By continuously removing part of the vapors and the equilibrium liquid solution and continuously adding feed, the process can be adjusted so that all concentrations and rates remain constant.

A continuous single vaporization of this type is known as a *simple equilibrium distillation*. Relationships between the concentrations and the flow rates can be obtained by the application of material balances. The total number of moles per

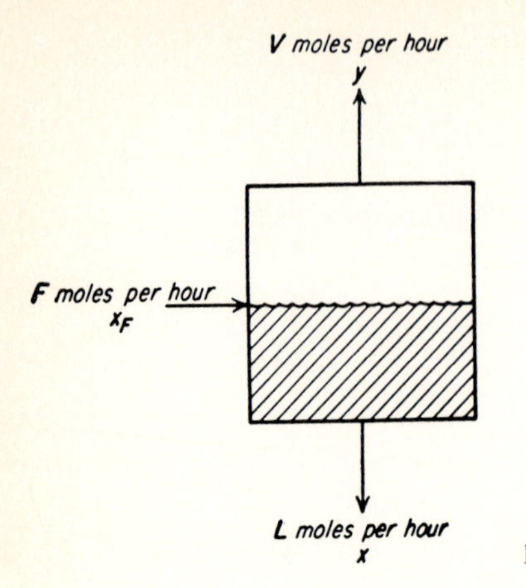


**Figure 8-2** Simple equilibrium distillation.

hour entering the system must equal the total number of moles per hour leaving the system, or

$$F = V + L \tag{8-14}$$

The amount of any one component entering the system in unit time must equal the amount of that component leaving the system in unit time. Since the liquid-and-vapor mole fractions all refer to the same component, a one-component material balance may be expressed as

$$Fx_F = Vy + Lx \tag{8-15}$$

or

$$Fx_F = Vy + (F - V)x \tag{8-16}$$

Equation (8-16) can be used to determine the rates at which the vapor and liquid should be removed in order to obtain a predetermined product composition from a given feed.

**Example 8-2: Determination of product withdrawal rate for a simple equilibrium distillation** A feed containing 0.482 mol fraction methyl alcohol and 0.518 mol fraction water is to be partially separated by a simple equilibrium distillation. How many moles of vapor should be removed per 100 mol of feed to give a vapor composition of 0.729 mol fraction methyl alcohol if the total vapor pressure is 760 mmHg?

SOLUTION From Fig. 8-1, the liquid concentration of a methyl alcohol–water mixture in equilibrium with a 0.729 mol fraction methyl alcohol vapor is 0.40 mol fraction methyl alcohol when the total vapor pressure is 760 mmHg.

Using Eq. (8.16) and a basis of 100 mol of feed,

$$Fx_F = Vy + (F - V)x$$
$$F = 100 \text{ mol}$$
$$y = 0.729$$
$$x_F = 0.482$$
$$x = 0.400$$

$$(100)(0.482) = V(0.729) + (100 - V)(0.400)$$

$$V = 24.9 \text{ mol}$$

If 24.9 mol of vapor is removed per 100 mol of feed, the vapor will contain 0.729 mol fraction methyl alcohol.

## RECTIFICATION

The preceding discussion has been limited to a consideration of simple distillations or distillations in which only one vaporization stage was involved. If a series of these simple distillations could be performed whereby the vapors from each step (or stage) were condensed and equilibrium vapors were evolved from the condensate, it would be possible to get a much greater separation than could be obtained from a single simple distillation.

For example, consider the case where a liquid mixture of methyl alcohol and water is subjected to a simple equilibrium distillation. If the total pressure is taken as 760 mmHg and the liquid concentration is 0.10 mol fraction methyl alcohol, the equilibrium vapors evolved must have a methyl alcohol mole fraction of 0.418. If these vapors were condensed and subjected to another simple equilibrium distillation, the evolved vapors would have a concentration of 0.74 mol fraction methyl alcohol. These steps could be continued until practically pure methyl alcohol would be obtained as the evolved vapors.

The stepwise operation just described is a simplified form of what actually happens in a practical distillation column. The heat evolved from the condensation of the vapors is employed to boil out new vapors from the succeeding steps. A portion of the previously evolved vapors is condensed and acts as the heat receiver for the fresh condensing vapors. These condensed vapors, which are returned in the individual steps, are called *reflux*. This type of distillation is known as *rectification*.

Rectification may be defined as a single-unit distillation operation in which vaporization occurs in repeated steps to give a much greater overall separation than could be obtained by one simple distillation. Figure 8-3 presents the comparison between a series of simple distillations and a rectification process. The term *fractionation* is synonymous with rectification, and the two expressions are used interchangeably.

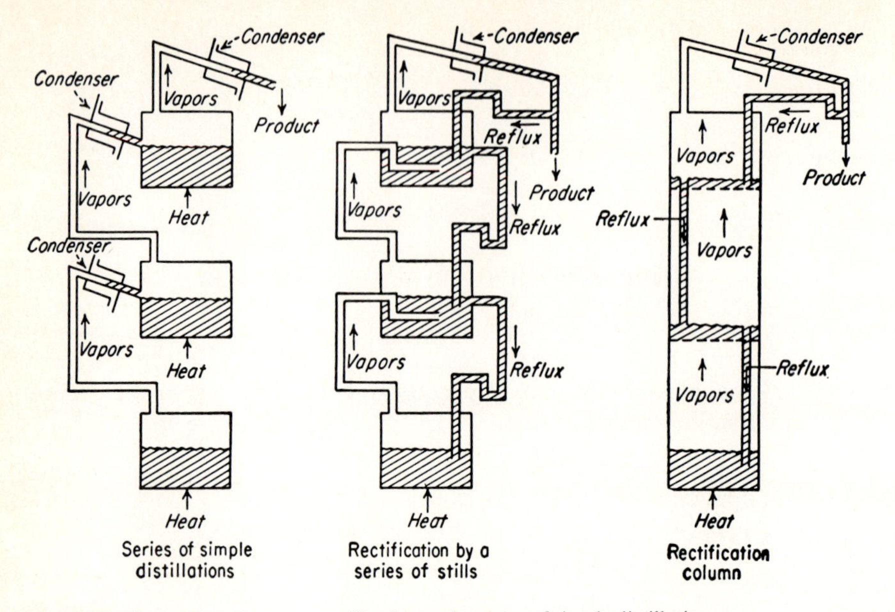


**Figure 8-3** Comparison between rectification and a series of simple distillations.

## Distillation Columns

Various methods are employed for obtaining the liquid and vapor contact necessary for rectification. Sieve, valve, or bubble-cap plate towers are widely used in industry. Distillation columns of this type consist of a series of plates as shown in the rectification column of Fig. 8-3 and in Fig. 4-19. There are a number of openings in each plate through which the vapors rise and pass into the liquid on the plate. The vapors bubble through the liquid where condensation and vaporization occur.

Packed distillation columns, as illustrated in Fig. 4-22, consist of a hollow shell filled with some material which offers a large amount of exposed surface area. Irregularly shaped rocks are occasionally used as the packing material in packed distillation columns, and even shoe eyelets have been used as the packing in small columns. A typical packing commonly used industrially is Raschig rings. This packing is ordinarily constructed from a ceramic material and consists of a large number of hollow cylinders with the outside diameter equal to the height. The Raschig rings may be dumped at random into the column to form the packing.

A still pot[1] is located at the bottom of the distillation column, and the heat required for the process is introduced into the unit at this point. Ordinarily, steam condensing on the inside of tubes is used to supply the necessary heat.

The final vapors evolved from the top of the distillation column are condensed, and part of the condensate is returned to the column as reflux and part is withdrawn as product. Continuous distillation columns are operated with the original

[1] In the terminology of the petroleum industry, a still pot is often referred to as a "reboiler." It is also occasionally called a "calandria."

feed being introduced at an intermediate point on the column. The section of the distillation column above the point where the feed enters is known as the *enriching section*, and the portion below the point where the feed enters is known as the *stripping section.*

### Reflux Ratio

The *reflux ratio*, as used in this book, may be defined as the moles of final distillate returned to the top of the distillation column per unit time divided by the moles of distillate withdrawn as product per unit time.

$$\text{Reflux ratio} = R = \frac{L}{D} \tag{8-17}$$

When the distillation unit is operating so that no product is withdrawn and all the top condensate is returned to the column, the reflux ratio is infinite. Under these conditions, the column is said to be operating at *total reflux*.

## METHODS OF CALCULATION

In the practical application of distillation operations, it is often necessary to determine the size of a distillation column required to give a certain separation of a known mixture. To determine the size of the unit, it is necessary to know how many theoretical stages or separate vaporizations are needed to give the required separation.

In distillation operations, a theoretical stage (or a theoretical plate) is defined as a single vaporization step in which the liquid and the vapors evolved from the liquid are in equilibrium. The evolved vapors rise into the next stage above, while the liquid in equilibrium with these vapors flows down into the next stage below. If calculations show that a given separation is attained when six repeated vaporizations and condensations occur, with equilibrium being attained in each of the six steps, a distillation column containing six theoretical stages would be necessary to produce the given separation. In the case of a distillation column containing bubble-cap plates, at least five plates in addition to the still pot would be required to make the separation. Actually, more than five plates would be needed, since perfect equilibrium is not attained on a real plate. This matter will be discussed later under the subject of distillation-column efficiency.

### McCabe-Thiele Method for Calculating the Number of Theoretical Stages with Binary (Two-Component) Mixtures

A graphical method for determining the number of stages theoretically necessary for any given separation has been developed by McCabe and Thiele.[2] The method

[2] W. McCabe and E. Thiele, *Ind Eng Chem*, **17**:605 (1925).

is based on material balances around certain parts of the column and on several basic assumptions.

The essential assumption made in the McCabe-Thiele method is that there must be equimolal overflow throughout the distillation column between any points where fresh feed is added and product is removed. For a ten-plate column with feed being introduced on the fourth tray, this assumption means the same number of moles of liquid reflux per unit time are flowing onto the tenth tray as are flowing onto all the other trays above the fourth one. Similarly, the same number of moles of liquid reflux per unit time are flowing onto the third tray as are flowing back into the still pot.

The essential assumption of equimolal overflow may be broken down into the following generalized assumptions:

1. Sensible-heat changes throughout the distillation column are negligible in comparison with the heat of vaporization of the components being separated.
2. The heat of vaporization per mole is equal for all components.
3. The heat of mixing of the components is negligible.
4. Heat losses from the distillation column are negligible.

Figure 8-4 represents a continuous distillation column, with feed being introduced at an intermediate point on the column and product being withdrawn as distillate and bottoms. The nomenclature used in the McCabe-Thiele development is indicated on the diagram presented in Fig. 8-4.

**The enriching section.** An overall material balance around the entire unit requires that the moles of feed entering per unit time must equal the sum of the moles of distillate and bottom product leaving per unit time, or

$$F = D + B \tag{8-18}$$

A total material balance around the section indicated by the dotted lines, labeled $I$ in Fig. 8-4 (the enriching section), gives

$$V_n = L_n + D \tag{8-19}$$

A material balance on the more volatile component over section $I$ gives

$$V_n y_{n+1} = L_n x_n + D x_D \tag{8-20}$$

Combining Eq. (8-19) with Eq. (8-20) and rearranging,

$$y_{n+1} = \frac{L_n}{L_n + D} x_n + \frac{D}{L_n + D} x_D \tag{8-21}$$

Since, by Eq. (8-17), the reflux ratio ($R$) equals $L_n/D$, Eq. (8-21) may be expressed as

$$y_{n+1} = \frac{R}{R + 1} x_n + \frac{x_D}{R + 1} \tag{8-22}$$

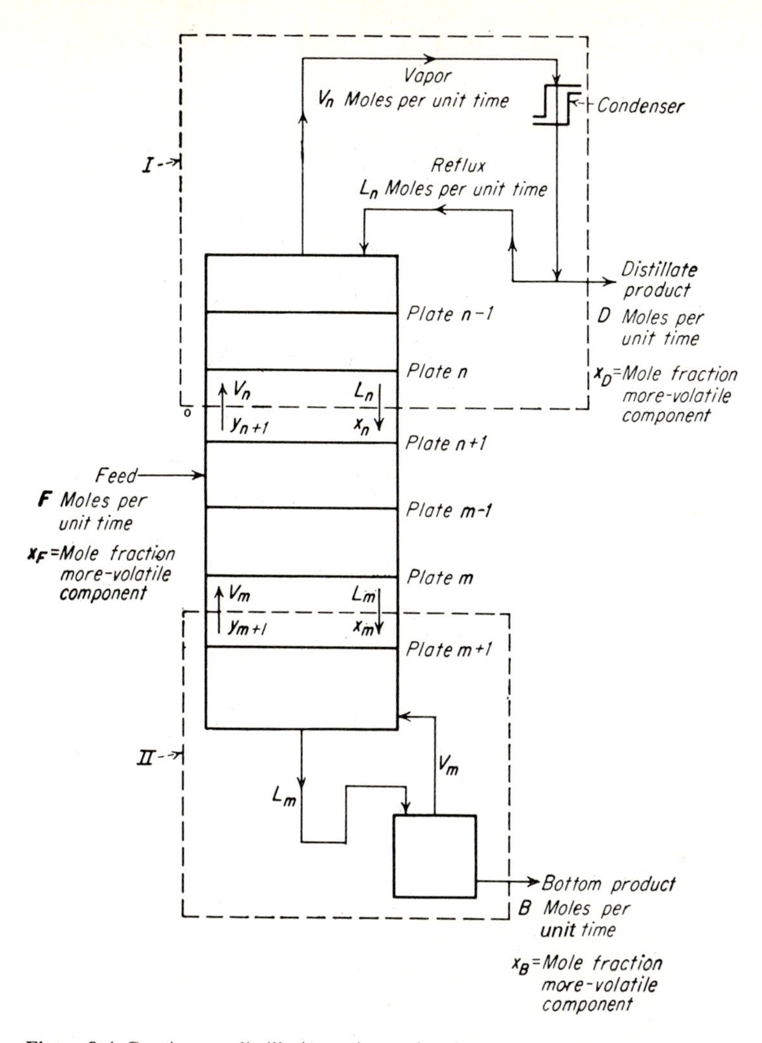


**Figure 8-4** Continuous distillation column showing material balance sections used in McCabe-Thiele method.

Equation (8-22) represents a relationship between the composition of the vapors rising into plate $n$ and the liquid composition leaving plate $n$. This equation applies to the entire enriching section and is known as *the equation for the enriching operating line.*

The number of theoretical separation stages required in the enriching section of a distillation column can be obtained graphically from an equilibrium $y$ vs. $x$ plot and a plot of $y_{n+1}$ vs. $x_n$ based on Eq. (8-22). This graphic method is best explained by consideration of an actual example. A distillate containing 0.80 mol fraction benzene and 0.20 mol fraction toluene is to be obtained from a bottoms containing 0.30 mol fraction benzene and 0.70 mol fraction toluene. Assume, for

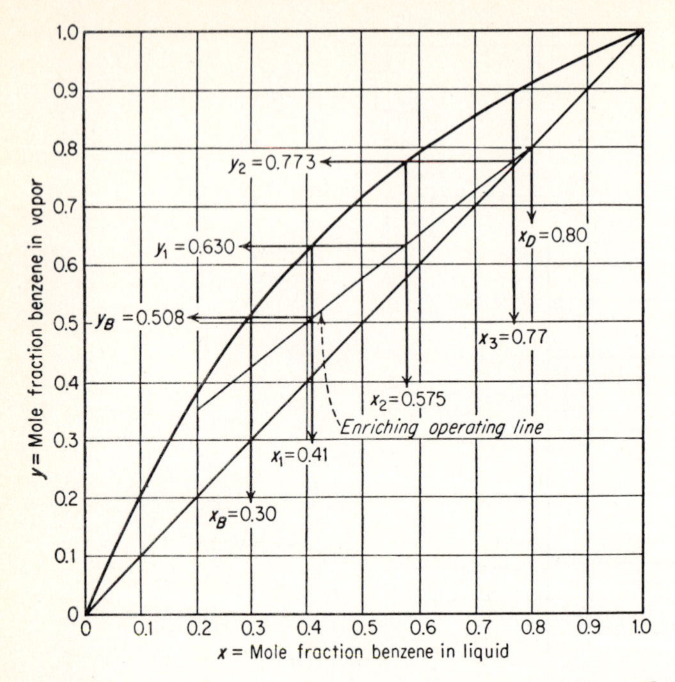


**Figure 8-5** Equilibrium $y$ vs. $x$ diagram for benzene-toluene at 760 mmHg.

this example, that the feed is introduced into the still pot and that the entire column acts as the enriching section. The distillation is to be carried out at atmospheric pressure. One mole of distillate product is to be withdrawn for every 3 mol of distillate returned to the column as reflux. This means the reflux ratio will be $\frac{3}{1} = 3$. Assuming constant molal overflow, the reflux ratio will be constant throughout the entire column.

An equilibrium $y$ vs. $x$ diagram for benzene-toluene mixtures at 760 mmHg is presented in Fig. 8-5. The enriching operating line, based on Eq. (8-22), is also plotted on this figure. Since $R/(R + 1)$ and $x_D/(R + 1)$ are constants for this case, the line based on Eq. (8-22) must be straight with a slope of $3/(3 + 1) = 0.75$. At the top of the column, the composition of the vapor leaving equals the composition of the condensed distillate returned as reflux. Therefore, when the top plate is considered as number $n + 1$, $y_{n+1} = x_D = x_n = 0.8$. This gives one point on the enriching operating line. Since the slope of this line is known, it is now possible to draw the entire line on the $y$ vs. $x$ plot without any further calculations.

The composition in the still pot is 0.30 mol fraction benzene. The vapors leaving the still pot are assumed to be in equilibrium with the liquid, and the equilibrium $y$ vs. $x$ plot indicates that these evolved vapors have a composition of 0.508 mol fraction benzene.

According to the enriching operating line [or Eq. (8-22)], the liquid leaving the first tray above the still pot must contain 0.41 mol fraction benzene if the vapors rising into this plate have a composition of 0.508 mol fraction benzene. This means the liquid composition on the first theoretical tray is 0.41 mol fraction benzene. Thus, the completion of one theoretical stage from the still pot to the first tray has resulted in an increase in the benzene liquid concentration from a mole fraction of 0.30 to a mole fraction of 0.41.

Assume the vapors evolved from the liquid on the first tray are in equilibrium with this liquid. Since the liquid composition on the first tray is 0.41 mol fraction benzene, the equilibrium vapors must have a composition of 0.63 mol fraction benzene, as shown in Fig. 8-5. Therefore, the vapors rising into the second tray contain 0.63 mol fraction benzene, and, according to the enriching operating line of Eq. (8-22), the liquid on the second tray must have a composition of 0.575 mol fraction benzene. Reference to Fig. 8-5 indicates that the reasoning process presented for the first two stages results in merely stepping off from the operating line to the equilibrium $y$ vs. $x$ curve over the liquid composition range between the bottoms and the distillate.

Completion of the steps in Fig. 8-5 shows that slightly more than three theoretical stages are needed to obtain the distillate composition of 0.8 mol fraction benzene. Therefore, if the liquid and the evolved vapors on each tray and in the still pot are in equilibrium, an efficient distillation column containing three actual bubble-cap trays in addition to the still pot should be capable of performing the required separation.

**The stripping section.** By making material balances over the section labeled *II* in Fig. 8-4, the following equations can be obtained which are applicable to the stripping section of a distillation column:

$$V_m = L_m - B \tag{8-23}$$

$$V_m y_{m+1} = L_m x_m - B x_B \tag{8-24}$$

$$y_{m+1} = \frac{L_m}{L_m - B} x_m - \frac{B x_B}{L_m - B} \tag{8-25}$$

Equation (8-25) is known as *the equation for the stripping operating line.* It is used for determining the number of stages in the stripping section exactly as Eq. (8-22) was used for determining the number of stages in the enriching section.

## EFFECT OF FEED CONDITIONS

The condition of the entering feed determines the relationship between the vapor rate in the enriching section $V_n$ and the vapor rate in the stripping section $V_m$. The feed condition also determines the relationship between $L_n$ and $L_m$. Thus, if the feed is introduced at a temperature below its boiling point, there will be more moles of liquid running down the stripping section per unit time than down the enriching section, and $L_m$ will be greater than $L_n$.

The symbol $q$ has been chosen to represent the condition of the feed. The quantity $q$ may be defined as the total heat actually required to vaporize 1 mol of the feed divided by the heat required to vaporize 1 mol of the liquid feed if it were at its boiling point.

$$q = \frac{\text{heat required to vaporize 1 mol of feed at entering conditions}}{\text{molal heat of vaporization of feed}} \tag{8-26}$$

If the feed enters at its boiling temperature, $q$ is equal to 1, and the mole rate of liquid flowing in the stripping section will equal the mole rate of liquid flowing in the enriching section plus the mole rate of feed introduced. If the feed were introduced as a saturated vapor, $q$ would be equal to zero, and the same number of moles of liquid per unit time would flow in both the enriching anu stripping sections.

From the definition of $q$, the relationship between the mole rate of liquid flowing in the stripping and enriching sections may be expressed as

$$L_m = L_n + qF \tag{8-27}$$

Similarly,

$$V_m = V_n + (q - 1)F \tag{8-28}$$

## The $q$ Line

The enriching operating line and the stripping operating line must intersect somewhere on a line which has the following equation:

$$y = \frac{q}{q-1}x - \frac{x_F}{q-1} \tag{8-29}$$

Equation (8-29) is known as *the equation for the q line*, and it may be derived from a combination of Eqs. (8-18), (8-21), (8-25), (8-27), and an overall volatile-component balance.

One point on a plot of $y$ vs. $x$ based on Eq. (8-29) is always found at $y = x = x_F$. Therefore, if $x_F$ is known, one point on the $q$ line is known. If the value of $q$ can be calculated, the slope of the $q$ line can be obtained from $q/(q - 1)$, and the entire straight line for Eq. (8-29) can be drawn.

The use of a $q$ line simplifies the graphical determination of the number of theoretical stages in a distillation column when both enriching and stripping sections are present. For example, if the reflux ratio and distillate composition are known, the enriching operating line can immediately be drawn on the $y$ vs. $x$ plot. If the feed composition and condition are known, the $q$ line can be drawn on the $y$ vs. $x$ plot. The point where the enriching operating line and the $q$ line intersect must be the top point of the stripping operating line. The correct stripping operating line can then be drawn on the graph by simply connecting this intersection point by a straight line with the bottoms point where $y = x = x_B$. This method gives exactly the same stripping operating line as would be obtained from a plot of Eq. (8-25). Example 8-3 presents a more detailed explanation of the $q$-line method for determining enriching and stripping operating lines.

## SUMMARY OF McCABE-THIELE METHOD OF CALCULATION

The McCable-Thiele method for determining the total number of theoretical stages in a distillation operation assumes constant molal overflow. This is equivalent to assuming straight operating lines in both the enriching and the stripping sections of the column. The important relationships and equations that have been developed are summarized in the following:

$$\text{Reflux ratio} = R = \frac{L_n}{D} = \frac{V_n - D}{D}$$

Equation for the enriching operating line:

$$y_{n+1} = \frac{R}{R+1} x_n + \frac{x_D}{R+1}$$

$$\text{Slope of enriching operating line} = \frac{L_n}{V_n} = \frac{R}{R+1}$$

Equation for the stripping operating line:

$$y_{m+1} = \frac{L_m}{L_m - B} x_m - \frac{B x_B}{L_m - B}$$

$$\text{Slope of stripping operating line} = \frac{L_m}{V_m}$$

$$q = \frac{\text{heat required to vaporize 1 mol of feed at entering conditions}}{\text{molal heat of vaporization of feed}}$$

$$L_m = L_n + qF$$

$$V_m = V_n + (q - 1)F$$

Equation for the $q$ line:

$$y = \frac{q}{q-1} x - \frac{x_F}{q-1}$$

$$\text{Slope of the } q \text{ line} = \frac{q}{q-1}$$

**Example 8-3: Determination of the number of theoretical stages for a given distillation operation** A mixture of benzene and toluene is fed to an intermediate point of a continuous fractionating column. The feed contains 45 mol % benzene and 55 mol % toluene. A distillate containing 92 mol % benzene and 8 mol % toluene is obtained. The bottoms contain 15 mol % benzene and 85 mol % toluene. One mole of distillate product is removed for every 4 mol returned to the unit as reflux. The feed enters at 131°F, and the

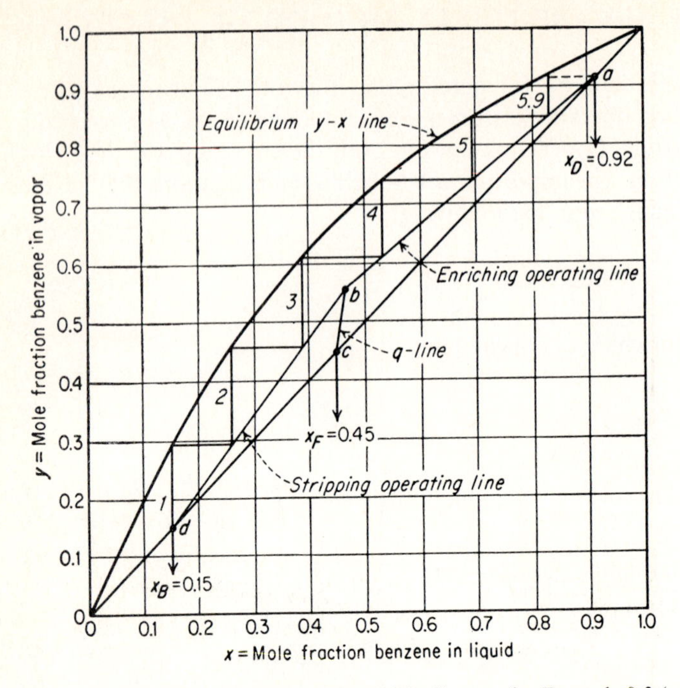


**Figure 8-6** Benzene-toluene McCabe–Thiele diagram for Example 8-3 (pressure = 760 mmHg).

unit operates at atmospheric pressure. How many theoretical stages (or plates) are present under these conditions?

The following data for benzene-toluene mixtures at 1 atm may be used.

The molal heat capacity of liquid mixtures of benzene and toluene in all proportions may be taken as 40 Btu/(lb mol)(°F). The molal heat of vaporization of benzene and toluene may be taken as 13,700 Btu/lb mol.

The atmospheric boiling point of a mixture containing 0.45 mol fraction benzene and 0.55 mol fraction toluene is 210°F.

The equilibrium $y$ vs. $x$ plot for benzene-toluene mixtures at 1 atm is presented in Fig. 8-6.

SOLUTION

The reflux ratio $= R = \frac{4}{1} = 4$.

Slope of enriching operating line $= R/R + 1 = \frac{4}{5} = 0.8$.

Top point of enriching operating line $= x_D$ on the 45° line in Fig. 8-6 = 0.92.

The enriching operating line is indicated in Fig. 8-6 by line *ab*.

$q = 13{,}700 + (40)(201 - 131)/13{,}700 = 1.204$

Slope of $q$ line $= q/q - 1 = 1.204/0.204 = 5.90$

One point on the $q$ line is $x_F$, on the 45° line in Fig. 8-6, equal to 0.45.

The $q$ line is indicated in Fig. 8-6 by line *cb*.

Since the $q$ line and the enriching operating line intersect at point $b$, the stripping operating line may be drawn by connecting point $b$ with $x_B = 0.15$ on the 45° line. The stripping operating line is indicated in Fig. 8-6 by line $db$.

By stepping off from the bottoms composition to the distillate composition, using the equilibrium line and the operating lines as the limits, the total number of stages obtained in Fig. 8-6 is 5.9.

Therefore, 5.9 theoretical stages (or plates) are present under the existing operating conditions.

## EFFICIENCY OF DISTILLATION UNITS

### Overall Plate Efficiency

In the preceding illustrative example, the total number of stages theoretically required for a given separation was determined. If a fractionating tower were made up of sieve plates and each of the plates and the still pot were 100 percent efficient, the given separation would be obtained if the number of actual plates in the tower were one less than the number of theoretical stages necessary. The still pot would act as one theoretical stage, and each of the plates would act as another theoretical stage.

In actual practice, plate columns are seldom 100 percent efficient, since complete equilibrium is not established between the liquid and evolved vapors on each tray. The *overall plate efficiency* of a plate tower is defined as the number of theoretical stages the complete unit is capable of producing minus one, divided by the actual number of individual plates in the distillation tower. This definition assumes the actual still pot is equivalent to one theoretical stage.

As an example, consider the case of an industrial distillation column containing 15 sieve trays in addition to the still pot. The unit has been tested and has been found to be capable of producing a total separation equivalent to 11.5 theoretical stages. Assuming the still pot as one theoretical stage, the other 10.5 stages must be in the actual column containing the plates. Therefore, the overall plate efficiency for this column, expressed as percent, is $(10.5/15)(100) = 70\%$. Overall plate efficiencies for industrial plate columns of the sieve, valve, or bubble-cap type usually are in the range of 50 to 80 percent.

### H.E.T.P.

The concept of an overall plate efficiency cannot be applied to packed columns since units of this type do not contain individual plates. The efficiency of packed columns is commonly expressed as the *height equivalent to a theoretical plate* (*H.E.T.P.*). This may be defined as the total packed height of the column divided by the number of theoretical stages (or plates) in the column. Thus, if the separation from the still pot to the distillate condenser were equivalent to 14 theoretical stages using a column with a packed height of 10 ft, the H.E.T.P. for the packed column would be (10)(12)/(14 − 1) = 9.2 in.

## H.T.U.

Chilton and Colburn[3] have proposed a method for evaluating the efficiency of packed towers based on a differential treatment of the separating process occurring in a distillation column. They express the efficiency in terms of so-called "transfer units." The total number of transfer units above the feed inlet may be obtained by use of the following equation:

$$\text{N.T.U.} = \int_{y_{f'}}^{y_T} \frac{dy}{y^* - y} \tag{8-30}$$

where $y$ = vapor composition at any point in tower above feed inlet
$y^*$ = vapor composition which would be in equilibrium with liquid at point where $y$ was determined
$y_T$ = vapor composition leaving top of tower
$y_{f'}$ = vapor composition at feed inlet point

The number of transfer units below the feed inlet can be obtained by an equation similar to Eq. (8-30) but employing different integration limits.

The *height of a transfer unit* (*H.T.U.*) is defined as the total packed height of the tower divided by the number of transfer units which the packed section is capable of producing. The advantage of the H.T.U. over the H.E.T.P. is that the H.T.U. is obtained on the basis of the differential process actually occurring in the tower, while the H.E.T.P. is based on a hypothetical stepwise operation.

## Murphree Plate Efficiency

The efficiencies of individual plates in a distillation tower may be reported as *Murphree plate efficiencies*.[4] This efficiency is defined as the actual vapor enrichment over one plate divided by the theoretical vapor enrichment which would have been obtained if the liquid on the plate and the vapors leaving the plate had reached equilibrium. Thus, if a vapor of composition $y_n$ enters plate $n - 1$, the actual vapors evolved have a composition of $y_{n-1}$. If the evolved vapors had left the liquid on plate $n - 1$ under equilibrium conditions, the composition of the vapors would have been $y^*_{n-1}$. The actual vapor enrichment is then $y_{n-1} - y_n$, and the theoretical vapor enrichment is $y^*_{n-1} - y_n$. Therefore

$$\text{Murphree plate efficiency for plate } n - 1 = \frac{y_{n-1} - y_n}{y^*_{n-1} - y_n} \tag{8-31}$$

[3] T. H. Chilton and A. P. Colburn, *Ind Eng Chem*, **27**:255, 904 (1935).
[4] E. Murphree, *Ind Eng Chem*, **17**:747 (1925).

## FEED PLATE LOCATION

The feed to a continuous distillation column should enter the column at the point where the distilling liquid has the same composition as the feed. If the feed is introduced at a point where it must mix with a material of different concentration, the approach to equilibrium in the column will be disturbed, and the overall fractionation efficiency of the unit will be reduced. For this reason, in designing a continuous distillation column, the approximate location of the feed-inlet tray should be determined. Inlets should be placed on this tray as well as on three or four of the trays immediately above and below the calculated feed tray. With the additional feed inlets available, the location of the feed tray can easily be changed to permit the column to operate at its maximum efficiency.

## MINIMUM REFLUX RATIO

The *minimum reflux ratio* may be defined as the reflux ratio that would be used to obtain a given separation if an infinite number of theoretical separation stages were used. This can be visualized by consideration of the equilibrium $y$ vs. $x$ plot presented in Fig. 8-7. If a mixture were to be separated from a bottoms composition of $x_B$ to a distillate composition of $x_D$ with no intermediate feed, the line *ab* in Fig. 8-7 would represent one possible enriching operating line. However, with this operating line, an infinite number of stepwise stages would be required to change the concentration from $x_B$ to $x_D$, owing to the pinch encountered between the equilibrium and operating lines at point *b*.

If the slope of the line *ab* were decreased, it would be impossible to make the required separation even with an infinite number of theoretical stages. If the slope of line *ab* were increased, the required separation could be made with a finite number of theoretical stages. Thus, the line *ab*, as shown on Fig. 8-7, is the only

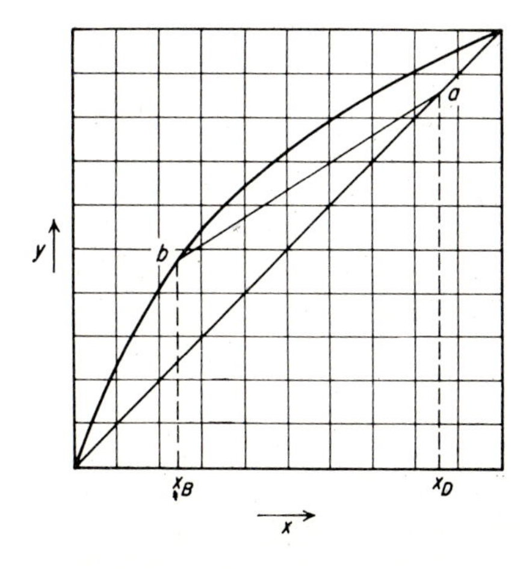


**Figure 8-7** Equilibrium $y$ vs. $x$ plot showing operating line for minimum reflux ratio.

operating line which meets the requirements of the minimum-reflux-ratio definition. Since the slope of an enriching operating line is $R/R + 1$, the minimum reflux ratio ($R_{min}$) can be obtained by determining the slope of the line *ab*, setting $R_{min}/R_{min} + 1$ equal to this slope and solving for $R_{min}$.

## TOTAL REFLUX

When a fractionating tower is operated with all of the condensed distillate being returned to the top of the tower as reflux, the reflux ratio has an infinite value and the unit is said to be operating at *total reflux*. Distillation operations carried out at total reflux do not permit the removal of any product. However, runs at total reflux are often made to determine the number of theoretical stages available in a given tower.

If a fractionation is carried out at total reflux using a binary mixture with a constant relative volatility, the following equation must apply:

$$\alpha^{N_p + 1} = \frac{(1 - x_B)(x_D)}{(x_B)(1 - x_D)} \tag{8-32}$$

where $\alpha$ = constant relative volatility
$N_p$ = number of theoretical plates in tower (not including still pot)
$x_B$ = mole fraction of more volatile component in bottoms (i.e., in still pot)
$x_D$ = mole fraction of more volatile component in distillate.

Equation (8-32) is known as the *Fenske equation*.[5] It can be derived by setting up the relative-volatility equation for each plate and the still pot. At total reflux, the composition of the vapors rising into a plate must equal the composition of the liquid on that plate. Therefore, when the product of all the relative volatilities is taken, all compositions cancel out except those for the bottoms and the distillate, and the final equation, as expressed by Eq. (8-32), is obtained.

The Fenske equation is often used in the experimental determination of the number of theoretical plates in distillation columns. For example, assume it is desired to determine the H.E.T.P. of a new distillation packing that has just come on the market. A tower is packed to a height of 10 ft with the new packing, and a distillation is performed using the packed tower operating at total reflux. A test mixture composed of *n*-heptane and methylcyclohexane is used, and the column is operated at atmospheric pressure. The relative volatility of the mixture, *n*-heptane–methylcyclohexane, is constant at a value of 1.070 when the operating pressure is 1 atm. In the experimental distillation, analyses indicate that the distillate composition is 0.88 mol fraction *n*-heptane and the still-pot composition is 0.15 mol fraction *n*-heptane. Since the unit is operating at total reflux with a mixture

[5] M. R. Fenske, *Ind Eng Chem*, **24**:482 (1932).

having a constant relative volatility, the Fenske equation can be used to calculate the total number of theoretical plates. Therefore

$$1.070^{N_p+1} = \frac{(1 - 0.15)(0.98)}{(0.15)(1 - 0.88)} = 41.6$$

and $N_p + 1 = 55.1$. Assuming the still pot as one theoretical plate, the total number of theoretical plates in the packed tower $= N_p = 54.1$. The H.E.T.P. for the new packing is $(10)(12)/54.1 = 2.2$ in.

The H.E.T.P. value for a given packing changes somewhat with changes in column diameter, packed height, boil-up rate, and test mixture. Therefore, these variables should be indicated when values of H.E.T.P. are reported.

## METHOD OF CALCULATION WHEN OPERATING LINE IS NOT STRAIGHT

The simplifying assumptions, giving straight operating lines in the McCabe-Thiele method for determining the number of theoretical stages, do not always apply. Ponchon[6] and Savarit[7] have developed a graphical method for determining the number of theoretical stages which does not assume constant molal overflow. This method is based on material and heat balances over the distillation unit and requires liquid-and-vapor enthalpy data and equilibrium $y$ vs. $x$ data for the mixture being fractionated.

## STEAM DISTILLATION

High-boiling liquids often cannot be purified by distillation at atmospheric pressure, since the components of the liquid tend to decompose at the high temperatures required. If the high-boiling substances are not soluble in water, it is possible to obtain a separation at lower temperatures by a process known as *steam distillation.*

Consider an example where live steam is injected directly into a liquid mixture which ordinarily boils at atmospheric pressure when the temperature is 500°F. If the water formed from the steam condensation is not mutually soluble with the original liquid mixture, the water layer and the liquid-mixture layer will exert their own vapor pressures as if each one were present alone. Thus, if the liquid temperature were 200°F, the vapor pressure due to the water would be 11.53 lb/in$^2$, absolute. Assume the original liquid mixture exerts a vapor pressure of 3.17 lb/in$^2$, absolute, when its temperature is 200°F. This means the total pressure over the two liquid layers at 200°F is $11.53 + 3.17 = 14.7$ lb/in$^2$, absolute $= 1$ atm. Therefore, the vapors can be removed and condensed at atmospheric pressure.

[6] M. Ponchon, *Tech mod*, **13**:20 (1921).

[7] P. Savarit, *Arts et métiers*, pp. 65, 142, 178, 241, 266, 307 (1922).

The condensed vapors separate into a water layer and a layer containing some of the original liquid mixture. These layers can be separated physically. In this manner, a distillation operation can be used to obtain a partial removal and recovery of the original mixture at atmospheric pressure, even though the actual temperature never exceeds 200°F.

The preceding example illustrates the process occurring in a steam distillation. An equation relating the weight of water in the vapors to the weight of the original liquid mixture in the vapors may be obtained as follows:

The total weight of water in the vapors divided by the molecular weight of water equals the moles of water in the vapors, or

$$\frac{W_a}{M_a} = n_a \tag{8-33}$$

Similarly, if $W_b$ is the weight of the original material in the vapors and $M_b$ is the molecular weight of the original material,

$$\frac{W_b}{M_b} = n_b \tag{8-34}$$

According to Dalton's law,

$$\frac{n_a}{n_a + n_b} P = p_a \tag{8-35}$$

and

$$\frac{n_b}{n_a + n_b} P = p_b \tag{8-36}$$

where $P$ is the total vapor pressure and $p_a$ and $p_b$ are the partial pressures of components $a$ and $b$ in the vapor.

Combining Eqs. (8-33) to (8-36),

$$\frac{W_a}{W_b} = \frac{M_a}{M_b}\left(\frac{p_a}{p_b}\right) \tag{8-37}$$

Equation (8-37) is the basic steam-distillation equation, and it may be used to calculate the ratio of the weights of the two phases obtained as distillate when a steam-distillation occurs.

## AZEOTROPIC AND EXTRACTIVE DISTILLATION

Many liquid mixtures cannot be separated by distillation because the volatilities of the components are equal. This means the relative volatility of the mixture is 1.0, and no fractionation can occur. Such a mixture is known as an *azeotrope*. A common example of an azeotrope is an 89.43 mol % mixture of ethyl alcohol and water at atmospheric pressure. The relative volatility of this mixture is 1.0, and no further purification can be obtained by conventional distillation methods.

It is often possible to change the relative volatility of a liquid mixture by adding a third component. For example, when benzene is added to the azeotropic mixture of water and ethyl alcohol, a new azeotropic mixture, consisting of ethyl alcohol, water, and benzene, is formed. This new ternary azeotrope has a relative volatility, referred to as ethyl alcohol, which is greater than 1. Therefore, the ternary azeotrope can be distilled off the top of a fractionating column, and practically pure ethyl alcohol can be recovered at the bottom of the column. A separating process of this type, where an azeotropic mixture is formed and removed from the top of the fractionating tower, is termed *azeotropic distillation.*

Addition of a third component to a binary mixture will sometimes lower the vapor pressure of the less volatile component of the binary mixture more than it lowers the vapor pressure of the more volatile component. An example of this is the addition of glycerol to a water–ethyl alcohol azeotrope. The vapor pressure of the water is lowered by the glycerol to such an extent that practically pure ethyl alcohol can be obtained as the distillate from a fractionating tower, while a mixture of glycerol, water, and ethyl alcohol is obtained at the bottom of the tower.

This type of distillation, where a mixture is separated by addition of a third component and this component is withdrawn at the base of the fractionating tower, is called *extractive distillation.*

## MULTICOMPONENT DISTILLATION

When a liquid mixture contains more than two components in significant quantities, it is called a *multicomponent mixture.* Distillation operations in the petroleum industry are ordinarily carried out with multicomponent mixtures.

Many different methods have been presented for calculating the number of theoretical stages necessary for a given multicomponent distillation. All of the methods require vapor-liquid equilibrium data, and it is often convenient to present these data for each component in the following form:

$$y = Kx \tag{8-38}$$

where $y$ = mole fraction of component in vapor phase
$x$ = mole fraction of component in liquid phase in equilibrium with $y$
$K$ = an empirical constant for each component at each temperature and pressure

Values of $K$ have been evaluated for many of the hydrocarbons encountered in the petroleum industry. With petroleum mixtures, the value of $K$ is essentially independent of the other components in the mixture. Although $K$ may vary with the concentration of the particular component, it is ordinarily permissible to use a constant average value of $K$ over the entire concentration range at each temperature and pressure.

One common method for making multicomponent distillation calculations is based on the same simplifying assumptions as those made in the McCabe-Thiele method for binary mixtures. Equations for the enriching and stripping lines are developed, and these equations are used in analytical plate-to-plate calculations, assuming the vapors evolved on each plate are in equilibrium with the liquid on that plate.[8]

## DISTILLATION COLUMN DESIGN

If a distillation column is designed with a large number of plates, it can handle a given separation with a lower reflux ratio than the same column could if it were designed with fewer plates. The column with more plates would have a higher initial cost than the one with fewer plates. However, the steam consumption (or rate of heat input to the still pot) would be greater in the column with fewer plates since higher boil-up-and-reflux rates would be required. Therefore, in designing distillation columns, an economic balance should be made between the added cost of putting more plates on the column and the money which can be saved by a decrease in steam or heat consumption. This economic balance determines the optimum number of plates for the column.

If the vapor velocity in a distillation column exceeds a certain limit, the liquid reflux will be unable to descend through the column, and a condition known as *flooding* will occur. When this limiting vapor velocity is exceeded, liquid starts to build up on the top trays of the column and may eventually flood over into the condenser. Therefore, in designing distillation columns, the diameter of the column must be large enough to keep the vapor velocity safely below the flooding velocity. Excessively high vapor velocities may also cause vapor-entrained droplets of liquid to be carried from one plate up to the next.

Design methods for predicting maximum allowable velocities in distillation columns are presented in Chap. 14 (Chemical Engineering Economics and Plant Design).

## APPLYING THE PRICIPLES OF DISTILLATION

One obvious application of the principles presented in this chapter is in the design of distillation equipment. Actually, the design of distillation equipment is seldom required of a chemical engineer, but he or she is often involved in the operation of the equipment. Therefore, it is particularly important for the chemical engineer to understand the physical aspects of distillation. Such things as the effect of change in reflux ratio or the possibility of increasing the capacity of a unit by insulating the distillation tower are usually of much more interest to the practical engineer than the number of trays in an existing column.

The equations developed in this chapter can be used for design purposes; however, the true value of these equations lies in their ability to indicate the physical

[8] For a detailed treatment of calculation methods in multicomponent distillations, see any of the numerous modern books specializing in distillation.

relationships among the different variables. The rest of this chapter deals with the practical applications of distillation principles and shows how a complete understanding of distillation requires a full realization of the interdependence of the crucial variables. The following discussion concerns continuous-distillation units in which feed is added continuously and distillate and bottoms are withdrawn continuously. Many of the remarks are also applicable to batch-distillation units.

## Qualitative Effect of Change in Variables

A continuous-distillation unit separates a feed material containing several volatile components into a distillate product and a bottoms product. Certain limits are ordinarily set on the concentrations of key materials in the distillate and in the bottoms, and the equipment usually must handle a set amount of feed per unit time. For example, the specified conditions for a distillation operation might require a unit to handle 5000 kg/h of a feed containing 10 wt% methanol and 90 wt% water to yield a distillate containing at least 99 wt% methanol and a bottoms containing less than 2 wt% methanol. These specifications set the values for five of the variables in the distillation operation. They are as follows:

1. The feed rate
2. The distillate composition
3. The bottoms composition
4. The rate of delivery for the distillate product (by material balances)
5. The rate of delivery for the bottoms product (by material balances)

There are two crucial variables which the operator can control to obtain the desired operating conditions. These are (1) the reflux ratio and (2) the amount of heat input to the still pot. Other factors affecting the overall operation of the unit are (1) the temperature of the feed, (2) the temperature of the reflux returned to the column, and (3) the location of the feed plate.

**Reflux ratio.** The reflux ratio indicates a relative measure of the amount of heat removed from the column, and it is the essential factor for setting the concentration of the distillate and the bottoms. An increase in the reflux ratio causes an increase in the concentration of the more volatile component in the distillate and a decrease in the concentration of the same component in the bottoms. The maximum concentration of the more volatile component in the distillate would be obtained at an infinite reflux ratio. However, under these conditions, no product could be withdrawn from the unit, and the operation would be of no value. A finite reflux ratio must be used, and the operator can change the value of the reflux ratio to control the concentration of the products.

It is possible to change the reflux ratio manually by means of a control valve on the line leading from the reflux condenser to the top of the distillation column. However, commercial reflux splitters are usually employed for this purpose. With these devices, the operator can change the reflux ratio by merely changing the instrument setting.

**Rate of heat supplied to still pot.** The rate at which the distillate product is removed depends on the amount of vapor condensing in the reflux condenser per unit time, which, in turn, depends on the rate of heat input to the still pot. For example, if 10 kg of reflux was returned to the top of a column per kilogram of distillate product withdrawn, it would be necessary for 11 kg of vapor to pass into the reflux condenser for each kilogram of distillate product. A constant amount of heat must be put into the still pot in order to cause 11 kg of vapor to enter the reflux condenser. Therefore, increasing the rate of heat input to the still pot increases the rate at which the distillate product may be removed.

The reflux ratio is the essential factor in setting the concentrations of the distillate and bottoms, while the rate of heat input to the still pot is the main variable affecting the rate of product delivery.

## Distillation-Column Control

The control of continuous-distillation units requires the maintenance of the proper balance between reflux ratio, rate of heat input to the still pot, and feed rate. For example, consider a continuous-distillation column operating under steady conditions such that all rates and concentrations remain constant. If the reflux ratio is increased, the concentration of the more volatile component in the distillate will increase, but the distillate product rate will decrease. Therefore, to maintain the desired percent removal of the more volatile component from the feed, it is necessary to decrease the feed rate or to increase the rate of heat input to the still pot.

Manual methods may be used for the control of distillation columns, but automatic controls are much more satisfactory. The results of a change in one of the distillation-unit variables are seldom immediately apparent, since there is a certain time lag caused by the presence of liquid (or holdup) in the column. Operators tend to overcorrect because they want immediate results, while automatic controllers can be adjusted to keep a continuous control with no overcorrections.

## Nonadiabaticity

Even if a distillation column is insulated, a certain amount of heat is lost from the walls of the column, and this results in a *nonadiabatic operation.* Usually, the effects of nonadiabaticity are not large, but in some cases the effects are appreciable. These heat losses cause increased condensation inside the column and give an increasing ratio of liquid flow rate to vapor flow rate from the top to the bottom of the column. Accordingly, the reflux ratio at the top of the column cannot always be used as a definite criterion for estimating the possible extent of separation which can be obtained in a distillation column.

Distillation columns located outdoors may appear to be more efficient in winter than in summer. This can be explained in two ways: (1) The effect of nonadiabaticity is more pronounced when the air around the column is cold. Consequently, at the same distillate reflux ratio, a better separation can be achieved in winter than in summer. (2) The liquid returned to the top of the column as reflux is colder in winter than in summer. This causes the same effect as heat losses from the column.

It may appear as if it would be advantageous to operate with considerable heat losses from the column walls, since this seems to increase the efficiency of the unit. However, this is not the case, for more heat must be supplied to the still pot in the nonadiabatic runs in order to maintain the same distillate product rate as in the adiabatic runs.

Nonadiabaticity also partly explains the reason why some columns appear to become more efficient as the rate of heat input to the still pot is decreased. The actual amount of heat lost from the column walls is essentially constant at all vapor rates, since the rate of heat transfer is proportional to the temperature difference across the wall. Therefore, the effects of nonadiabaticity become more pronounced at lower vapor rates or at lower rates of heat input to the still pot.

### Concentration and Temperature Distribution

With binary mixtures, the concentration of the more volatile component increases smoothly from the bottom to the top of a distillation column. The maximum temperature always occurs at the bottom of the column. With binary mixtures, the temperature decreases steadily for points progressively higher in the column until the minimum temperature is reached at the top of the column.

Concentration and temperature distributions over multicomponent distillation units do not show steady changes and are widely different from those found with binary mixtures. The concentration of certain components may actually reach a maximum value on an intermediate plate of a continuous column. The rate of temperature change with multicomponent mixtures varies considerably over the length of the column and may be much larger at one part of the column than at another.

## PROBLEMS

**8-1** The vapor pressure of pure benzene is 986 mmHg at a temperature of 89°C. The vapor pressure of pure toluene is 392 mmHg at a temperature of 89°C. What will the mole fraction of benzene be in a binary liquid mixture of benzene and toluene when the temperature is 89°C and the total equilibrium vapor pressure of benzene and toluene is 720 mmHg?

**8-2** A liquid mixture of ethyl alcohol and water is in equilibrium with a vapor containing ethyl alcohol and water at a total pressure of 760 mmHg. A sample of the vapor indicates that it contains 3.3 mol of ethyl alcohol for every 1.7 mol of water. If the liquid has a mol fraction of 0.52 ethyl alcohol, what is the relative volatility for the mixture?

**8-3** A liquid mixture of methyl alcohol and water contains 0.41 mol fraction methyl alcohol. This mixture is fed continuously to a unit where a simple equilibrium distillation occurs at atmospheric pressure. What will the concentration of the distillate be if 30 mol of distillate and 70 mol of the equilibrium liquid are withdrawn per 100 mol of feed?

**8-4** A mixture containing 0.70 mol fraction $a$ and 0.30 mol fraction $b$ is subjected to a simple batch distillation until the instantaneous composition of the vapors leaving becomes 0.60 mol fraction $a$. If the relative volatility for this mixture is constant at $\alpha_{ab} = 1.5$, estimate the average composition of the total distillate collected.

**8-5** A liquid mixture containing 0.4 mol fraction methyl alcohol and 0.6 mol fraction water is fed to an intermediate point on a continuous-distillation column containing eight actual sieve trays. The distillate composition is 0.93 mol fraction methyl alcohol, and the bottoms composition (from the still pot) is 0.10 mol fraction methyl alcohol when the reflux ratio is 2.0. The column is operated at atmospheric pressure and the feed enters at its boiling temperature. If the McCabe-Thiele simplifying assumptions may be made, what is the overall plate efficiency?

**8-6** What is the minimum reflux ratio that can be used in an enriching distillation column to obtain a distillate composition of 0.90 mol fraction benzene from a bottoms containing 0.20 mol fraction benzene and 0.80 mol fraction toluene? The column is to operate at atmospheric pressure.

Equilibrium $y$ vs. $x$ data for benzene-toluene mixtures at atmospheric pressure are given below.

| $x$ mol fraction benzene | $y$ mol fraction benzene | Boiling point at 760 mmHg, °C |
|---|---|---|
| 0.00 | 0.00 | 110.4 |
| 0.05 | 0.111 | 108.3 |
| 0.10 | 0.209 | 106.2 |
| 0.20 | 0.375 | 102.2 |
| 0.30 | 0.508 | 98.5 |
| 0.40 | 0.621 | 95.3 |
| 0.50 | 0.714 | 92.2 |
| 0.60 | 0.791 | 89.5 |
| 0.70 | 0.855 | 87.0 |
| 0.80 | 0.912 | 84.6 |
| 0.90 | 0.959 | 82.3 |
| 0.95 | 0.981 | 81.3 |
| 1.00 | 1.000 | 80.3 |

**8-7** The H.E.T.P. of a certain packed tower may be taken as 0.203 m. A binary mixture with a constant relative volatility of 1.120 is to be separated in this unit. The tower is used as an enriching section, and the liquid concentration of the more volatile component in the still pot is maintained at a mole fraction of 0.15. If the tower is packed to a height of 4.57 m, what is the maximum composition of distillate obtainable under any conditions?

**8-8** Experimental tests on a single bubble-cap plate using a methyl alcohol–water test mixture indicate the liquid composition on the table is 0.40 mol fraction methyl alcohol. The actual composition of the vapors leaving the plate is 0.68 mol fraction methyl alcohol, and the vapor composition entering the plate is 0.51 mol fraction methyl alcohol. What is the Murphree plate efficiency for this case if the total pressure is 760 mmHg?

**8-9** The vapor pressures of dimethylaniline at various temperatures are as follows:

| Temperature, °C | Vapor pressure, mmHg | Temperature, °C | Vapor pressure, mmHg |
|---|---|---|---|
| 70.0 | 10 | 125.8 | 100 |
| 84.8 | 20 | 146.5 | 200 |
| 101.6 | 40 | 169.2 | 400 |
| 111.9 | 60 | 193.1 | 760 |

Considering water and dimethylaniline as completely insoluble, calculate the following.

(*a*) The temperature and composition of the vapor when a mixture of dimethylaniline and water boils at a total pressure of 730 mmHg.

(*b*) The kilograms of dimethylaniline obtained as distillate in a steam distillation for every 100 kg of water obtained in the distillate when the total pressure is 730 mmHg. (The molecular weight of dimethylaniline is 121.2.)

**8-10** Twenty thousand pounds per hour of a liquid feed containing 45 wt % benzene and 55 wt % toluene is fed to an intermediate point in a continuous distillation column. The distillate composition is 92 wt % benzene, and the bottoms composition is 10 wt % benzene. The column contains valve plates, and its overall plate efficiency may be taken as 70 percent. A reflux ratio of two times the minimum reflux ratio is used. The liquid feed may be assumed as entering at its boiling point. The column operates at a total pressure of 760 mmHg, and the equilibrium $y$ vs. $x$ data for benzene-toluene mixtures, as given in Prob. 8-6, may be used.

Assuming the still pot actually acts as one theoretical stage, determine the following:

(*a*) The number of actual plates the column should contain.

(*b*) The recommended location for the feed plate on the actual column.

(*c*) The pounds of distillate product obtained per hour.

(*d*) The pounds of steam theoretically necessary per hour if saturated steam at 35 $lb/in^2$, absolute, is used in the heating coils. The heat of vaporization for 1 lb mol of the bottoms mixture may be taken as 13,700 Btu.

(*e*) The recommended diameter of the column if the allowable vapor velocity based on the total cross-sectional area is 2.0 ft/s. The perfect-gas law is applicable, and the bottoms temperature should be used.

(*f*) If the feed had been introduced as a cold liquid instead of at its boiling point, would the recommended diameter of the stripping section be larger or smaller than that of the enriching section?

**8-11** A mixture is being fed to an intermediate point in a continuous-distillation column at the rate of 160 kg mol/h. The reflux ratio $R$ is 8.0. Ninety kilogram moles per hour of distillate product is withdrawn, and 70 kg mol/h of bottoms product is withdrawn. The feed enters at 422 K, and the boiling point of the feed is 311 K. The heat capacity of the feed in the vapor state is 1256 J/(kg)(K), and the heat required to vaporize 1 kg of the feed at its boiling temperature is 335,000 J. Determine the number of kilogram moles of vapor per hour passing through the stripping section of the tower.

CHAPTER

# NINE

# ABSORPTION AND EXTRACTION

Absorption and extraction are unit operations involving the transfer of material between two phases. *Absorption*, as presented in this book, refers to an operation in which the transfer of material is from a gaseous phase to a liquid phase. For example, when ammonia in the gaseous state is dissolved by liquid water, there is a transfer of ammonia from the gas phase to the liquid phase, and the operation of absorption occurs.

The term *extraction* refers to an operation in which one or more components of a liquid or a solid are transferred to another liquid. When liquid–liquid extraction is used, at least one of the liquid components must be sufficiently insoluble so that two liquid phases are formed. One example of extraction is the separation of certain components of petroleum-base oils by the use of acetone. If acetone is added to the oil, some of the components of the oil dissolve in the acetone, and the remainder of the oil forms an immiscible (nonmixing) layer with the acetone solution. These two layers can be separated. In this process, the transfer of part of one liquid to another liquid has occurred, and the unit operation of extraction has been performed.

Many of the basic concepts and principles applied in the treatments of distillation and heat transfer are directly applicable to absorption and extraction operations. The chemical engineer should realize that the various unit operations are not separate and distinct from each other. Certain of the unit operations, although commonly presented under different headings, involve similar principles. For example, distillation, absorption, and extraction are all concerned with separating the components of a mixture by a process involving the presence of two separate phases. A comprehension of the theoretical treatments for all three of these unit operations requires a clear understanding of material-balance applications, equilibrium relationships, and the concept of theoretical separating stages.

In some cases, chemical engineering principles have been presented on the basis of generalized concepts rather than on the basis of the individual operations. However, the unit-operations presentation is much more suitable for beginning

**Table 9-1 Nomenclature for absorption and extraction**

$a$ = interfacial area per unit volume of absorbing apparatus, $m^2/m^3$ or $ft^2/ft^3$
$A$ = surface-contact area available for mass transfer, $m^2$ or $ft^2$
$c$ = concentration of a component in a liquid solution, kg mol of the component per $m^3$ of the liquid solution or lb mol/$ft^3$
$G_M$ = flow rate of inert gas through absorption tower, kg mol/(s)($m^2$) or lb mol/(h)($ft^2$)
$H$ = Henry's law constant, kg mol/($m^3$)(atm) or lb mol/($ft^3$)(atm)
$H_o$ = amount of nonextractable material in a mixture being treated by an extraction process, kg or lb
$k$ = absorption film coefficient, with subscript $G$, kg mol/(s)($m^2$)(atm) or lb mol/(h)($ft^2$)(atm); with subscript $L$, kg mol/(s)($m^2$)(kg mol/$m^3$) or lb mol/(h)($ft^2$)(lb mol/$ft^3$)
$K$ = overall absorption coefficient, with subscript $G$, kg mol/(s)($m^2$)(atm) or lb mol/(h)($ft^2$)(atm); with subscript $L$, kg mol/(s)($m^2$)(kg mol/$m^3$) or lb mol/(h)($ft^2$)(lb mol/$ft^3$)
$L_M$ = flow rate of pure absorbing liquid through absorption tower, kg mol/(s)($m^2$) or lb mol/(h)($ft^2$)
$N$ = rate of absorption, kg mol/s or lb mol/h
$p$ = partial pressure, atm
$S_o$ = amount of extracting solvent added in an extraction process, kg or lb
$V$ = volume of absorbing apparatus, $m^3$ or $ft^3$
$X$ = amount of material dissolved in a unit amount of a liquid solvent, kg mol/kg mol or kg/kg; or lb mol/lb mol or lb/lb
$Y$ = amount of one material present in a unit amount of a second material, kg mol/kg mol or kg/kg; or lb mol/lb mol or lb/lb

Subscripts

1 refers to conditions at gas entrance to an absorption tower; refers to equilibrium conditions in extraction
2 refers to conditions at gas exit from an absorption tower; refers to equilibrium conditions in extraction
$a$, $b$, and $c$ refer to components of a mixture or to plates in an absorption tower
$aG$ refers to main body of a gas
$ai$ refers to interface between liquid and gas films
$aL$ refers to main body of a liquid
$G$ refers to gas film
$L$ refers to liquid film
$o$ refers to original or starting conditions

chemical engineers as long as they realize the various unit operations are flexible and employ many of the same principles.

In this chapter, the similarity of certain absorption and extraction theories to other unit-operation theories will be indicated.

**American engineering units compared to SI.** The nomenclature shown in Table 9-1 gives standard units used in absorption and extraction operations for both the American engineering system and SI. Following are the factors used for converting from one system to another.

For $a$, 1 $ft^2/ft^3$ = 3.281 $m^2/m^3$
For $c$, 1 lb mol/$ft^3$ = 16.018 kg mol/$m^3$
For $G_M$ or $L_M$, 1 lb mol/(h)($ft^2$) = $1.356 \times 10^{-3}$ kg mol/(s)($m^2$)

For $H$, 1 lb mol/(ft$^3$)(atm) = 16.018 kg mol/(m$^3$)(atm)
For $k_G$ or $K_G$, 1 lb mol/(h)(ft$^2$)(atm) = $1.356 \times 10^{-3}$ kg mol/(s)(m$^2$)(atm)
For $k_G a$ or $K_G a$, 1 lb mol/(h)(ft$^3$)(atm) = $4.449 \times 10^{-3}$ kg mol/(s)(m$^3$)(atm)
For $k_L$ or $K_L$, 1 lb mol/(h)(ft$^2$)(lb mol/ft$^3$) = $8.467 \times 10^{-5}$ kg mol/(s)(m$^2$)(kg mol/m$^3$)
For $k_L a$ or $K_L a$, 1 lb mol/(h)(ft$^3$)(lb mol/ft$^3$) = $2.778 \times 10^{-4}$ kg mol/(s)(m$^3$)(kg mol/m$^3$)
For $N$, 1 lb mol/h = $1.260 \times 10^{-4}$ kg mol/s

To simplify the presentation of the material in this chapter, all of the units used in the discussion are in the American engineering system. With the conversions given in the preceding list, it should be a very easy operation to convert to SI units if desired.

## DIFFUSION

The transfer of material from one phase to another takes place by a process known as diffusion. *Diffusion* may here be defined as the spontaneous intermingling of the particles forming a solution. Thus, when ammonia vapor is absorbed in water, the ammonia particles or molecules must diffuse through the gas to the surface of the liquid water. These molecules must then enter the liquid solution and diffuse away from the water surface into the main body of the liquid.

## EQUIPMENT

Since material can be transferred from one phase to another only when the surfaces of the two phases are in contact, it is desirable to provide the largest possible surface-contact area in order to obtain the maximum amount of material transfer. This can be accomplished by breaking one of the phases into small segments and forcing these segments to pass through the other phase. In absorption operations, this is carried out by the use of plate towers where the gas, in the form of finely divided bubbles, is forced to rise through the liquid layer.

Another method commonly used for obtaining large areas of surface contact between gas and liquid phases is to break up the liquid phase into an extended surface and pass the gases over this surface. Various types of packed towers are used for this purpose.

Equipment for extraction operations should be designed to give the maximum possible contact between the two phases. Arrangements must be made to transfer the two phases from one stage to another if the extraction is performed in more than one step. In liquid–liquid extraction, contact between the two phases is commonly obtained by the use of packed-tower sections. Vibrating perforated plates and venturi injectors have also been suggested for use in obtaining efficient surface contact for liquid–liquid extraction.

## HENRY'S LAW[1]

Henry's law states that the concentration of a component in a liquid solution is directly proportional to the equilibrium partial pressure of that component over the liquid solution, or

$$c_a = Hp_a \tag{9-1}$$

where $c_a$ = concentration of component $a$ in a liquid solution, expressed, in this book, as kilogram (pound) moles of component $a$ per cubic meter (cubic foot) of the liquid solution, kg mol/m$^3$ or lb mol/ft$^3$

$H$ = Henry's law constant, kg mol/(m$^3$)(atm) or lb mol/(ft$^3$)(atm)

$p_a$ = partial pressure of component $a$ in *equilibrium* with the solution having a concentration of $c_a$, atm

Henry's law is a liquid law and is followed only under certain conditions. In general, when the concentration of a volatile component in a solution is very low, that component tends to follow Henry's law. Thus, when the concentration of ammonia in water is below approximately 5 mol %, Henry's law may be assumed as applicable to the ammonia. The constant $H$ depends on the components of the solution and on the temperature of the solution, and its value must be determined experimentally.

Henry's law is essentially a special case of Raoult's law, where $H$ would be equivalent to the reciprocal of the pure-component vapor pressure if $c_a$ were expressed as a mole fraction. Since many absorption operations deal with dilute solutions, Henry's law is of considerable use for determining equilibrium relationships in absorption calculations.

## METHODS OF CALCULATION FOR ABSORPTION COLUMNS

The number of theoretical stages required for a continuous-absorption operation may be determined by a graphical method which is analogous to the McCabe–Thiele method for determining theoretical stages in distillation columns.

Consider a continuous-absorption column, such as the one shown in Fig. 9-1, where only one component of the gas is absorbed. The following nomenclature applies to Fig. 9-1:

$G_M$ = kilogram or pound moles of inert gas passing through 1 m$^2$ or 1 ft$^2$ of cross section of the absorption tower in unit time, kg mol/(s)(m$^2$) or lb mol/(h)(ft$^2$)

$L_M$ = kilogram or pound moles of pure absorbing liquid passing through 1 m$^2$ or 1 ft$^2$ of cross section of the absorption tower in unit time, kg mol/(s)(m$^2$) or lb mol/(h)(ft$^2$)

[1] In many texts, the constant $H$ in Henry's law is defined as the reciprocal of the constant given in Eq. (9-1), and $c_a$ is often replaced by mole fraction of $a$.

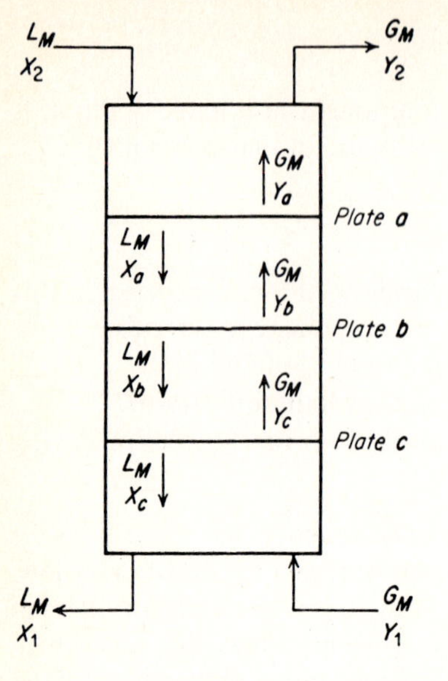


**Figure 9-1** Continuous-absorption column used for treatment of graphical theoretical-stages determination.

$X$ = kilogram or pound moles of dissolved absorbable material per kilogram or pound mole of pure absorbing liquid, kg mol/kg mol or lb mol/lb mol

$Y$ = kilogram or pound moles of gaseous absorbable material per kilogram or pound mole of inert gas, kg mol/kg mol or lb mol/lb mol

Subscripts 1 and 2 refer to the section of the column where gases enter and leave, respectively

Subscripts $a$, $b$, and $c$ refer to plates $a$, $b$, and $c$, respectively

A material balance around the entire column, based on the amount of absorbable material, gives

$$L_M(X_1 - X_2) = G_M(Y_1 - Y_2) \tag{9-2}$$

Equation (9-2) can be considered as an operating equation, since it may be made to represent the conditions at any section of the column by merely changing the $X$ and $Y$ subscripts. Thus, by an absorbable-component material balance around plate $b$,

$$L_M(X_b - X_a) = G_M(Y_c - Y_b) \tag{9-3}$$

If a plot is made of $Y$ vs. $X$, Eq. (9-2) or (9-3) can be represented on this plot as the operating line with a constant slope of $L_M/G_M$. Such a plot is presented in Fig. 9-2, and the operating line is designated by line $AD$. Curve $OE$, on this same plot, is a line representing the equilibrium data for the absorption operation.

When a plate is theoretically perfect, the liquid on the plate and the vapors leaving the plate have had sufficient contact to permit the establishment of equilibrium conditions. Under these conditions, $X_a$ and $Y_a$, as well as $X_b$ and $Y_b$, are equilibrium values; therefore, they represent points on the equilibrium curve.

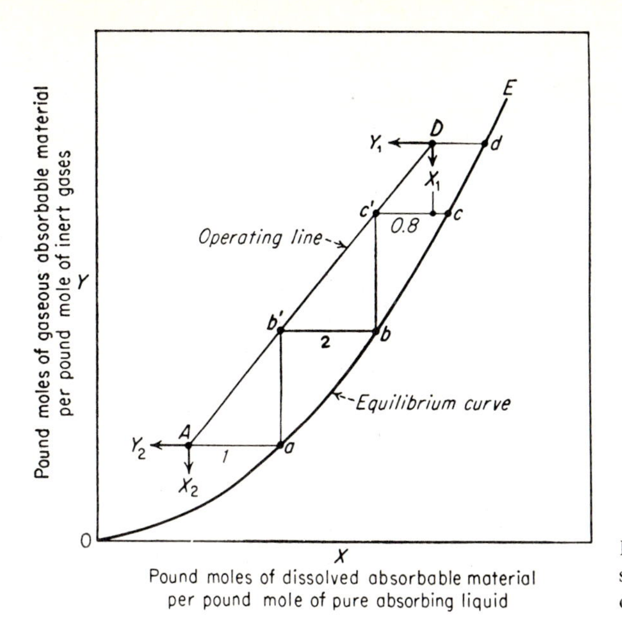


**Figure 9-2** Equilibrium diagram showing graphical determination of theoretical stages.

Point $A$ in Fig. 9-2 represents the conditions at the top of the absorption column. If plate $a$ (assumed as a theoretically perfect plate) is the top plate in the column, $Y_a$ equals $Y_2$, and $Y_2$ must be in equilibrium with $X_a$. Therefore, point $a$ in Fig. 9-2 represents the theoretical liquid concentration on tray $a$, and point $b'$ on the operating line represents the vapor composition rising into tray $a$. This completes one theoretical stage, and the vapor composition has changed from the $Y$ value represented by point $A$ (or $a$) to the $Y$ value represented by point $b'$.

This stepwise operation can be continued on down the column until point $D$, representing the conditions at the bottom of the column, has been reached. The absorption operation depicted in Fig. 9-2 would require approximately 2.8 theoretical stages. The reasoning behind this method for determining the number of theoretical stages is similar to that used in the McCabe–Thiele treatment of theoretical stages in distillation columns.

**Overall plate efficiency.** The overall plate efficiency of a bubble-cap absorption column is simply the number of theoretical stages (or plates) a given column can produce, divided by the actual number of plates in the column. Murphree individual plate efficiencies for absorption columns are determined in the same manner as for distillation columns.

## Minimum Ratio of Liquid to Gas

The minimum ratio of liquid to gas in absorption operations is defined as the value of $L_M/G_M$ which would produce a given gas separation if an infinite number of theoretical stages were used. In Fig. 9-2, the slope of a straight line between points $A$ and $d$ would represent the value of $L_M/G_M$ where an infinite number of stages would produce the required change in vapor concentration.

**Example 9-1: Graphical determination of the number of actual plates required for an absorption operation** A bubble-cap plate absorption column is to be used to absorb ammonia in water. A gaseous mixture containing 20 mol % $NH_3$ and 80 mol % air enters the bottom of the absorption tower. A total of 50 lb mol of gaseous $NH_3$ enters the tower per hour. Four thousand pounds of pure water enters the top of the absorption tower per hour, and the equipment is operated at atmospheric pressure with a constant temperature of 60°F. Assuming the overall plate efficiency for the tower as 60 percent, determine the actual number of bubble-cap plates required to absorb 90 percent of the entering ammonia. Neglect the effect of water vapor in the gases. Experimental equilibrium data for $NH_3$ gas and water solutions of $NH_3$ at 60°F and a pressure of 1 atm are as follows:

| | | | | | |
|---|---|---|---|---|---|
| Liquid concentration, moles $NH_3$/mol $H_2O$ | 0.000 | 0.053 | 0.111 | 0.177 | 0.250 |
| Partial pressure of $NH_3$, mmHg | 0.0 | 32.0 | 62.0 | 103.0 | 166.0 |
| Gas concentration, mol $NH_3$/mol inerts | 0.000 | 0.044 | 0.089 | 0.157 | 0.280 |

SOLUTION

**Basis**

1 h

Moles of gaseous $NH_3$ per mole of inert gas at entrance to tower = $Y_1$ = 20/80 = 0.25.

Total moles of $NH_3$ absorbed by water = (50)(0.90) = 45.

Moles of absorbed $NH_3$ per mole of water at liquid exit from tower = $X_1$ = (45)(18/4000) = 0.202.

Total moles of inert gases passing through tower = (80/20)(50) = 200.

Moles of gaseous $NH_3$ per mole of inert gases at gas exit from tower = $Y_2$ = (50)(0.10)/(200) = 0.025.

Moles of absorbed $NH_3$ per mole of water at liquid entrance to tower = $X_2$ = 0.000.

The graphical solution to this problem is presented in Fig. 9-3. The operating line is represented by a straight line between $Y_1, X_1$ and $Y_2, X_2$. The slope of this line is

$$\frac{L_M}{G_M} = \frac{4000}{(18)(200)} = 1.11$$

From Fig. 9-3, it can be seen that 4.4 theoretical stages are required to absorb 90 percent of the entering ammonia under the given conditions.

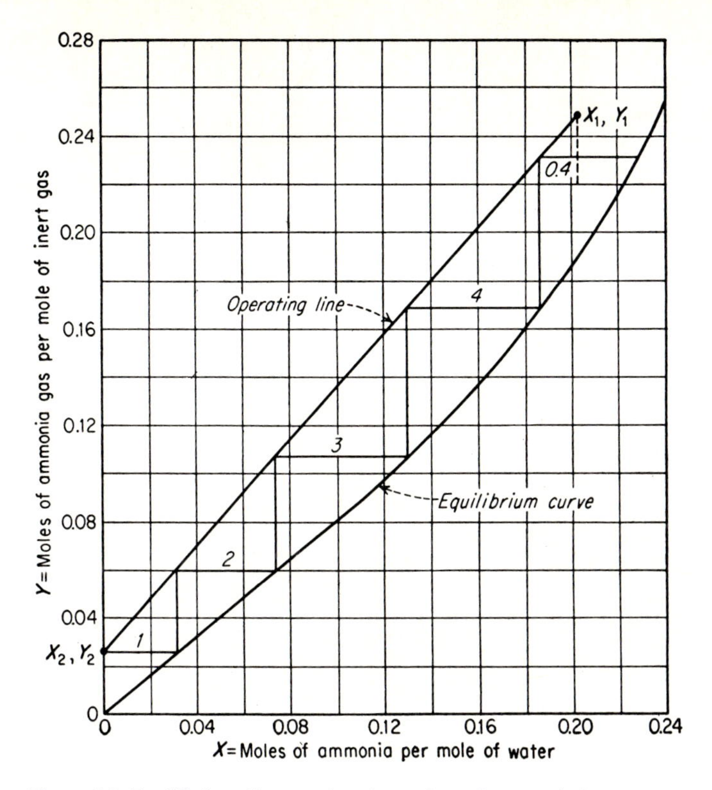


**Figure 9-3** Equilibrium diagram for absorption of ammonia in water at 60°F and a pressure of 1 atm (Example 9-1).

Since the overall plate efficiency for the absorption tower is 60 percent, the total number of actual bubble-cap plates necessary is $4.4/0.6 = 7.33$ plates. Therefore, the absorption tower should contain eight actual bubble-cap plates.

## APPLICATION OF ABSORPTION COEFFICIENTS

In distillation operations, the stepwise graphical determination of the total number of theoretical stages required for a certain operation gives results that are immediately applicable to practical design problems. This is due to the fact that the overall plate efficiency of a plate column and the H.E.T.P. of packed columns are not greatly affected by the type of mixture being distilled. Therefore, efficiencies determined with one mixture may generally be applied to other mixtures without the introduction of large errors. However, this is not the case in absorption operations.

The rate of absorption varies widely with different mixtures. Certain absorbing mixtures may indicate that a plate absorber has a high plate efficiency since the rate of absorption for the particular components is high. This high rate of absorption results in a rapid approach to equilibrium with a resultant high plate efficiency. On the other hand, if an absorbing mixture with a very low rate of absorption were separated in the same plate tower at the same throughput rates, the plate efficiency of the unit would be very low.

This difficulty may be overcome by methods of calculation involving absorption coefficients which take the rate of absorption into account. These methods are very useful and are applicable not only to absorption operations, but to many different types of mass-transfer operations.

Packed absorption towers are usually designed on the basis of available absorption coefficients, and plate towers are commonly designed using the theoretical-stage concept combined with experimental overall plate efficiencies.

## TWO-FILM THEORY OF ABSORPTION

The concept of liquid and gas films was introduced in the consideration of heat-transfer and fluid-flow principles. This same concept may be applied to gas absorption. When a component of a gas is absorbed by a liquid, the gaseous component must diffuse through the gas to the surface of the liquid. A thin stationary film of vapor may be assumed as existing at the boundary between the liquid and the gas. Just as in the theoretical consideration of heat-transfer films, the entire resistance to the flow of material from the body of the gas to the liquid surface may be assumed as existing in the gas film.

When the gaseous component enters the liquid phase, it must diffuse away from the surface into the body of the liquid. The resistance to this mass transfer may be assumed as existing in a fictitious liquid film. If it is assumed that there is no resistance at the interface between the gas and liquid films, the overall resistance to the mass transfer must be due to the two films: one gas film and one liquid film.

This theory is analogous to that postulated for the flow of heat between two confined materials. The only difference lies in the presence of an added resistance to the flow of heat due to the confining walls.

**Absorption where one film is controlling.** Because of the wide range in solubilities of various gases in liquids, there are certain cases where the resistance to mass transfer offered by either the gas film or the liquid film becomes negligible when the two resistances are compared. For example, if a gas, such as hydrogen chloride, which is very soluble in water, is absorbed by water, the gas will be sucked into the liquid as soon as it gets through the gas film and reaches the liquid surface. Thus, the resistance of the water film is practically negligible in comparison with the resistance offered by the gas film, and the gas film is said to be controlling. On the other hand, if a slightly soluble gas, such as oxygen, were being absorbed by water, the oxygen would pass through the water film very slowly in comparison with its possible rate through the gas film, and the water film would be controlling.

**Conditions at the interface.** Equilibrium conditions may be assumed to exist at the interface between the liquid and the gas films. For example, at a temperature of 80°F, a water solution of ammonia containing 5 mol% ammonia is in equilibrium with a gas in which the partial pressure of ammonia is 0.0708 atm. Therefore, if the liquid concentration at the interface is 5 mol% ammonia in water, the pressure of the ammonia in the gas at the interface is 0.0708 atm.

## Absorption Coefficients

The rate at which a component of a gas can be transferred through a gas film to the vapor-liquid interface is proportional to the surface area available for the transfer. The rate is also proportional to the force available for driving the gaseous component through the gas film. This driving force may be taken as the difference between the pressure of the component in the main body of the gas and the pressure of the component at the vapor-liquid interface. This may be expressed as

$$N = k_G A(p_{aG} - p_{ai}) \tag{9-4}$$

where $N$ = rate at which component is transferred through gas film, kg mol/s or lb mol/h

$k_G$ = gas-film absorption coefficient, kg mol/(s)(m$^2$)(atm) or lb mol/(h)(ft$^2$)(atm)

$A$ = surface-contact area available for mass transfer, m$^2$ or ft$^2$

$p_{aG}$ = partial pressure of component in main body of the gas, atm

$p_{ai}$ = partial pressure of component at vapor-liquid interface, atm

Equation (9-4) is analogous to the equation expressing the rate of heat transfer through fluid films.

The rate at which a component is transferred through a liquid film can be expressed in a similar manner as

$$N = k_L A(c_{ai} - c_{aL}) \tag{9-5}$$

where $k_L$ = liquid-film absorption coefficient, kg mol/(s)(m$^2$)(kg mol/m$^3$) or lb mol/(h)(ft$^2$)(lb mol/ft$^3$)

$c_{ai}$ = concentration of component in liquid at vapor-liquid interface, kg mol/m$^3$ or lb mol/ft$^3$

$c_{aL}$ = concentration of component in main body of liquid, kg mol/m$^3$ or lb mol/ft$^3$

Just as in heat-transfer considerations, an overall coefficient may be employed for calculations dealing with absorption operations. This overall coefficient must be a function of the individual film coefficients. Since mass transfer by diffusion is controlled by the two separate driving forces of gas-pressure difference and liquid-concentration difference, it is necessary to define two overall coefficients. The symbol $K_G$ represents the overall coefficient based on the same driving-force units as used for the gas-film coefficient, and $K_L$ represents the overall coefficient based on the same driving-force units as used for the liquid-film coefficient. The equations for these overall coefficients are

$$N = K_G A(p_{aG} - p_{aL}) \tag{9-6}$$

$$N = K_L A(c_{aG} - c_{aL}) \tag{9-7}$$

where $K_G$ = overall absorption coefficient based on same units as gas-film coefficient, kg mol/(s)(m$^2$)(atm) or lb mol/(h)(ft$^2$)(atm)

$K_L$ = overall absorption coefficient based on same units as liquid-film coefficient, kg mol/(s)(m$^2$)(kg mol/m$^3$) or lb mol/(h)(ft$^2$)(lb mol/ft$^3$)

$p_{aL}$ = partial pressure of component if it were in equilibrium with a solution having same concentration as main body of liquid, atm

$c_{aG}$ = concentration of component if it were in a solution in equilibrium with a gas having same component partial pressure as main body of gas, kg mol/m$^3$ or lb mol/ft$^3$

Since the gas film and the liquid film represent resistances in series, the rate of material passing through each film must be the same, or

$$\begin{aligned} N &= k_G A(p_{aG} - p_{ai}) = k_L A(c_{ai} - c_{aL}) \\ &= K_G A(p_{aG} - p_{aL}) = K_L A(c_{aG} - c_{aL}) \end{aligned} \tag{9-8}$$

When Henry's law is applicable to the component being absorbed, Eq. (9-1) may be written as

$$c_{ai} = Hp_{ai} \tag{9-9}$$

$$c_{aL} = Hp_{aL} \tag{9-10}$$

$$c_{aG} = Hp_{aG} \tag{9-11}$$

From Eqs. (9-8) to (9-11), the following expressions may be derived relating the overall coefficients to the individual film coefficients for the case where Henry's law applies:

$$\frac{1}{K_G} = \frac{1}{k_G} + \frac{1}{Hk_L} \tag{9-12}$$

$$\frac{1}{K_L} = \frac{1}{k_L} + \frac{H}{k_G} \tag{9-13}$$

Equations (9-12) and (9-13) are similar to the equations used for expressing resistance to heat transfer. Thus, in the case of mass transfer, the total resistance can be expressed as $1/K_G$ or $1/K_L$, and the individual film resistances can be expressed as $1/k_G$, $1/Hk_L$, $1/k_L$, or $H/k_G$. As an example, the fraction of the total mass-transfer resistance due to the gas film would be $(1/k_G)/(1/K_G)$ or $(H/k_G)/(1/K_L)$.

**Interfacial-area term.** In Eqs. (9-4) to (9-8), the symbol $A$ represents the interfacial surface-contact area available for mass transfer between the two phases. In most absorption equipment, such as packed towers, it is very difficult to determine the total interfacial area. Therefore, experimentally determined rates of mass transfer are commonly reported on the basis of a unit volume of the absorption equipment rather than on a basis of a unit area of active interface. This is accomplished by reporting the absorption coefficients as $K_G a$, $K_L a$, $k_G a$, or $k_L a$, where $a$ represents the interfacial area per unit volume of the absorbing apparatus. Thus the units of $K_G a$ become kilogram moles per second per square meter per atmosphere, or pound moles per hour per cubic foot per atmosphere.

Since

$$a = \frac{A}{V} \tag{9-14}$$

where $V$ equals the total volume of the absorption tower, Eq. (9-8) may be expressed as

$$\begin{aligned} N &= k_G a V(p_{aG} - p_{ai}) = k_L a V(c_{ai} - c_{aL}) \\ &= K_G a V(p_{aG} - p_{aL}) = K_L a V(c_{aG} - c_{aL}) \end{aligned} \tag{9-15}$$

Equations (9-12) and (9-13) may be expressed as

$$\frac{1}{K_G a} = \frac{1}{k_G a} + \frac{1}{Hk_L a} \tag{9-16}$$

$$\frac{1}{K_L a} = \frac{1}{k_L a} + \frac{H}{k_G a} \tag{9-17}$$

## Methods of Calculation Using Overall Absorption Coefficients

The design of packed absorption towers is usually based on the application of experimental values of absorption coefficients. In this treatment, overall gas-film coefficients will be used, although the same results would be obtained using overall liquid-film absorption coefficients. In order to simplify the method of calculation, only mixtures that follow Henry's law will be considered. Since a large portion of commercial absorption operations involve systems wherein the absorbed component is in a dilute solution and, therefore, follows Henry's law, this simplification is not impractical. Most absorption columns are operated with liquid and gas flowing countercurrently and continuously. Consequently, these conditions will be assumed in the following discussion of calculation methods.

According to Eq. (9-18),

$$N = K_G a V(p_{aG} - p_{aL}) \tag{9-18}$$

The value of $(p_{aG} - p_{aL})$ varies over the length of the absorption tower, and some form of a mean value must be employed if Eq. (9-18) is to be applied over the total length of the tower. When Henry's law applies, a log-mean value for $(p_{aG} - p_{aL})$ may be used, based on the pressure differences at the gas entrance and exit of the tower.

The symbol $N$ is defined as the rate at which the component being absorbed is transferred through the gas film. If the units of $N$ are taken as moles per hour, the value of $N$ corresponding to an overall log-mean driving force becomes

$N$ = moles of absorbable material entering absorption tower per hour minus moles of absorbable material leaving absorption tower per hour

Under these conditions, Eq. (9-18) can be used directly for absorption-tower calculations. If experimental values of $K_G a$ are available, the size of absorption tower necessary for a given amount of absorption of a material following Henry's law can be calculated as follows: Assume a stack gas containing 1 mol of $SO_2$ for every 100 mol of inert gas is to pass through an absorption tower countercurrently to a water–sodium sulfite solution under such conditions that the exit gases will contain 0.05 mol of $SO_2$ per 100 mol of inert gas. A total of 10,000 lb mol of inert gas is to pass through the tower per hour and the absorption unit is to operate at a pressure of essentially 1 atm. A grid-type packing will be used in the absorption tower, and empirical data indicate that the value of $K_G a$ for this packing is 11.3 lb mol/(h)(ft$^3$)(atm) for the given operating conditions. According to the available equilibrium data, the solubility of $SO_2$ in the liquid is so great that the pressure of $SO_2$ in equilibrium with the liquid may be taken as zero at all the concentrations encountered. If the available head space limits the height of the tower to 20 ft, the required diameter can be obtained by applying Eq. (9-15) as follows:

$$p_{aG} \text{ (at bottom of tower)} = \frac{1}{100 + 1}(1) = 0.00990 \text{ atm}$$

$$p_{aG} \text{ (at top of tower)} = \frac{0.05}{100 + 0.05}(1) = 0.00050 \text{ atm}$$

$$p_{aL} \text{ (at bottom and top of tower)} = 0.0000 \text{ atm}$$

$$(p_{aG} - p_{aL})_{\text{log mean}} = \frac{(0.00990 - 0.0) - (0.00050 - 0.0)}{2.3 \log \dfrac{(0.00990 - 0.0)}{(0.00050 - 0.0)}}$$

$$= 0.00315 \text{ atm}$$

$N = (0.01 - 0.0005)(10{,}000) = 95$ lb mol of $SO_2$ absorbed per hour. By Eq. (9-15),

$$V = \frac{N}{K_G a(p_{aG} - p_{aL})_{\text{log mean}}} = \frac{95}{(11.3)(0.00315)} = 2670 \text{ ft}^3$$

$$\text{Cross-sectional area of column} = 2670/20 = 133.5 \text{ ft}^2$$

$$\text{Required diameter of column} = \left[\frac{(133.5)(4)}{\pi}\right]^{1/2} = 13.0 \text{ ft}$$

The preceding example illustrates the manner in which absorption coefficients may be applied to the design of absorption equipment. Empirical values of the absorption coefficients must be available before this method can be applied. The methods of calculation for more complicated cases are presented in detail in many chemical engineering texts.

## Height of a Transfer Unit

The H.T.U. concept is often used in the treatment of absorption operations. The basic definitions applying to transfer units, as presented in Chap. 8 (Distillation), also apply to absorption operations.

**Example 9-2: Experimental determination of overall absorption coefficients** Two hundred pound moles per hour of a gaseous mixture containing 20 mol % $NH_3$ and 80 mol % air is introduced into the bottom of a packed absorption tower. The tower is 20 ft high and has a diameter of 4 ft. Pure water enters the top of the tower continuously. Experimental results indicate that 95 percent of the ammonia is absorbed by the water, and the liquid leaving the bottom of the tower contains 1 mol of ammonia per 5.9 $ft^3$ of solution. The tower is operated at a constant temperature of 70°F at atmospheric pressure. Under these conditions, ammonia follows Henry's law, and the Henry's law constant may be obtained from the fact that, at 70°F, the vapor pressure of ammonia is 0.056 atm when in equilibrium with a liquid solution containing 1 mol of ammonia per 5.9 $ft^3$ of solution.

Determine the values of $K_G a$ and $K_L a$ for the absorption tower operating under the given conditions.

SOLUTION

**Basis**

1 h

Total moles of ammonia absorbed = $N$ = (200)(0.20)(0.95) = 38 lb mol/h
Tower volume = $V$ = (20)(3.14)(2)(2) = 251.2 $ft^3$
Henry's law constant = $H$ = 1/(5.9)(0.056) = 3.03 lb mol/($ft^3$)(atm)
At the bottom of the tower,

$$p_{aG} = 0.20 \text{ atm}$$

$$p_{aL} = \frac{1}{(5.9)(3.03)} = 0.056 \text{ atm}$$

$$p_{aG} - p_{aL} = 0.20 - 0.056 = 0.144 \text{ atm}$$

At the top of the tower,

$$p_{aG} = \frac{(40 - 38)(1)}{(160 + 2)} = 0.01235 \text{ atm}$$

$$p_{aL} = 0.0 \text{ atm}$$

$$p_{aG} - p_{aL} = 0.01235 \text{ atm}$$

Log mean $(p_{aG} - p_{aL})$ over the column is

$$\frac{0.144 - 0.01235}{2.3 \log \dfrac{0.144}{0.01235}} = 0.0537 \text{ atm}$$

From Eq. (9-15),

$$K_G a = \frac{N}{V(p_{aG} - p_{aL})_{\text{log mean}}} = \frac{38}{(251.2)(0.0537)} = 2.82 \text{ lb mol/(h)(ft}^3\text{)(atm)}$$

At the bottom of the tower,

$$c_{aL} = \frac{1}{5.9} = 0.170 \text{ lb mol/ft}^3$$

$$c_{aG} = (3.03)(0.2) = 0.606 \text{ lb/mol/ft}^3$$

$$c_{aG} - c_{aL} = 0.606 - 0.170 = 0.436 \text{ lb mol/ft}^3$$

At the top of the tower,

$$c_{aL} = 0.00 \text{ lb mol/ft}^3$$

$$c_{aG} = (3.3)(0.01235) = 0.0374 \text{ lb mol/ft}^3$$

$$c_{aG} - c_{aL} = 0.0374 \text{ lb mol/ft}^3$$

Log mean $(c_{aG} - c_{aL})$ over the column is

$$\frac{0.436 - 0.0374}{2.3 \log \dfrac{0.436}{0.0374}} = 0.163 \text{ lb mol/ft}^3$$

From Eq. (9-15),

$$K_L a = \frac{N}{V(c_{aG} - c_{aL})_{\text{log mean}}} = \frac{38}{(251.2)(0.163)}$$
$$= 0.93 \text{ lb mol/(h)(ft}^3\text{)(lb mol/ft}^3\text{)}$$

CHECK When Henry's law applies, $HK_L a = K_G a$. Therefore, these results may be checked as follows:

$$HK_L a = (3.03)(0.93) = K_G a = 2.82 \text{ lb mol/(h)(ft}^3\text{)(atm)}$$

## METHODS OF CALCULATION FOR EXTRACTION OPERATIONS

Liquid–liquid extraction is often employed to separate two mutually soluble liquids. In operations of this type, a third liquid, having a preferential solubility for one of the components of the mixture, must be added to the solution, and a separate liquid phase must be formed.

The liquid which is added to the solution to bring about the extraction is known as the *solvent*. This solvent takes up part of the components of the original solution and forms an immiscible layer with the remaining solution. The solvent layer is called the *extract*, and the other layer, composed of the remainder of the original solution plus some of the solvent, is termed the *raffinate*.

## Case Where the Solvent Is Immiscible with One of the Solution Components

The basic principles of extraction can be presented clearly by consideration of the case where the solvent is completely immiscible with one of the components of the solution on which the extraction is being performed. Under these conditions, it is possible to set up simple material balances to give the total separation theoretically possible for each contacting stage.

Consider the example of a dilute solution of acetaldehyde in toluene, where part of the acetaldehyde is to be extracted using water as the extracting solvent. Toluene and water are practically immiscible in mixtures containing up to 15 percent by weight acetaldehyde. Therefore, when water is added to the solution, some of the acetaldehyde dissolves in this extracting solvent and two liquid layers are formed.

The extract (water) layer contains only acetaldehyde and water, while the raffinate layer contains only acetaldehyde and toluene. If the two layers are at equilibrium at any given temperature, there must be a definite relationship between the acetaldehyde concentration in the extract and in the raffinate. The equilibrium relationships are determined experimentally just as the equilibrium data used in distillation operations are determined experimentally.

Let $X_o$ represent the pounds of acetaldehyde (i.e., the extractable component) per pound of toluene in the initial solution, $S_o$ the pounds of water (i.e., the extracting solvent) added, $H_o$ the pounds of toluene in the initial solution, and $Y_1$ and $X_1$ the equilibrium concentrations of acetaldehyde in the extract and raffinate as pounds per pound of water and pounds per pound of toluene, respectively.

At equilibrium conditions, the following overall material balance on the acetaldehyde may be set up:

$$X_o H_o = H_o X_1 + S_o Y_1 \tag{9-19}$$

or

$$H_o = S_o \left( \frac{Y_1}{X_o - X_1} \right) \tag{9-20}$$

The values of $H_o$, $S_o$, and $X_o$ are usually known for a one-stage extraction, and the equilibrium relationships between $X_1$ and $Y_1$ at the temperature of operation are also known. Therefore, the values of $X_1$ and $Y_1$ for the single extraction can be determined and the amount of separation can be found.

Ordinarily, the values of $X_1$ and $Y_1$ are related in such a manner that a trial-and-error or a graphical solution must be used. However, at low concentrations, $X_1$ and $Y_1$ are often related by simple mathematical equations, and a direct analytical solution of the extraction material-balance equation can be made.

**Example 9-3: Determination of amount of separation in a single extraction** A solution of acetaldehyde and toluene containing 11 kg of acetaldehyde and 100 kg of toluene is to be extracted by 80 kg of water. The extraction is to be performed at 17°C, where the equilibrium relationship between the acetaldehyde concentration in the extract and raffinate may be expressed as $Y_1 = 2.2X_1$.

$Y_1$ = kg acetaldehyde per kg of water in the extract phase

$X_1$ = kg acetaldehyde per kg of toluene in the raffinate phase

If equilibrium is obtained between the phases in a single (one-stage) extraction process, how many kilograms of acetaldehyde will be removed by the water? Toluene and water may be assumed as completely immiscible under these conditions.

SOLUTION Using Eq. (9-18),

$$\frac{(11)(100)}{100} = 100X_1 + 80Y_1 \qquad (A)$$

Also,

$$Y_1 = 2.2X_1 \qquad (B)$$

Equations (*A*) and (*B*) can be solved simultaneously to give $Y_1 = 0.0877$ kg acetaldehyde per kilogram of water. Therefore, the total kilograms of acetaldehyde extracted by the water if equilibrium conditions are attained is $(0.0877)(80) = 7.02$ kg.

## Extraction Employing More Than One Stage

The preceding discussion has been limited to a consideration of a single extraction stage. It is possible to combine a series of these stages so that each stage gives additional separation.

## Extraction When Materials Are Not Completely Immiscible

In most industrial applications of extraction, the solvent is present to some extent in both phases. The calculations of the separation per theoretical stage for cases of this type are usually carried out graphically, employing a triangular equilibrium diagram relating the concentrations of all three components in the various phases.[2]

## PROBLEMS

**9-1** An aqueous solution of ammonia containing 1 lb mol of $NH_3$ per 5.9 $ft^3$ of solution is in equilibrium with a gaseous mixture of $NH_3$ and air at 90°F when the partial pressure of the $NH_3$ is 0.0926 atm. Assuming Henry's law is applicable to the $NH_3$ over the range of concentration involved, what will be the equilibrium partial pressure of the $NH_3$ when the solution contains 1 mol % $NH_3$ and 99 mol % water if the temperature is 90°F? The density of an aqueous solution of $NH_3$ containing 1 mol % $NH_3$ at 90°F may be taken as 61.9 lb/$ft^3$.

[2] For details see any of the numerous books on mass transfer by extraction.

**9-2** A continuous-absorption tower is used for absorbing HCl in water. Two hundred pound moles per hour of a gaseous mixture containing 5 mol of HCl per 2 mol of air enters the bottom of the tower, and 15,000 lb of pure water enters the top of the tower per hour. The aqueous HCl solution leaving the bottom of the tower contains 1 mol of HCl for every 7 mol of water. Assuming no water is vaporized in the tower, determine the moles of HCl per mole of air in the exit gas stream.

**9-3** The equilibrium data for a certain absorption operation may be represented by the equation.

$$Y = 5.2X^2$$

where $X$ = kg mol of dissolved absorbable material per kg mol of pure absorbing liquid
$Y$ = kg mol of gaseous absorbable material per kg mol of inert gas

The unit is operated as a continuous countercurrent absorption. The entering gas contains 5 mol of absorbable material per 20 mol of inert gas. The absorbing liquid enters the column as a pure material. Determine the minimum molal ratio of liquid to gas ($L_M/G_M$) if the gases leaving the tower contain 1 mol of absorbable material per 50 mol of inert gas.

**9-4** How many theoretical absorption stages would be required to make the separation indicated in Prob. 9-3 if a molal liquid-to-gas ratio ($L_M/G_M$) of 1.3 times the minimum value were used?

**9-5** Ammonia is being absorbed in water at 70°F under such conditions that Henry's law applies. The value of the Henry's law constant $H$ is 3.03 lb mol/(ft$^3$)(atm), and the value of $K_G a$ for the absorption is 3.0 lb mol/(h)(ft$^3$)(atm). Determine the percent of the total resistance to mass transfer due to the gas film if the value of $k_L a$ is 5.1 lb mol/(h)(ft$^3$)(lb mol/ft$^3$).

**9-6** A packed tower with an inside volume of 8.495 m$^3$ is to be used for an ammonia absorption under such conditions that $K_G a$ = 0.01246 kg mol/(s)(m$^3$)(atm) and Henry's law is applicable to the ammonia solution. The pressure-difference driving force at the top of the column ($p_{aG} - p_{aL}$) is 0.009 atm, and the pressure-difference driving force at the bottom of the column is 0.090 atm. If 249.5 kg of ammonia, in addition to the inert gases, enters the tower per hour, what percent of the entering ammonia will be absorbed?

**9-7** A liquid mixture of acetaldehyde and toluene contains 8 lb of acetaldehyde and 90 lb of toluene. Part of the acetaldehyde in this solution is to be extracted, using pure water as the extracting agent. The extraction is to be performed in two stages, using 25 lb of fresh water for each stage. The raffinate layer from the first stage is treated by fresh water in the second stage. The extraction takes place at 17°C, and the equilibrium data given in Example 9-3 may be employed. Assuming toluene and water as immiscible, what would be the weight percent of acetaldehyde in a mixture of the extracts from both stages if each of the extractions were theoretically perfect?

**9-8** An aqueous solution of acetic acid is to be extracted with isopropyl ether. The solution contains 24.6 kg of acetic acid and 80 kg of water. If 100 kg of isopropyl ether is added to the solution, what weight of acetic acid will be extracted by the isopropyl ether if equilibrium conditions are attained? Water and isopropyl ether may be considered as completely immiscible under these conditions. At the temperature of the extraction, the following equilibrium data apply:

| kg acetic acid/kg isopropyl ether | 0.030 | 0.046 | 0.063 | 0.070 | 0.078 | 0.086 | 0.106 |
|---|---|---|---|---|---|---|---|
| kg acetic acid/kg water | 0.10 | 0.15 | 0.20 | 0.22 | 0.24 | 0.26 | 0.30 |

CHAPTER
# TEN

# HUMIDIFICATION AND DEHUMIDIFICATION

Air-conditioning units in homes and business establishments are practical examples of applications involving the unit operations of humidification and dehumidification. Standard methods have been developed for dealing with these operations, and the chemical engineer should become familiar with the special terminology involved.

When dry air is brought into contact with water, some of the water vaporizes into the air. If the air and liquid water are kept in contact, the liquid will continue to vaporize until the vapor pressure of the water equals the equilibrium vapor pressure of water at the temperature of the mixture. Under these conditions, the air is said to be *saturated*.

The unit operation of *humidification* involves the transfer of water from the liquid phase into a gaseous mixture of air and water vapor. *Dehumidification* deals with a decrease in the moisture content of air by an operation whereby water is transferred from the vapor state to the liquid state.

While humidification and dehumidification are not necessarily limited to water-and-air mixtures, the practical application of these operations almost always involves this mixture. Therefore, this treatment will deal primarily with the humidification and dehumidification of air containing water vapor, although the reader should realize that the principles are also applicable to other mixtures.

Since the transfer of water between liquid and gaseous phases involves mass transfer, many of the principles applied to evaporation, distillation, absorption, and extraction are also applicable to humidification and dehumidification. Actually, the operation of humidification or dehumidification is merely a special case of absorption, and it is treated as a separate operation because of its extended practical applications.

The evaporation of a liquid involves an absorption of heat. Therefore, the unit operation of heat transfer must be combined with mass transfer in carrying out humidification or dehumidification operations.

**Table 10-1 Nomenclature for humidification and dehumidification**

$c_s$ = humid heat, amount of heat necessary to raise temperature of unit mass of dry air plus any contained moisture by 1 degree, J/(kg dry air)(K) or Btu/(lb dry air)(°F)
$H$ = humidity, kg water vapor per kg dry air or lb water vapor per lb dry air
$H_P$ = percentage humidity, 100 times ratio of actual humidity of air to humidity if air were saturated at same temperature and total pressure
$H_R$ = relative humidity, 100 times ratio of actual vapor pressure of water in air to water-vapor pressure if air were saturated at same temperature
$H_S$ = humidity of air saturated with water vapor at adiabatic saturation temperature, kg water vapor per kg dry air or lb water vapor per lb dry air.
$p_a$ = actual partial pressure of water vapor, mmHg
$p_S$ = equilibrium (or saturation) partial pressure of water, mmHg
$P$ = total pressure, mmHg
$t$ = temperature, °C, K, or °F
$t_S$ = adiabatic saturation temperature, °C, K, or °F

Greek symbols

$\lambda_S$ = lambda, latent heat of vaporization of unit mass of water at adiabatic saturation temperature, J/kg or Btu/lb

Subscripts

1 refers to initial or entering conditions

## DEFINITIONS

**Humidity ($H$).** The number of kilograms (or pounds) of water vapor carried by 1 kg (or 1 lb) of dry air is known as the *humidity* (sometimes called *absolute humidity*) of the air. Thus, if a mixture of air and water vapor contains 0.01 lb of water vapor for every pound of dry air, the humidity of the air is 0.01. The symbol $H$ is used to represent humidity.

**Percentage humidity ($H_P$).** It is convenient to relate the actual humidity of air to the humidity of the same air if it were saturated with water vapor. *Percentage humidity* may be defined as 100 times the ratio of the actual humidity of air to the humidity if the air were saturated at the same temperature and total pressure. At a temperature of 80°F and a pressure of 760 mmHg, air saturated with water vapor contains 0.0233 lb of water vapor per pound of dry air. If a mixture of air and water vapor at 80°F and 760 mmHg pressure has a humidity of 0.0100, the percentage humidity of the mixture is $(0.0100/0.0233)(100) = H_P = 42.9\%$.

**Percentage relative humidity ($H_R$).** The extent of saturation of an air and water-vapor mixture is sometimes expressed on the basis of water-vapor partial pressures. The degree of saturation, expressed as 100 times the actual partial pressure of water vapor in the air divided by the saturation vapor pressure of pure water at the same temperature, is known as the *percentage relative humidity*. The symbol $H_R$ is used to designate percentage relative humidity.

**Relationship between $H_P$ and $H_R$.** The molecular weight of dry air may be taken as 29.0, and the molecular weight of water is 18.0. Assuming that Dalton's law is applicable, $p_a/(P - p_a)$ or $p_s/(P - p_s)$ is the number of moles of water vapor per mole of dry air in the air-water mixture, and the percentage humidity of any given gaseous mixture of air and water vapor may be expressed as follows:

$$H_P = \frac{H_{\text{actual}}}{H_{\text{saturated}}}(100) = \frac{\dfrac{18p_a}{29(P - p_a)}}{\dfrac{18p_S}{29(P - p_S)}}(100) = \frac{p_a(P - p_S)}{p_S(P - p_a)}(100) \qquad (10\text{-}1)$$

where $p_a$ = actual partial pressure of water vapor in gaseous mixture, mmHg
$p_S$ = equilibrium (or saturation) partial pressure of liquid water at same temperature as original gaseous mixture, mmHg
$P$ = total pressure of dry air and water vapor in gaseous mixture, mmHg

The percentage relative humidity may be expressed as

$$H_R = \frac{p_a}{p_S}(100) \qquad (10\text{-}2)$$

Equations (10-1) and (10-2) may be combined to give the following relationship between the percentage humidity and the percentage relative humidity for any given mixture:

$$H_P = H_R\left(\frac{P - p_S}{P - p_a}\right) \qquad (10\text{-}3)$$

**Example 10-1: Determination of percentage humidity from vapor-pressure data** The air in a room at a temperature of 90°F and a total pressure of 760 mmHg contains 0.021 lb of water vapor per pound of dry air. Determine the percentage humidity and the percentage relative humidity of the air if data from the steam tables indicate that the equilibrium vapor pressure of liquid water at 90°F is 0.698 lb/in$^2$, absolute.

SOLUTION

$$0.698 \text{ lb/in}^2\text{, absolute} = (0.698)\left(\frac{760}{14.7}\right) = 36.1 \text{ mmHg}$$

If the air were saturated at 90°F and 760 mmHg, the humidity would be

$$\frac{(18)(36.1)}{(29)(760 - 36.1)} = 0.03095 \text{ lb water/lb dry air}$$

Actual humidity of the air = 0.021 lb water/lb dry air
Percentage humidity = $H_P$ = (0.021/0.03095)(100) = 67.9 %

The actual partial pressure of the water vapor may be determined from the equation

$$0.021 = \frac{18(p_a)}{29(760 - p_a)}$$

$$p_a = 24.9 \text{ mmHg}$$

$$\text{Percentage relative humidity} = H_R = \frac{p_a}{p_S}(100) = \frac{24.9}{36.1}(100)$$

$$= 69.0\ \%$$

A quick check on these results may be made by the use of Eq. (10-3).

$$H_P = H_R\left(\frac{P - p_S}{P - p_a}\right) \tag{10-3}$$

$$67.9 = 69.0\left(\frac{760 - 36.1}{760 - 24.9}\right) = 67.9$$

**Dew point.** The temperature at which a given air and water-vapor mixture would be saturated is termed the *dew point.* For example, since the saturation vapor pressure of liquid water at 90°F is 36.1 mmHg, if the partial pressure of water in air is 36.1 mmHg, the dew point for this mixture must be 90°F. The dew point is independent of the actual (or dry-bulb) temperature of the mixture.

**Humid heat** ($c_s$). *Humid heat* may be defined as the amount of heat required to raise the temperature of unit mass of dry air plus any contained moisture by 1 degree. Using American engineering units with heat expressed as Btu, unit mass as 1 lb, and temperature as 1°F, the heat capacity of dry air and water vapor may be assumed constant at 0.24 Btu/(lb)(°F) and 0.45 Btu/(lb)(°F), respectively, over the range of temperatures usually encountered. Therefore, the value of the humid heat for air and water-vapor mixtures may be obtained from the following equation:

$$\text{Humid heat} = c_s = 0.24 + 0.45\,H \text{ Btu/(lb dry air)(°F)} \tag{10-4}$$

With SI units, the expression for humid heat becomes 1004.8 + 1884.1 $H$ J/(kg dry air)(K).

**Saturated volume and humid volume.** The volume in cubic meters (or cubic feet) occupied by 1 kg (or 1 lb) of dry air at any specified temperature and pressure, plus sufficient water vapor to saturate the air, is defined as the *saturated volume.* In a similar manner, the *humid volume* is defined as the volume occupied by unit mass of dry air plus any moisture the air may contain. Saturated volumes and humid volumes at atmospheric pressure can be determined by direct application of the perfect-gas law.

**Dry-bulb temperature.** If a dry thermometer, suitably shielded with respect to any thermal radiation, is placed in air, the thermometer will give a temperature reading which is referred to as the *dry-bulb temperature.*

## ADIABATIC CONDITIONS

In an *adiabatic* operation there is no gain or loss of heat in the region surrounding the system. Since many humidification and dehumidification operations essentially meet this condition, it is important to consider the relationships involved in adiabatic evaporations.

**Wet-bulb temperature.** If a drop of liquid water is brought into contact with a large amount of unsaturated air, some of the liquid will evaporate into the air. As the water evaporates, it absorbs heat and the liquid tends to cool. However, when the liquid becomes colder than the air, there will be a transfer of sensible heat from the air to the liquid. In this manner, a dynamic equilibrium is quickly obtained where the amount of heat absorbed by the evaporation exactly equals the amount of heat gained by the liquid from the surrounding air. Under these conditions, there is no net gain or loss of heat in the surroundings, and an adiabatic evaporation is occurring. At these adiabatic conditions, the temperature of the liquid has a definite equilibrium value which is lower than the air temperature but higher than the dew point of air.

If the surface of a thermometer bulb is kept wet by means of a wick or cloth thoroughly saturated with water, the same type of adiabatic evaporation will occur as with a drop of water. The thermometer with the wetted bulb indicates the equilibrium temperature attained in the adiabatic vaporization. The equilibrium vaporization temperature determined in this manner, under conditions where a sufficiently high air velocity past the wetted bulb is maintained, is called the *wet-bulb temperature.*

### Adiabatic Humidifying Equation

When equilibrium conditions have been attained in an adiabatic evaporation process, it is possible to set up an overall heat balance for the process. Consider the case in which 1 lb of dry air with a humidity of $H_1$ and a temperature in degrees Fahrenheit of $t_1$ enters an adiabatic humidifier of the type shown in Fig. 10-1. The liquid water in the humidifier is at the adiabatic evaporation temperature, and the air and liquid water have sufficient contact so that the air leaves the humidifier saturated with water vapor at the adiabatic evaporation (or saturation) temperature. The adiabatic saturation temperature in degrees Fahrenheit is designated by $t_s$, and the humidity of the leaving air is indicated by $H_S$.

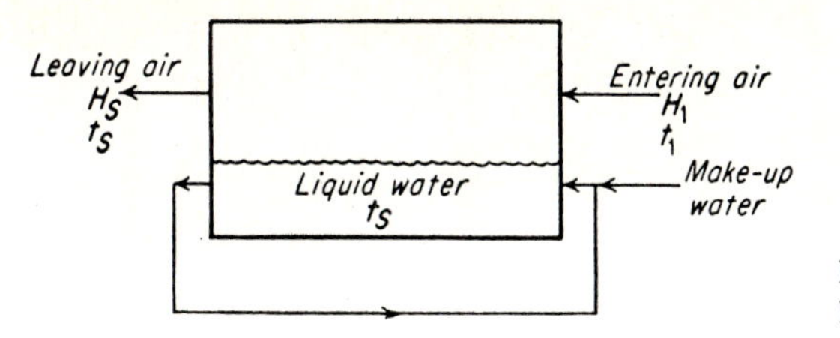


**Figure 10-1** Drawing of an adiabatic humidifier.

By using a zero heat-energy level at the datum temperature of $t_S$ and a basis of 1 lb of dry air, it is possible to make the following heat balance around the adiabatic humidification unit shown in Fig. 10-1:

Heat input:

$$\text{Sensible heat in entering air} = c_{s_1}(t_1 - t_S) = (0.24 + 0.45H_1)(t_1 - t_S)$$

$$\text{Latent heat in entering air} = \lambda_S(H_1)$$

where $\lambda_S$ equals the heat of vaporization in British thermal units of 1 lb of water at the adiabatic saturation temperature. (The heat of vaporization is assumed as constant over the temperature range involved.)

Heat output:

$$\text{Sensible heat in leaving air} = 0.0$$

since the zero heat-energy level is taken at $t_S$, which is the temperature of the leaving air.

$$\text{Latent heat in leaving air} = \lambda_S(H_S)$$

Since the heat input equals the heat output,

$$-\frac{c_{s_1}}{\lambda_S} = \frac{H_S - H_1}{t_S - t_1} \tag{10-5}$$

and

$$t_1 = \frac{\lambda_S(H_S - H_1)}{0.24 + 0.45H_1} + t_S \tag{10-6}$$

If $t_S$ is fixed, the values of $\lambda_S$ and $H_S$ are also fixed, and the only variables in Eq. (10-6) are $t_1$ and $H_1$. Thus, for any given adiabatic saturation temperature, Eq. (10-6) may be represented by a line on a plot of $H$ vs. $t$. These adiabatic humidification lines are included on a general plot known as a humidity chart.

**Relationship between wet-bulb temperature and adiabatic saturation temperature.** The wet-bulb temperature is obtained when the heat required to vaporize a small amount of water into a large volume of air exactly equals the sensible heat transferred from the air to the liquid water. The amount of heat necessary to vaporize the water depends on the rate at which the water can escape from the liquid surface.

This rate depends upon the diffusion coefficient. The sensible heat transferred to the water depends upon the heat-transfer coefficient between the air and the liquid water.

For water and air systems, the relationship between the diffusion coefficient and the heat-transfer coefficient is such that the wet-bulb temperature and the adiabatic saturation temperature are almost equal. Therefore, for all practical purposes, the adiabatic saturation temperature $t_S$ as shown in Eq. (10-6) is the same as the wet-bulb temperature. This equality between wet-bulb and adiabatic saturation temperatures applies only to mixtures of air and water vapor. For other mixtures, the values of the diffusion and heat-transfer coefficients must be obtained and other equations must be applied.

## THE HUMIDITY CHART

Figure 10-2 presents the common humidity chart for mixtures of water vapor and air at a total pressure of 760 mmHg. The information given in Fig. 10-2 applies only at a total pressure of 760 mmHg, and corrections must be made before data from the chart can be applied to give accurate results at other total pressures. For the accuracy required in ordinary engineering calculations, values determined from Fig. 10-2 may be employed over a total-pressure range of 730 to 780 mmHg without applying correction factors.

### Humidity Lines

The humidity chart presents lines relating air temperature to humidity at different values of percentage humidity. These lines are represented by a family of curves starting at the lower left-hand corner of the plot and sloping upward to the right. Each member of the family of curves is labeled according to its percentage humidity.

The 100 percent humidity line represents the humidities of air at various temperatures when the air is saturated with water vapor. For example, when air at 120°F and atmospheric pressure is saturated with water vapor, the humidity is 0.081 lb of water vapor per pound of dry air. If air at 120°F has 50 percent humidity, the pounds of water vapor per pound of dry air should be (0.5)(0.081) = 0.0405. From Fig. 10-2, it can be seen that air at 120°F and 50 percent humidity does contain 0.0405 lb of water vapor per pound of dry air.

### Humid-Heat Line

Equation (10-4) is represented by a straight line on the humidity chart giving the relationship between humidity and humid heat.

### Adiabatic Humidification Lines

The adiabatic humidification lines are based on Eq. (10-6), and they present the relationship between temperature $t_1$ and humidity $H_1$ at different values of the

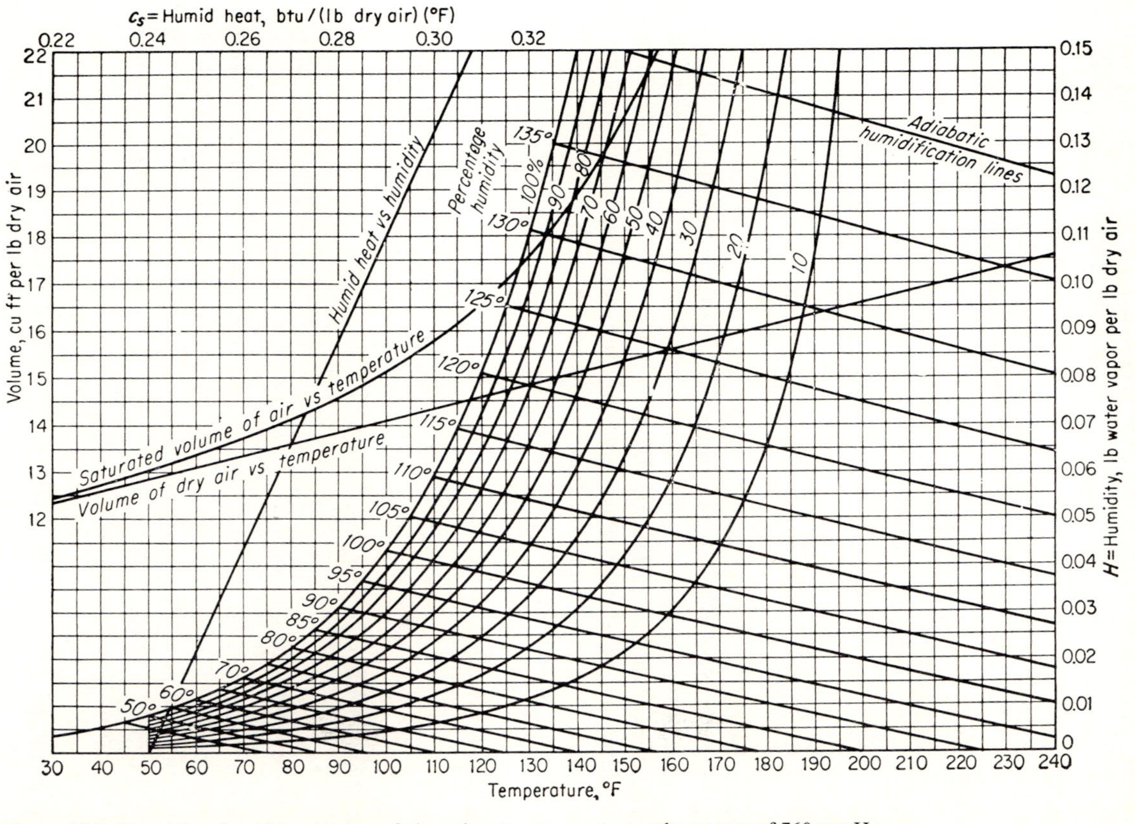


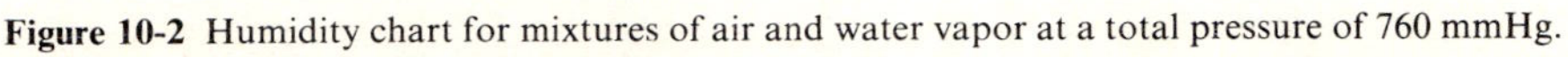

**Figure 10-2** Humidity chart for mixtures of air and water vapor at a total pressure of 760 mmHg.

adiabatic saturation temperature. Since for air and water-vapor mixtures the adiabatic saturation temperature and the wet-bulb temperature can be taken as identical, the adiabatic humidification lines show the relation between air humidity and air temperature for different wet-bulb temperatures. Thus, if the dry-bulb (or actual) temperature of air is 120°F and the humidity of the air is 0.038 lb of water vapor per pound of dry air, the wet-bulb temperature, according to Fig. 10-2, must be 100°F.

These lines are represented on the humidity chart by a family of curves, with each line starting at the 100 percent humidity line and sloping downward to the right. Each member of the family of curves is labeled according to its adiabatic saturation temperature or, for the case of air and water-vapor mixtures, the wet-bulb temperature.

### Air-Volume Lines

The saturated volume of air versus temperature is presented as a curve on the humidity chart. Just beneath the saturated-volume curve, a line is presented relating the volume of dry air to the temperature.

The humid volume of air may be determined from a knowledge of the air humidity by interpolating between the saturated-volume and dry-volume lines. For example, Fig. 10-2 indicates that the volume of 1 lb of dry air at 120°F is 14.6 $ft^3$, and the volume of saturated air at 120°F is 16.6 $ft^3$/lb of dry air. The humid volume of air having 40 percent humidity at 120°F would then be $14.6 + 40/100(16.6 - 14.6) = 15.4$ $ft^3$/lb of dry air.

## DETERMINATION OF HUMIDITY

### Wet-Bulb Temperature Method

The humidity of air may be determined from the wet-bulb and dry-bulb temperatures of the air. The wet-bulb temperature can be obtained by determining the equilibrium temperature when air is passed over a wetted thermometer bulb. The bulb of the thermometer is kept wet by enclosing it in a wick or cloth which is thoroughly saturated with water.

If the wet-bulb and dry-bulb temperatures of a given air are known, the humidity can be found from the adiabatic humidification lines in Fig. 10-2. Since the adiabatic humidification lines are based on Eq. (10-6), for any one set of wet-bulb and dry-bulb temperatures there can be only one value for the air humidity. For example, if the dry-bulb temperature of atmospheric air is 120°F and the wet-bulb temperature is 100°F, the point where the 100°F adiabatic humidification line intersects the 120°F temperature line on Fig. 10-2 must represent the humidity of the air. In this example, the two lines intersect at a humidity of 0.038 lb of water vapor per pound of dry air.

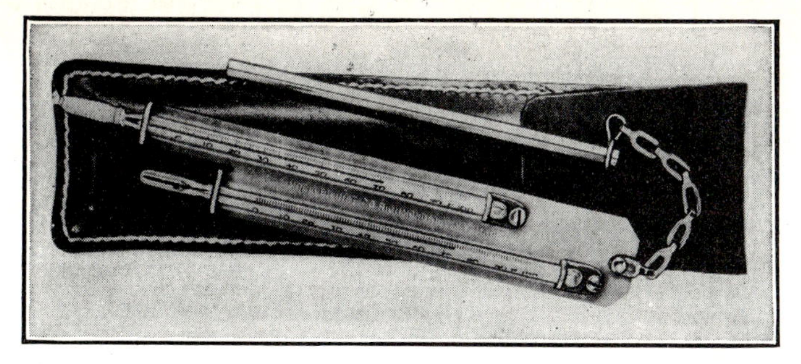

**Figure 10-3** Sling psychrometer. (*Courtesy of Moeller Instrument Company.*)

The method for determining the humidity of air at atmospheric pressure from wet-bulb and dry-bulb temperatures may be outlined as follows:

1. On the humidity chart (Fig. 10-2), locate the adiabatic humidification line corresponding to the wet-bulb temperature. Interpolate between lines if necessary.
2. Locate the vertical temperature line representing the dry-bulb temperature.
3. The humidity corresponding to the intersection point of the two lines located in steps 1 and 2 represents the humidity of the air.

### The Sling Psychrometer

The sling psychrometer, as shown in Fig. 10-3, is commonly used for measuring the wet-bulb and dry-bulb temperatures of atmospheric air. The adiabatic humidification temperature and the wet-bulb temperature of air and water-vapor mixtures may be assumed equal if the air velocity past the wetted bulb is high enough to minimize the effect of radiation. The air velocity past the wet bulb should be about 4.6 m/s or 15 ft/s.

Both the wet-bulb and the dry-bulb thermometers are located in the sling psychrometer. The psychrometer can be rotated by hand to obtain the desired air velocity past the wet bulb. The rotation should be continued until the wet-bulb temperature has become constant; however, it is essential that the temperature readings be taken before the wet-bulb wick becomes dry.

## AIR CONDITIONING

Many curing processes require careful control of the humidity of air contacting the materials. Air-conditioning units are in widespread use for controlling the humidity and temperature in homes and business establishments. As a result of the large demand for equipment, the field of air conditioning has been developed extensively, and the general design methods have become standardized.

**Humidifiers.** In a humidifier, the temperature of the liquid water must be higher than the dew point of the air. Air may be humidified by spraying water or steam into the air or by the use of a packed or bubble-cap tower where the air and liquid water may come into intimate contact.

If the water temperature is higher than the wet-bulb temperature of the air, the water will be cooled and will approach the wet-bulb temperature of the air. In an adiabatic humidifier, the water is recirculated, and the temperature of the water, as well as the temperature of the exit gases, becomes approximately the same as the wet-bulb temperature of the entering air.

**Dehumidifiers.** If water is brought into contact with air under conditions such that the water temperature is less than the dew point, some of the water vapor will condense out of the air. This is the process which takes place in a dehumidifier.

### Water Coolers

When air is humidified by contact with liquid water having a temperature higher than the wet-bulb temperature of the air, the water is cooled. This process is frequently used for the primary purpose of cooling the water rather than humidifying the air.

Water-cooling towers are often constructed of wood slats placed in a large wooden frame so that the slats act as a type of tower packing. Water runs down over the slats and contacts air which rises up through the tower. The air may rise through the tower owing to a natural draft caused by the decreased density of the warmer air at the top of the tower, or the air may be forced through the tower by a blower.

## METHODS OF CALCULATION

Humidification and dehumidification calculations are based on material and energy balances and on rate coefficients for heat and mass transfer. Humidity charts can be used to simplify the calculations and reduce the amount of time required for the calculations.

### Material Balances

On the basis of unit time, the total amount of water vapor contained by air leaving a humidifying unit must be equal to the amount of water vapor in the air entering the unit plus any moisture gained from the liquid water flowing through the humidifier. Similarly, for a dehumidifying unit, the amount of water vapor in the exit air plus the weight of moisture condensed out of the air must be equal to the total amount of water vapor contained by the air entering the unit.

The application of humidification or dehumidification material balances is greatly simplified if a given mass of dry air is chosen as a basis. With this basis, the values from the humidity chart can be used directly. For example, consider the case in which air, having a humidity of 0.02 kg of water vapor per kilogram of dry air, enters a dehumidifier and the leaving air has a humidity of 0.015. On a basis of 1 kg of dry air, the amount of water vapor condensed out of the air in the dehumidifier must be $0.02 - 0.015 = 0.005$ kg.

## Energy Balances

The total amount of heat energy entering a humidifier or a dehumidifier must equal the total heat energy coming out. Where the heat values are based on some convenient temperature (such as 0°C or 32°F) as the zero energy level, the heat input to the system consists of the following items:

1. Sensible heat in the entering air and water vapor
2. Latent heat of the entering water vapor
3. Sensible heat in the entering liquid water

The heat leaving the system is found in the following places:

1. Sensible heat in the leaving air and water vapor
2. Latent heat of the leaving water vapor
3. Sensible heat in the leaving liquid water
4. Unaccounted-for losses as radiation or convection from the outside surface of the unit.

## Heat- and Mass-Transfer Coefficients

Experimental values for the rate of mass transfer in certain types of humidification or dehumidification equipment are often expressed as diffusion coefficients. These coefficients are analogous to the absorption coefficients discussed in Chap. 9, with the one exception that the driving force is expressed as a humidity difference rather than a partial pressure or concentration difference.

Since the transfer of sensible heat between the liquid and gas phases is of considerable importance in humidification and dehumidification operations, experimental values of heat-transfer coefficients are often determined for the various types of equipment. The mass- and heat-transfer coefficients can only be evaluated experimentally. They are of considerable value for design purposes when standard equipment is used, since previous experience permits a close approximation of the actual values of the coefficients.

# PROBLEMS

**10-1** A mixture of air and water vapor weighs 80.0 kg. Determine the humidity of the air if there is 1.4 kg of water vapor in the mixture.

**10-2** A mixture of air and water vapor contains 5 lb of water vapor for every 200 lb of dry air. If the temperature of the mixture is 120°F and the total pressure is 760 mmHg, determine the following: (*a*) actual vapor pressure of water in the air; (*b*) dew point of the mixture; (*c*) percentage humidity of the air; (*d*) relative humidity of the air.

**10-3** The partial pressure of water vapor is 50 mmHg in a mixture of air and water vapor at 110°F and 1 atm pressure. How many Btu of heat energy must be added to 25 lb of this mixture to raise the temperature to 125°F?

**10-4** By use of the humidity chart, determine the following for air at 90°F and 1 atm when the percentage humidity is 50 percent: (*a*) the humidity of the air; (*b*) the humid heat of the air; (*c*) the volume occupied by 1 kg of dry air plus the contained moisture; (*d*) the adiabatic saturation temperature.

**10-5** An adiabatic humidifier operates at atmospheric pressure under such conditions that the air leaving the unit is saturated at the adiabatic saturation temperature. If the air enters the humidifier at 80°F with a humidity of 0.005 lb of water vapor per pound of dry air, what will be the humidity of the air leaving the unit?

**10-6** By the use of a sling psychrometer, it is found that the air in a room at atmospheric pressure has a wet-bulb temperature of 65°F and a dry-bulb temperature of 90°F.

(*a*) What is the humidity of the air?

(*b*) What is the percentage humidity of the air?

**10-7** A compressor delivers air at a total pressure of 40 lb/in$^2$, absolute, and a temperature of 90°F. If this air is saturated with water vapor, what is the humidity of the air? Assume Dalton's law is applicable to this mixture.

**10-8** Air at 760 mmHg pressure, having a wet-bulb temperature of 50°F and a dry-bulb temperature of 70°F, enters a humidifier. If the air leaving the unit has a humidity of 0.007 kg of water vapor per kilogram of dry air, how many kilograms of water vapor are evaporated into the air for every 100 kg of dry air entering the unit?

**10-9** Fresh air is introduced into a room at such a rate that the air in the room is changed every 5 min. The room is 30 by 30 by 10 ft and the air in the room is at 70°F with a percentage humidity of 60 percent. Outside air at a temperature of 60°F is used as the air supply. The outside air is heated to the room temperature by a heat exchanger. How many Btu per hour must be supplied by the heat exchanger? All total pressures may be taken as 760 mmHg.

CHAPTER
# ELEVEN
# DRYING

*Drying* may be defined as the removal of water from materials containing a relatively small amount of water. Drying operations are carried out by thermal means, and they are thus differentiated from purely physical water-and-solid separations. Evaporation ordinarily refers to the removal of water vapor in a pure state from a mixture containing a relatively high amount of water, while drying operations usually involve the removal of small amounts of water by passing air or some other gas over a material having a relatively low water content.

The term "drying" is not necessarily limited to the removal of small amounts of water, since liquids other than water may also be removed from mixtures to yield "dry" residues. However, water is by far the most common liquid involved in drying operations. While the principles presented in this chapter are applicable to other liquids, the discussion will be limited to the removal of water.

When water is removed from a substance by thermal means, the transfer of mass and the transfer of heat are involved. Heat must be supplied to vaporize the water, and the water must diffuse through the various resistances in order to escape into the free vapor state. Thus, the magnitudes of the diffusion coefficients and the heat-transfer coefficients have a large effect on the drying rate.

## METHODS OF DRYING

A material may be dried by passing hot gases over its surface. Heat is transferred from the hot gas to the drying material, where the heat causes water to vaporize into the gas.

Drying may also be accomplished by indirect heating of the substance containing the water, using a stream of cool gas to carry away the evolved water vapor. Air is ordinarily used as the gas for carrying away the water vapor, and the air humidity must be sufficiently low to permit the gas to pick up additional water vapor without becoming saturated.

**Table 11-1 Nomenclature for drying**

$Q$ = total weight of dry solid, kg or lb
$R$ = drying rate, kg water removed/h or lb water removed/h
$R_c$ = rate of drying during the constant-rate period, kg water removed/h or lb water removed/h
$R_o$ = rate of drying at start of the operation, kg water removed/h or lb water removed/h
$R_2$ = rate of drying at end of first (or $B$) falling-rate period, kg water removed/h or lb water removed/h
$T_E$ = equilibrium total moisture content, kg water/kg dry solid or lb water/lb dry solid
$W$ = free moisture content, kg water/kg dry solid or lb water/lb dry solid
$W_C$ = critical free moisture content, kg water/kg dry solid or lb water/lb dry solid
$W_F$ = free moisture content at end of drying, kg water/kg dry solid or lb water/lb dry solid
$W_o$ = free moisture content at start of drying, kg water/kg dry solid or lb water/lb dry solid
$W_1$ = free moisture content at start of constant-rate period, kg water/kg dry solid or lb water/lb dry solid
$W_2$ = free moisture content at end of first (or $B$) falling-rate period, kg water/kg dry solid or lb water/lb dry solid

Greek symbols

$\theta_c$ = theta, time of drying in constant-rate period, h
$\theta_f$ = theta, time of drying in falling-rate period, h

Many different types of driers are in common use, and the final choice of a drier depends on the amount and kind of material to be dried, the type of final product desired, available facilities, and the economics of the process. Certain materials must be dried at low temperatures to keep away from discoloration or decomposition, and special types of equipment such as vacuum driers may be used for this purpose.

Drum driers and some types of rotary driers supply the necessary transfer of heat by the use of hot surfaces, while tray driers, spray driers, and some rotary driers obtain heat transfer by contact with hot gases.

## Tray Driers

A tray drier contains a number of trays in which the material to be dried is placed. Hot air passes over these trays and carries away the evolved water vapor. When the material becomes sufficiently dry, the trays and the dry tray contents are removed. Some tray driers are operated batchwise, and some are operated continuously with the trays containing the wet material entering one end of the drier and passing slowly through the unit until they emerge at the outlet end with most of the water removed.

## Drum Driers

Certain types of materials may be dried by the use of a drum drier, which consists essentially of a revolving drum with heat supplied to the inside of the drum. If the drying material is a liquid, it will cling to the outside of the revolving surface in a thin film and be dried in one revolution. Thus, the system can be arranged so that the revolving drum dips into the liquid and the dry material is removed by a knife blade before the surface dips into the liquid at the end of the revolution.

### Rotary Driers

A rotary drier consists of a horizontal cylindrical shell rotating around its horizontal axis. The wet feed material enters one end of the cylindrical shell, and hot gases may be forced into the other end. The rotating action lifts the wet material and permits it to drop through the hot gas, thus bringing about a rapid rate of drying. The dry-solids exit end of the cylinder is slightly lower than the entrance end to permit the feed to move through the drier by gravity.

The movement of the feed is often aided by the presence of a stationary blade scraper located inside the cylinder. The pitch of the blades is set to force the solid material toward the exit. These blades also serve to keep thick cakes from building up on the cylinder walls.

In some rotary driers, the cylinder wall is stationary and the raising and dropping of the feed particles is accomplished by an internal rotating blade. Many rotary driers of this type have jacketed cylinder walls so that heat may be supplied through the walls of the cylinder rather than by the passage of hot gases.

### Spray Driers

Liquid mixtures may be dried very effectively by spraying the mixture into a stream of hot gases. The small droplets fall through the hot gases, and the large amount of heat-transfer area exposed permits a rapid rate of drying.

### Vacuum Driers

By carrying out a drying operation under a vacuum, it is possible to obtain high rates of water removal at low temperatures. Therefore, heat-sensitive materials are often dried with the various types of drying equipment operated at reduced pressures rather than at atmospheric pressure.

## PRINCIPLES OF DRYING

A drying operation involves the simultaneous transfer of mass and heat, and the rate of water removal is determined by the rate at which mass and heat may be transferred. Consider the case where water is being removed from a solid under such conditions that the water is present as a liquid on the surface of the solid. If hot air passes over the surface and acts as the heat source, the air and water will be in direct contact, and the resistance to heat and mass transfer will be due only to the liquid and gas films. However, if liquid water is not present at the surface of the solid or only covers part of the surface, evaporation will occur beneath the surface or on a reduced surface area, and added heat- and mass-transfer resistances will be encountered. Thus, in a batch process, the rate of drying should be greatest at the beginning of the operation when water covers the entire surface of the solid. As the drying proceeds, a time will be reached when water is no longer available at the surface, and the rate of drying will decrease.

**Equilibrium moisture content** ($T_E$). In a given mixture the extent of drying which can be accomplished by air at any set value of humidity and temperature is limited. The *equilibrium moisture content* may be defined as the number of kilograms (or pounds) of water per kilogram (or pound) of dry solid when the drying limit has been attained by use of air at any given temperature and humidity. The equilibrium moisture content is determined by the properties of the solid material and by the humidity and temperature of the air.

**Dry-weight basis.** Drying calculations are simplified if water conditions are expressed on the basis of 1 kg (or 1 lb) of dry solid. Using this basis, the amount of water removed can be determined very easily by obtaining the difference between the water contents at any two times.

**Free moisture content** ($W$). The free moisture content represents the amount of water that can be removed from a wet solid under any given conditions of air humidity and temperature.

Using a dry-weight basis, the *free moisture content* may be defined as the total kilograms (or pounds) of water contained by 1 kg (or 1 lb) of dry solid minus the equilibrium moisture content.

## Rate of Drying

Figure 11-1 represents a typical drying-rate curve for a batch process in which the gas passes over the wet solid at a velocity sufficient to keep the temperature and humidity of the gas essentially constant. The rate of drying, expressed as pounds of water removed per hour, is plotted against the free moisture content of the wet solid. The free moisture content is expressed as pounds of water per pound of dry solid. The symbol $R$ is used to represent the rate of drying as pounds of water removed per hour.

Drying-rate curves are based on experimental data of free moisture content at different times during the course of a drying operation. For example, consider the case in which a known weight of dry solid is mixed with water and submitted to a batch experimental drying. If the total weight of the wet solid is noted at regular intervals while the material is drying, a plot can be made of the total weight versus the time. The slope of this curve at any point must represent the ratio of change in wet-solid weight to change in time. Therefore, the slope represents the instantaneous rate of drying corresponding to one particular wet-solid weight.

If the experimental drying process is continued until the weight of the solid remains constant, the equilibrium moisture content can be determined. Since the total weight of dry solid is known, the free moisture content corresponding to any

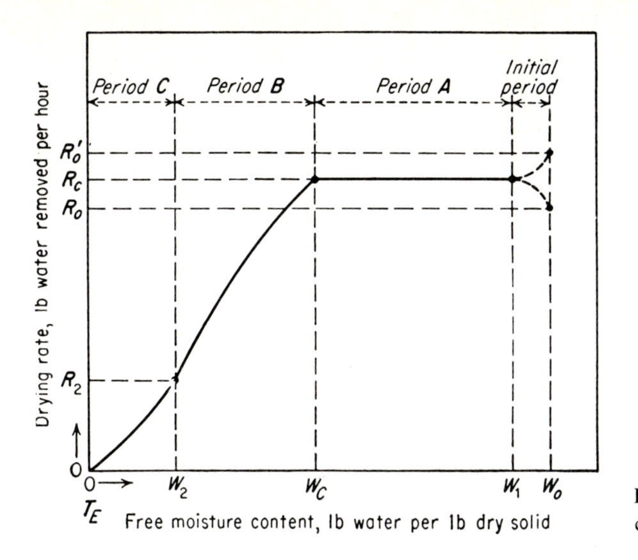


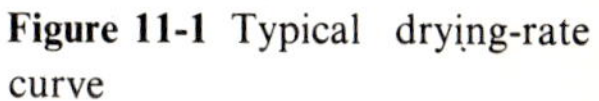
**Figure 11-1** Typical drying-rate curve

wet weight can be determined. These experimental data of instantaneous drying rate at various free moisture contents may be used to prepare a curve of the type presented in Fig. 11-1.

Typical drying-rate curves may be divided into different periods as the water content of the solid decreases.

**Initial period.** Referring to Fig. 11-1, there is an initial period at the start of the drying when the rate of water removal is adjusting to the conditions of the drying apparatus. This period is relatively short and, in many cases, cannot be detected experimentally.

**Period *A* (constant-rate period).** As long as water covers the entire surface of the solid, the rate of drying is constant. This period is represented on a typical drying-rate curve by a straight horizontal line showing no change in rate of water removal as the moisture content of the solid is reduced. During this period, water is diffusing through the solid at a rate sufficient to keep the entire surface wet. As the water content of the solid decreases, the amount of liquid diffusing to the surface becomes less, until a moisture content is finally reached where the diffusion rate is not sufficient to keep the entire surface wet.

**Period *B* (falling-rate period).** When a part of the solid surface is no longer wetted by the liquid, the water area available for direct heat and mass transfer decreases. This causes a decrease in the drying rate, as indicated in Fig. 11-1.

The free moisture content of the solid at the end of the constant-rate period is designated as the *critical free moisture content*. Critical free moisture content is ordinarily expressed as kilograms of water per kilogram of dry solid (or pounds of water per pound of dry solid), and it must be determined experimentally for each material and for each drying condition.

During period *B*, most of the water is still escaping by vaporizing at the surface of the solid. However, as the free moisture content decreases, the available wetted surface area decreases resulting in a steady reduction of the drying rate.

**Period *C* (falling-rate period).** In ordinary cases, it is very difficult to distinguish between period *B* and period *C*. Theoretically, period *C* represents the part of the drying cycle where the water vaporization has become entirely subsurface, with water no longer diffusing to the surface before vaporizing.

The lines representing periods *B* and *C* on the drying-rate curve are, ordinarily, approximately straight. In many cases, a single straight line may be drawn to represent periods *B* and *C* combined as one falling-rate period. The rate of drying becomes zero at the equilibrium moisture content. Therefore, the drying-rate curve must end at a zero rate and a total moisture content of $T_E$ (i.e., zero free moisture content).

## METHODS OF CALCULATION

One of the most important factors to be determined in drying calculations is the length of time required to produce a given amount of drying. Accurate calculations require a reasonably complete set of data giving the drying rates for the material under known conditions.

### Constant-Rate Period

Various empirical methods have been proposed for estimating the rate of drying during the constant-rate period. However, the best methods are based on actual experimental data obtained under conditions where the feed material and relative exposed surface area, as well as gas velocity, temperature, and humidity, are substantially the same as those used in the final drier. The time required for the constant-rate period can be determined directly from these data.

The basic method is best illustrated by an example. Assume a small-scale batch experiment has been carried out on a porous solid. The results of the experiment indicate that the rate of drying during the constant-rate period is 3.0 lb of water removed per hour when the initial charge contains water mixed with 6.0 lb of dry solid. The results also show that the critical free moisture content is 0.15 lb of water per pound of dry solid and that the equilibrium moisture content is 0.03 lb of water per pound of dry solid. A larger drier is to be operated with the same material under the same conditions of air velocity, humidity, and temperature as the small-scale drier. The exposed surface area per pound of dry solid will be the same in each drier.

If 1000 lb of dry solid containing 400 lb of water is to be dried per hour, the experimental data can be used to calculate the time of drying in the constant-rate period. The constant-rate period will continue until the total moisture contained by the solid is (0.15 + 0.03)(1000) = 180 lb of water. During this period, 400 − 180 = 220 lb of water will have been removed. From the experimental data, the rate of drying is 3.0/6.0 = 0.5 lb of water per hour per pound of dry solid. Therefore, the time for the constant-rate period in the large drier must be 220/(1000)(0.5) = 0.44 h.

## Falling-Rate Period

If experimental data are available giving the drying-rate curve during the falling-rate period, the drying time during this period can be determined by a direct graphical integration.[1]

If these data are not available, an approximate method is often employed. It is assumed that the drying-rate curve during the falling-rate period may be represented by a straight line between the critical free moisture content and the zero free moisture content on a plot such as the one presented in Fig. 11-1. When this assumption is made, the drying time during the falling-rate period may be determined from the following equation:[2]

$$\theta_f = \frac{2.3 W_C Q}{R_c} \log \frac{W_C}{W_F} \tag{11-1}$$

where $\theta_f$ = time of drying in falling-rate period, h

$Q$ = total pounds of dry solid, kg or lb

$R_c$ = rate of drying during the constant-rate period, kg water removed/h or lb water removed/h

$W_C$ = critical free moisture content, kg water/kg dry solid or lb water/lb dry solid

$W_F$ = free moisture content at end of drying, kg water/kg dry solid or lb water/lb dry solid

**Example 11-1: Determination of total drying time in a batch-drying operation** A solid is dried in a batch operation under such conditions that the rate of drying during the constant-rate period is 50 kg of water removed per hour and the critical free moisture content is 0.30 kg of water per kilogram of dry solid. The curve of drying rate versus moisture content may be assumed as a straight line during the falling-rate period, and the equilibrium moisture content of the material is 0.05 kg of water per kilogram of dry solid. If the material contains 500 kg of dry solid and 300 kg of water at the start of the drying, estimate the time required to obtain a final product containing 0.1 kg of moisture per kilogram of dry solid.

[1] See App. B for graphical-integration method for determining drying time during falling-rate period.

[2] See App. B for derivation of Eq. (11-1).

SOLUTION

Constant-rate period:

Weight of water removed = 300 − (500)(0.30 + 0.05) = 125 kg

Drying time in constant-rate period = $\theta_c$ = 125/50 = 2.5 h

Falling-rate period:

$Q$ = 500 kg of dry solid

$R_c$ = 50 kg water removed/h

$W_C$ = 0.30 kg water/kg dry solid

$W_F$ = 0.10 − 0.05 = 0.05 kg water/kg dry solid

From Eq. (11-1), the drying time in the falling-rate period is

$$\theta_f = \frac{(2.3)(0.30)(500)}{50} \log \frac{0.3}{0.05} = 5.37 \text{ h}$$

The total drying time required to obtain a final product containing 0.1 kg of water per kilogram of dry solid is

$$\theta_c + \theta_f = 2.5 + 5.37 = 7.9 \text{ h}$$

## PROBLEMS

**11-1** Under conditions where a certain sand has reached its equilibrium moisture content, 50 kg of the sand and water contains 48.7 kg of dry sand. What is the equilibrium moisture content $T_E$ of the sand under these conditions?

**11-2** From the following batch-drying data obtained in a tray drier, determine the rate of drying as pounds of water removed per hour when the free moisture content is 0.11 lb of water per pound of dry Filter-Cel.

Material being dried: mixture containing water and 12.5 lb of dry Filter-Cel
Dry-bulb temperature of air: 230°F
Wet-bulb temperature of air: 110°F
Equilibrium moisture content: 0.04 lb water/lb dry Filter-Cel

| Time, h | 0.0 | 0.2 | 0.4 | 0.6 | 0.7 | 0.9 | 1.1 | 1.3 | 1.4 |
|---|---|---|---|---|---|---|---|---|---|
| Weight of wet material, lb | 40.0 | 32.9 | 25.8 | 18.7 | 16.4 | 14.3 | 13.4 | 13.0 | 13.0 |

**11-3** Using the data of Prob. 11-2, estimate the critical free moisture content of the Filter-Cel under the given conditions.

**11-4** Determine the rate of drying during the constant-rate period for the data given in Prob. 11-2. Express this rate as pounds of water removed per hour per square foot of exposed surface area, assuming the total exposed surface area as 10 ft$^2$.

**11-5** A batch drier removes water from a solid material at a rate of 30 lb/h during the constant-rate period. Under the operating conditions, the critical free-moisture content is 0.5 lb of water per pound of dry solid, and the equilibrium-moisture content is 0.04 lb of water per pound of dry solid. The curve of drying rate versus moisture content may be assumed as a straight line during the entire falling-rate period. Three hundred pounds of dry solid, containing 200 lb of water, enters the drier. How long will the total drying require if the final product contains 0.08 lb of water per pound of dry solid?

**11-6** Using the data of Prob. 11-5, determine how many cubic feet of air per hour, entering at 760 mmHg 120°F, must be blown through the drier during the constant-rate period if the entering air has a humidity of 0.03 lb water vapor per pound of dry air and the exit-air humidity is 0.05.

**11-7** A soap drier reduces the water content of a given soap from 30% by weight to 5% by weight. It is estimated that the cost of drying is $10/500 kg of bone-dry soap in the mixture passing through the drier. Soap containing 5 wt% water sells at $10/100 kg at the factory. How much should the factory charge for soap containing 30 wt% water? Express the answer as dollars per 100 kg of the moist soap containing 30 wt% water.

**11-8** A tray drier is to be used for drying a solid material. Each tray is 3 ft wide by 4 ft long, and the wet material has a depth of 2 in. The density of the wet material is 50 lb/ft$^3$. The wet material contains 3 lb of water per pound of dry solid. How many trays are necessary to obtain 3000 lb of a product containing 1 lb of water per pound of dry solid?

CHAPTER

# TWELVE

## FILTRATION

Solid material may be separated from a liquid by forcing the mixture through a porous medium which retains the solid. The operation carried out in a physical separation of this type is known as *filtration.* The solids are removed from the liquid and remain on the porous filtering medium as a cake.

The development of this field has been based on practical experience with very little consideration given to theory. The result has been a wide variety of different types of filtration equipment and operational methods. The present filtration theory is based largely on experimental data, and most calculations require specific experimental information on the particular slurry (or solid-liquid mixture) involved.

A driving force is required to push a mixture of solids and liquids into the filtering unit and through the resistances offered by the cake and filtering medium. This driving force may be obtained by pressure on the entering side of the unit or by vacuum on the liquid-exit end. Certain types of filters obtain the required pressure by use of a liquid head so that gravity forces the liquid through the filtering medium, while other types employ positive or vacuum pumps to supply the necessary driving force.

A strong fabric is commonly used as a filtering medium. However, the cloth itself is not necessarily the true filtering medium, since the average size of the solid particles in a liquid slurry is often smaller than the average opening between the fibers making up the cloth. The fabric causes the solid particles to become entangled on the surface of the cloth. This results in the formation of a layer of solid material, and this layer is the true filtering medium. Thus, it is extremely important for this initial layer to be formed at the beginning of the operation.

**Table 12-1 Nomenclature for filtration**

$A$ = total filtering area, $m^2$ or $ft^2$
$C$ = proportionality constant in basic filtration equation, (Pa)(h)/$m^2$ or (lbf)(h)/$ft^4$
$K$ = a constant for each filtering operation, $m^6$/h or $ft^6$/h
$L$ = thickness of cake, m or ft
$P$ = pressure drop across cake, Pa or lbf/$ft^2$
$s$ = exponent of compressibility for filtration cake, dimensionless
$V$ = total volume of filtrate up to any time $\theta$, $m^3$ or $ft^3$
$V_c$ = a fictitious volume of filtrate necessary to lay down a cake with resistance equal to that of press and filter cloth, $m^3$ or $ft^3$

Greek symbols

$\alpha$ = alpha, specific cake resistance, (Pa)(h)(m)/kg or (lbf)(h)/(lb)(ft)
$\alpha'$ = alpha, a constant dependent on materials being filtered
$\theta$ = theta, total filtering time, h
$\theta_c$ = theta, time required to lay down a fictitious cake equivalent to press and frame resistance, h
$\rho_c$ = rho, density of cake expressed as mass of dry-cake solids per unit volume of wet filter cake, kg/$m^3$ or lb/$ft^3$

# TYPES OF FILTERS

## Gravity Filters

A gravity filter ordinarily consists of a tank with a perforated floor containing a porous material such as sand. Solids can be filtered out of liquid mixtures by letting the mixture run through the porous bed, using gravity as the driving force. Gravity filters are commonly used to process large amounts of liquids containing small quantities of solids relative to the total amount of mixture. This type of filter is widely used for clarifying water in water-purification plants.

## Plate-and-Frame Filters

A plate-and-frame filter press consists of a series of plates arranged alternately with separating frames. Each side of the plates is covered with a filter cloth, and the whole assembly is held together by a mechanical force.

The frames separating the plates form a hollow space into which the slurry is forced. The liquid passes through the filter cloth on the plates, leaving the suspended solids as a cake on the surface of the cloth. The liquid passes into corrugations on the plate and from there to a duct which carries the filtrate (i.e., the filtered liquid) out of the unit. When the space in the frames becomes filled with cake, the filtration must be stopped and the cake must be removed. Thus, the operation must be carried out batchwise, and considerable time and labor are required for the cleaning operations.

The plates and frames are constructed of metal, wood, or rubber, and pumps are required to supply the pressure for forcing the slurry into and through the unit.

Owing to the intermittent operation and high cost of labor, plate-and-frame filter presses are generally used only when the value of the cake is high and the quantity of cake is relatively small.

### Rotary Filters

Rotary filters may be operated continuously. For this reason, most large-scale filtration operations now use rotary filters instead of plate-and-frame presses.

A rotary filter consists, essentially, of a cylinder covered by a filter cloth, with a vacuum exerted on the inside of the cylinder so that liquid is pulled through the cloth. The cylinder rotates about its horizontal axis, and the surface dips into the slurry solution. The vacuum sucks liquid through the cloth, causing a cake to form on the outside area. After the cake passes out of the slurry, wash water may be sprayed onto the cake, and, as the rotation continues, air can be sucked through the cake. Thus, in the course of one revolution, the cake is formed, washed, and dried. Just before the cake is ready to dip into the slurry again at the completion of the revolution, a knife blade removes the cake and drops it into storage or onto a conveying receiver.

The equipment can be arranged mechanically to give separate streams of filtrate and wash water at the vacuum exit from the unit (see Chap. 4, Industrial Chemical Engineering Equipment).

## OPERATION OF FILTERS

### Precoating and Filter Aids

Since the filter cloth requires a layer of cake on it before it becomes an effective filtering medium, it is essential to obtain a *precoat* at the start of the filtering operation. In some cases, this initial layer may be obtained by passing the slurry through the filter at a slow rate. This permits a thin cake to form on the cloth, since there is not sufficient force to drive the particles on through the cloth. As soon as the initial layer is formed, the throughput rate may be increased.

Certain types of slurries tend to plug the pores of the filter cloth, resulting in a very high resistance to the passage of the liquid. This may be overcome by laying down an initial layer of a special material which does not penetrate and plug up the filter cloth. A material of this type is known as a *filter aid*. It permits the quick formation of the initial filtering layer without increasing the resistance of the cloth, and it also permits the cake to be removed easily at the end of the operation.

A filter aid may be used by mixing it directly with the slurry during the entire operation. This results in a more porous and less compressible cake giving decreased resistance to the passage of the liquid. A filter aid should have a low density and be chemically inert toward the mixture being filtered. The most common filter aid is kieselguhr, or diatomaceous earth. This material is practically pure silica and has a very complex skeletal structure. Charcoal, sawdust, asbestos, and other granular materials are occasionally used as filter aids.

### Constant-Pressure Filtration

A filter press may be operated by maintaining a constant pressure on the mixture entering the filtering unit. Under these conditions, the pressure drop across the cake and filtering medium remains constant throughout the entire run. As the cake thickness increases, the resistance to the passage of liquid also increases. Therefore, if the pressure-drop driving force is held constant, the rate of slurry and filtrate flow will decrease during the course of the run.

### Constant-Rate Filtration

The pressure on the solution entering a filtering unit may be controlled to maintain a constant flow rate of filtrate. Most filter presses combine constant-rate and constant-pressure operations in the individual runs. Thus the unit is operated at a low constant rate until a layer of filtering cake has been formed. The rate is then increased, and the operation during the remainder of the run is at constant pressure.

### Compressible and Noncompressible Cakes

When pressure is applied to a filter cake, there is a tendency for the solid particles to squeeze more closely together, resulting in a compression of the cake. If the pressure is sufficiently great, the cake will become very dense and offer a high resistance to the passage of any additional liquid. Practically all solids encountered in filtration operations are compressible to some extent. The precipitated hydroxides of aluminum and iron are examples of highly compressible sludges, while calcium carbonate, sand, and other coarse granular materials are examples of slightly compressible sludges.

For most engineering calculations, the slightly compressible sludges are considered as noncompressible. The addition of filter aids serves to increase the ability of the cake to resist compression, and these aids are often used to approach the condition of noncompressible cakes.

## METHODS OF CALCULATION

The rate at which filtrate is obtained in a filtering operation is governed by the materials making up the slurry and by the physical conditions of the operation. If the resistance of the filtering cloth is assumed as negligible in comparison with the resistance of the cake, the rate of filtration with a noncompressible cake is directly proportional to the available filtering area and to the pressure-difference driving force. The resistance of the cake may be assumed to be directly proportional to the cake thickness; therefore, the rate of filtration must be inversely proportional to the thickness of the cake. These statements may be expressed mathematically as

$$\frac{dV}{d\theta} = \frac{PA}{CL} \tag{12-1}$$

where $dV/d\theta$ = rate of filtrate flow, $m^3/h$ or $ft^3/h$
$V$ = total volume of filtrate up to any time $\theta$, $m^3$ or $ft^3$
$\theta$ = total filtering time, h
$P$ = pressure drop across cake, Pa or $lbf/ft^2$
$A$ = total filtering area, $m^2$ or $ft^2$
$L$ = thickness of cake, m or ft
$C$ = the proportionality constant, (Pa)(h)/$m^2$ or (lbf)(h)/$ft^4$

The thickness of the cake may be expressed as

$$L = \frac{wV}{\rho_c A} \tag{12-2}$$

where $w$ = mass of dry-cake solids per unit volume of filtrate, $kg/m^3$ or $lb/ft^3$
$\rho_c$ = density of cake expressed as mass of dry-cake solids per unit volume of wet filter cake, $kg/m^3$ or $lb/ft^3$

Equations (12-1) and (12-2) may be combined to give

$$\frac{dV}{d\theta} = \frac{PA^2}{\alpha w V} \tag{12-3}$$

where $\alpha$ equals $C/\rho_c$ and has the dimensions of (Pa)(h)(m)/kg or (lbf)(h)/(lb)(ft). The constant $\alpha$ is known as the specific cake resistance, and its value must be determined experimentally for each slurry.

The viscosity of the filtrate has an effect on the rate of filtration. In this treatment, the effect of viscosity is included in the specific-cake-resistance constant.

For a constant-pressure filtration, Eq. (12-3) may be integrated between the limits of zero and $V$ to give

$$V^2 = \frac{2PA^2}{\alpha w}\theta \tag{12-4}$$

In the usual range of operating conditions, the value of the specific cake resistance for compressible cakes may be related to the pressure difference by the empirical dimensional equation.

$$\alpha = \alpha' P^s \tag{12-5}$$

where $\alpha'$ is a constant dependent on the materials in the slurry and $s$ is a constant known as the *compressibility exponent of the cake.* The value of $s$ would be zero for a perfectly noncompressible cake and 1 for a completely compressible cake. For commercial slurries, the value of $s$ is usually between 0.1 and 0.8.

**Example 12-1: Estimation of filtering area required for a given filtration operation** A slurry containing 3.0 lb of dry solid per cubic foot of solid-free liquid is to be passed through a filter press at a constant pressure difference of 2 lbf/in$^2$. The specific cake resistance for the slurry may be represented by the equation

$$\alpha = 1.31P^{0.2}\ \text{(lbf)(h)/(lb)(ft), with } P = \text{lbf/ft}^2$$

Under these conditions, estimate the total filtering area required for one filtration run to deliver 400 ft$^3$ of filtrate in 2 h, assuming 1 ft$^3$ of filtrate is obtained for every cubic foot of solid-free liquid entering the unit.

The resistance of the press and cloth may be neglected, and the filtrate may be assumed as completely free of all solid material. Work the problem with American engineering units and with SI units.

SOLUTION With American engineering units:

$P = (2\ \text{lbf/in}^2)(144\ \text{in}^2/\text{ft}^2) = 288\ \text{lbf/ft}$
$\alpha = 1.31(288)^{0.2} = 4.07\ \text{(lbf)(h)/(lb)(ft)}$
$w = 3.0\ \text{lb/ft}^3$
$\theta = 2\ \text{h}$
$V = 400\ \text{ft}^3$

Under these conditions, Eq. (12-4) is applicable.

$$A = V\left(\frac{\alpha w}{2P\theta}\right)^{1/2} = (400\ \text{ft}^3)\left[\frac{4.07\ \text{(lbf)(h)}}{\text{(lb)(ft)}}\,\bigg|\,\frac{3\ \text{lb}}{\text{ft}^3}\,\bigg|\,\frac{}{2}\,\bigg|\,\frac{\text{ft}^2}{288\ \text{lbf}}\,\bigg|\,\frac{}{2\ \text{h}}\right]^{1/2}$$

$$= 41.2\ \text{ft}^2$$

Therefore, a total filtration area of 41.2 ft$^2$ would permit the delivery of 400 ft$^3$ of filtrate in 2 h with the given slurry mixture and operating conditions.

With SI units:

Conversion factors from App. A:

1 lbf/ft$^2$ = 47.88 Pa  1 lbf = 4.448 N  1 lb = 0.4536 kg

1 ft = 0.3048 m  1 lb/ft$^3$ = 16.018 kg/m$^3$  1 ft$^3$ = 0.02832 m$^3$

1 ft$^2$ = 0.09290 m$^2$

$$P = (2)(144) = 288\ \text{lbf/ft}^2 = \frac{288\ \text{lbf/ft}^2}{}\,\bigg|\,\frac{47.88\ \text{Pa}}{\text{lbf/ft}^2} = 13{,}790\ \text{Pa}$$

$\alpha = 1.31\ P^{0.2}$ is an empirical dimensional equation that only holds for P in units of pound-force/per square foot and results in $\alpha$ units of (lbf)(h)/(lb)(ft).

Therefore,

$$\alpha = 1.31(288)^{0.2} = 4.07\ \text{(lbf)(h)/(lb)(ft)}$$

For SI, the units of $\alpha$ must be converted to pascal-hour-meters per kilogram or to newton-hours per kilogram per meter since 1 Pa = 1 N/m². Therefore,

$$\alpha = \frac{4.07\ (\text{lbf})(\text{h})}{(\text{lb})(\text{ft})}\left|\frac{4.448\ \text{N}}{\text{lbf}}\right|\frac{\text{lb}}{0.4536\ \text{kg}}\left|\frac{\text{ft}}{0.3048\ \text{m}}\right. = 130.94\ \frac{(\text{N})(\text{h})}{(\text{kg})(\text{m})}$$

$$= 130.94\ \frac{(\text{Pa})(\text{h})(\text{m})}{\text{kg}}$$

$$w = \frac{3.0\ \text{lb/ft}^3}{}\left|\frac{16.018\ \text{kg/m}^3}{\text{lb/ft}^3}\right. = 48.05\ \text{kgm}^3$$

$$\theta = 2\ \text{h}$$

$$V = \frac{400\ \text{ft}^3}{}\left|\frac{0.02832\ \text{m}^3}{\text{ft}^3}\right. = 11.33\ \text{m}^3$$

Using Eq. (12-4),

$$A = V\left(\frac{\alpha w}{2P\theta}\right)^{1/2}$$

$$= (11.33\ \text{m}^3)\left[\frac{130.94\ (\text{Pa})(\text{h})(\text{m})}{\text{kg}}\left|\frac{48.05\ \text{kg}}{\text{m}^3}\right|\frac{}{2}\left|\frac{}{13{,}790\ \text{Pa}}\right|\frac{}{2\ \text{h}}\right]^{1/2}$$

$$= 3.83\ \text{m}^2 = \frac{3.83\ \text{m}^2}{}\left|\frac{\text{ft}^2}{0.09290\ \text{m}^2}\right. = 41.2\ \text{ft}^2$$

Thus, the SI solution gives the same result as the American engineering solution, and a total filtration area of 41.2 ft² or 3.83 m² would permit the delivery of 400 ft³ or 11.33 m³ of filtrate in 2 h with the given slurry mixture and operating conditions.

Equation (12-4) is often expressed in a simplified form including terms for the resistance due to the press and filter cloth. The form generally used is

$$(V + V_c)^2 = K(\theta + \theta_c) \tag{12-6}$$

where $V_c$ = a fictitious volume of filtrate necessary to lay down a cake with resistance equal to that of press and filter cloth, m³ or ft³

$\theta_c$ = time required to lay down the fictitious cake equivalent to press and filter cloth resistance, h

$K$ = a constant for each filtering operation, equivalent to $2PA^2/\alpha w$, m⁶/h or ft⁶/h

## Optimum Cleaning Cycle in Batch Filtrations

Equation (12-4) or (12-6) may be used to determine the time a filtering unit, operating at constant pressure, should be kept on stream to give the maximum capacity. For example, consider the case of a filtration where the constant value of

$2PA^2/\alpha w$ is known and the resistance of the press and filter cloth is negligible. The experimental value of $2PA^2/\alpha w$ has been found to be 75.5 $ft^6/h$, and the time the unit must be shut down for each cleaning is 2 h. The length of time to operate between cleanings to give the maximum capacity can be calculated by using Eq. (12-4).

In this case, $V^2 = 75.5\theta$. Letting $\theta$ be the filtering time for one operating cycle and $V$ the amount of filtrate collected in one cycle, the total number of cycles per 24 h must be $24/(\theta + 2)$. The total volume of filtrate collected in 24 h is $V[24/(\theta + 2)]$, or the total volume of filtrate per 24 h equals $(75.5\theta)^{1/2}[24/(\theta + 2)]$.

The value of $\theta$ where $(75.5\theta)^{1/2}[24/(\theta + 2)]$ is a maximum represents the optimum length of the operating cycle. The simplest way to find the optimum value of $\theta$ is by trial and error. In this example, the maximum volume of filtrate is obtained when $\theta$ is between 1.8 and 2.2 h. Since the filtrate volume is almost constant over this range of operating time, the particular time chosen for the optimum operation cycle would be the value to give the fewest number of shutdowns while still permitting a maximum filtrate delivery.

## PROBLEMS*

**12-1*** A plate-and-frame filter press containing 20 frames is used to filter a slurry made up to 10 lb of dry solids per 100 lb of liquid-and-solid mixture. The inside dimensions of each frame are 2 ft by 2 ft by 1 in thick. The cake formed in the filtration is noncompressible and contains 0.7 lb of dry solids per pound of cake. How many pounds of solid-free filtrate can be delivered before the press is filled with a wet cake having a density of 90 $lb/ft^3$?

**12-2*** A rotary filter turns at the rate of 2.0 rpm. The fraction of total filtering area immersed in the slurry is 0.20. It has been observed that 1.5 $ft^3$ of filtrate is delivered per minute per square foot of submerged area with a given slurry under these operating conditions. If 6.0 $ft^3$ of the filtrate is delivered per revolution by this rotary filter, what is the total area of the filter cloth on the drum?

**12-3** A filtration is carried out in a plate-and-frame filter press at a constant pressure difference of 3.0 $lbf/in^2$. The total filtering area is 80 $ft^2$, and the specific cake resistance is 3.5 (lbf)(h)/(lb)(ft). How many cubic feet of filtrate will be obtained in 1 h if the filter cake contains 10 lb of dry solids per cubic foot of filtrate? The resistance of the frame and cloth may be assumed as negligible, and the cake may be considered as noncompressible.

**12-4** Experimental filtration data for a compressible cake indicate that the specific cake resistance is 3.1 (lbf)(h)/(lb)(ft) when the pressure-difference driving force is 2 $lbf/in^2$ and 3.9 (lbf)(h)/(lb)(ft) when the pressure difference is 5 $lbf/in^2$. Estimate the value of the compressibility exponent for the cake.

**12-5** A certain plate-and-frame filter press is operated 24 h per day. The unit must be shut down periodically in order to remove the accumulated filter cake. The performance during one filtration run can be represented by

$$(V + 5)^2 = 78(\theta + 0.2) \qquad V = ft^3.\ \theta = h$$

How many hours should each filtration run last between shutdowns in order to obtain the maximum amount of filtrate per day if the total time required for each cleaning shutdown is 1 h?

* The problems with asterisks should be solved using American engineering units and also SI units.

CHAPTER
# THIRTEEN

# CHEMICAL TECHNOLOGY

Chemical technology is that part of chemical engineering dealing with chemical changes and physical conditions involved in industrial processes. The physical changes which take place in chemical engineering operations have been treated in the discussions of various unit operations. To obtain an overall concept of the full chemical plant and to understand the complete physical process, it is necessary to supplement the study of the unit operations with a consideration of chemical technology.

The backwoods distiller may be satisfied by merely setting up and operating equipment that yields a potable fluid. However, a true chemical engineer would want to know the chemical reactions occurring in the process and the reasons why the equipment functions properly, along with the best temperatures, pressures, and operating conditions to achieve the desired products. With a complete understanding of the physical processes and the chemical changes, the optimum product and most efficient operation can be obtained.

There are many volumes of books dealing with the various chemical processes used in industry, and detailed treatments are available in the literature.[1] Consequently, the purpose of this chapter is not to present a comprehensive coverage of all chemical technology, which would be impossible in one chapter or even in one book. Instead, this chapter presents examples of chemical technology for a few of the classic chemical processes that have been of historic importance in industry. The treatment emphasizes the broad principles involved in chemical technology that are generally applicable to many different industries and is intended to give the reader a general understanding of the type of information included under the general subject of chemical technology.

[1] For example, see the various editions of *Chemical Process Industries*, written by R. Norris Shreve and published by McGraw-Hill.

## FUNDAMENTAL CONSIDERATIONS

### Safety Considerations

Equipment for chemical processes should be set up to permit safe operation. This means the entire process, including chemical reactions, necessary controls, and equipment, should function in a manner involving little danger to personnel and minimum fire hazards.

Consideration must be given to possible emergencies which might arise, and adequate precautions should be taken. Fire hazards and exposure of personnel to potential danger spots should be eliminated or reduced to a minimum.

### Instrumentation

The modern chemical industry relies largely on instrument and computer control rather than on hand control by individual operators. Many processes involving chemical reactions must be operated with careful temperature regulation, and the reactions must be stopped when a certain degree of completion is attained. By proper instrumentation, the temperatures can be controlled and the reactions stopped automatically. This automatic operation ensures a uniform product and does not require the constant attention of a skilled operator.

Instruments are widely used to control temperatures, pressures, and the progress of chemical reactions. Recording instruments are of great value for indicating the progressive changes of important variables during the course of an operation and serve to provide necessary information for computer controls.

### Materials of Construction

Each of the many chemical processes employs certain materials of construction which are best suited to the particular conditions encountered. A complete treatment of the chemical technology for any given process should consider the materials which can be used for construction of the equipment.

### Special Equipment

Different types of specialized equipment, such as reacting vessels, autoclaves, filters, and furnaces, are required in the chemical industry. These items must have special features which are determined by the process involved.

Many manufacturing firms provide catalogs which serve as excellent guides for kinds of chemical equipment available. These catalogs are normally put out annually and represent a valuable addition to any chemical engineer's library.

### Chemical Control

While automatic instruments may be used to regulate the important variables in a chemical operation, it is always necessary to make control tests on the materials involved. Careful tests are made on the final product to determine the quality of the material. Chemical tests are also carried out on the intermediate materials to make certain the process is proceeding correctly. These chemical control tests may be carried out by highly skilled technicians in their laboratories, by process operators in the plant, or directly by analytical units installed as part of the operating process.

## PRODUCTION OF SULFURIC ACID

Sulfuric acid has historically been produced by two methods: the contact process and the chamber process. Both of these methods involve the formation of sulfur dioxide, which is oxidized to sulfur trioxide and absorbed in water to give sulfuric acid.

The contact process has essentially replaced the old-time chamber process. Chamber plants produce weak sulfuric acid (approximately 70% by weight) and have operating problems that are eliminated in the contact process. Contact plants are capable of producing 100% acid and even stronger grades containing dissolved $SO_3$ in excess of the amount necessary to react with the available water. Sulfuric acid containing excess $SO_3$ is called *oleum.* The oleums are marketed on the basis of the percentage of unreacted $SO_3$ dissolved in the 100% sulfuric acid solution. Thus, a 20% oleum contains 20 kg of $SO_3$ for every 80 kg of $H_2SO_4$; that is, the free $SO_3$ content is 20 percent of the total weight.

### The Contact Process

Figure 13-1 shows a flow sheet for a conventional sulfuric-acid contact process. The essential materials of construction are stainless steel, lead, duriron, acid brick, and steel.

The reactions involved in the process are

$$S + O_2 \longrightarrow SO_2$$

$$SO_2 + \tfrac{1}{2}O_2 \longrightarrow SO_3$$

$$SO_3 + H_2O \longrightarrow H_2SO_4 \qquad \text{(in presence of } H_2SO_4\text{)}$$

All these reactions evolve heat, and a large amount of cooling water is required to keep the temperatures under control. Most commercial plants obtain $SO_2$ by direct oxidation of molten sulfur, but some installations produce the $SO_2$ by oxidation of iron pyrites according to the following reaction:

$$4FeS_2 + 11O_2 \longrightarrow 2Fe_2O_3 + 8SO_2$$

The reaction $SO_2 + \frac{1}{2}O_2 \rightarrow SO_3$ requires careful temperature control, and a special catalyst must be used. At a temperature of 450°C, the reaction proceeds to 97 percent of completion. At lower temperatures the percent conversion falls off, and at higher temperatures the reaction reverses, causing the $SO_3$ to

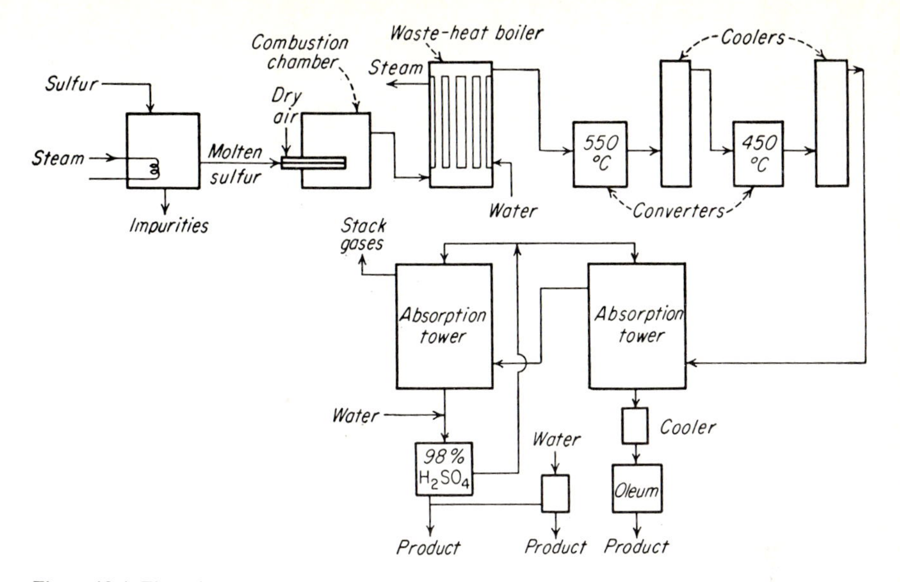


**Figure 13-1** Flow sheet showing manufacture of sulfuric acid by the contact process.

decompose into $SO_2$ and $O_2$. Two converters are often used for oxidizing the $SO_2$ to $SO_3$. The first converter operates at a temperature of 550°C and rapidly oxidizes 80 percent of the $SO_2$ to $SO_3$. The second converter operates at the optimum temperature of 450°C, where the oxidation rate is slower but the conversion approaches 97 percent.

Vanadium pentoxide is generally used as the catalyst for the oxidation of $SO_2$ to $SO_3$. The vanadium catalyst is usually in the form of pellets containing diatomaceous earth as a carrier.

Platinum dispersed on a carrier of asbestos or silica gel has also been used as a catalyst. The platinum catalyst loses its activity with use and poisons more rapidly than the vanadium catalyst.

As shown in Fig. 13-1, the $SO_3$ from the last converter is cooled and passes into absorption towers where 98 percent by weight sulfuric acid and oleum are produced. The absorption towers are usually constructed with steel walls lined with acid-proof brick and packed with chemical stoneware.

Since 98% acid has been found to be the most efficient absorbing agent for $SO_3$, acid of this strength is introduced into the tops of the absorption towers. Water is added to the product acid to give grades of lower concentrations. An electrostatic precipitator is often used to remove the last traces of acid mist from the gases that are leaving the final absorption tower.

**Modern alterations in the contact process.** Several improvements on the conventional contact process have been put into commercial use in recent years. The cooling-water requirements have been reduced by as much as 90 percent by eliminating the coolers following the converters. Temperature control is obtained by direct injection of cool air into the gas stream between converter stages. In place of two separate converters, a single converter is used. This converter is divided into a series of stages permitting 99 percent of the entering $SO_2$ to be converted into $SO_3$.

In place of packed absorption columns, the more recent type of contact process employs a horizontal cylindrical vessel which gives efficient absorption of the hot $SO_3$ by direct contact of gas and liquid. No packing is used in the absorption system, and cooling is obtained by evaporation of water from the absorbing acid.

The acid mist evolved from the absorption system is removed by means of a venturi scrubber followed by a cyclone separator. The advantages of this more recent type of process over the conventional process are a large saving on cooling water and simplified design and operation. The disadvantage is that the more recent type of process is not well adapted for producing sulfuric acid of strengths greater than 95 wt %.

## PRODUCTION OF NITRIC ACID

Since about 1930, the standard method for producing nitric acid has been by water absorption of nitrogen oxides produced from the oxidation of ammonia. Prior to 1930, nitric acid was obtained by treatment sodium nitrate with sulfuric acid in a heated vessel and condensing and absorbing the evolved gases. The reaction for the production of nitric acid from sodium nitrate is

$$NaNO_3 + H_2SO_4 \longrightarrow NaHSO_4 + HNO_3$$

When synthetic ammonia became available in large quantities, the most economical method for producing nitric acid soon became the oxidation of ammonia followed by water or acid absorption of the nitrogen oxides produced. This method has been developed until now practically all commercial nitric acid is produced by the ammonia-oxidation process.

Owing to the highly corrosive action of nitric acid, stainless steel is used almost exclusively as the material of construction when any acid is present. Steel pipes and tanks may be used for the ammonia.

The reactions for the production of nitric acid by the ammonia-oxidation process may be represented as follows:

$$4NH_3 + 5O_2 \longrightarrow 4NO + 6H_2O$$

$$2NO + O_2 \longrightarrow 2NO_2$$

$$3NO_2 + H_2O \longrightarrow 2HNO_3 + NO$$

Figure 13-2 presents a flow diagram for an ammonia-oxidation nitric acid plant. The ammonia is oxidized at a temperature of 940°C, using a fine-mesh platinum screen as a catalyst. The catalyst contains 95 percent platinum and 5 percent rhodium. The platinum serves as the catalyst for the reaction, and the

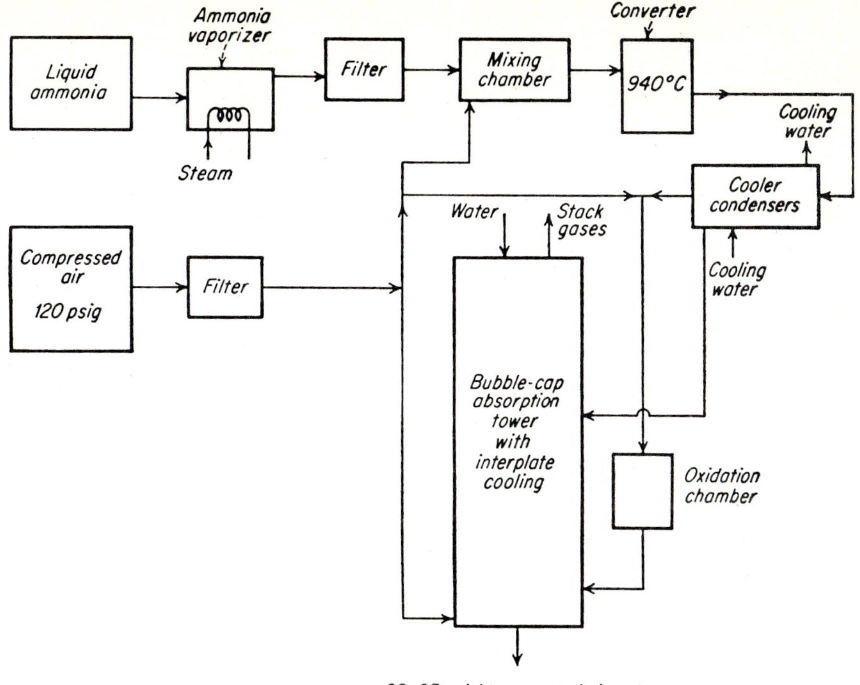


**Figure 13-2** Flow sheet showing manufacture of nitric acid by the ammonia-oxidation process.

rhodium acts primarily as an accelerator for the initial reaction when the unit is put into operation. Approximately 96 percent of the ammonia is converted to NO, the remainder going to nitrogen or other oxides of nitrogen.

Some of the NO is oxidized to $NO_2$ and reacts with water to form nitric acid before the gases enter the absorption tower. This condensed acid has a strength of about 40 wt % nitric acid. It is added to the absorption tower at an intermediate point.

In the absorption tower, the $NO_2$ reacts with water to form nitric acid and NO. This NO is oxidized to $NO_2$ which, in turn, reacts with water to yield more nitric acid and NO. Since the oxidation of NO continues throughout the entire absorption tower, it is necessary to add excess air to the entering gases. Enough excess air is added to maintain an oxygen content of approximately 3% by volume in the gases leaving the top of the tower.

Absorption towers are operated at pressures between 120 lb/in$^2$, gage, and atmospheric pressure. Most units in the United States run at the elevated pressures, while a large proportion of plants in foreign countries operate at atmospheric pressure. Bubble-cap plate towers are commonly used for the absorption operation.

The reaction of $NO_2$ with water evolves heat, and the kinetics of the reactions require that the absorption temperature be maintained at 85°F or less. Therefore, it is necessary to cool the acid in the absorption tower. This is ordinarily done by passing cold water through cooling coils located on each tray.

Product acid, having a maximum strength of 68% $HNO_3$ by weight, can be obtained at operating pressures of 120 lb/in$^2$, gage, using ordinary cooling water. The maximum strength of product acid when the absorption is carried out at atmospheric pressure is approximately 55 wt %.

Concentrated nitric acid (95 to 98 wt %) may be obtained from the weaker ammonia-oxidation acid by employing the dehydrating action of sulfuric acid to remove the water.

## FUEL GASES

The combustion of fuel gases gives off heat energy which finds many applications for industrial and home use. Because of handling convenience, cleanliness, and efficiency, the use of fuel gases has become universal. Producer gas, water gas, coal gas, and natural gas are examples of typical fuel gases. The quantities of these materials are ordinarily expressed on a volume basis as cubic feet at 60°F and 1 atm pressure with the gas saturated with water vapor.

### Producer Gas

This gas is obtained by passing air through a thick bed of coke. The following reaction occurs where the air enters the bed:

$$C + O_2 \longrightarrow CO_2$$

As the $CO_2$ passes through the hot coke, the $CO_2$ reacts with carbon to yield CO as follows:

$$CO_2 + C \longrightarrow 2CO$$

Steam is usually introduced with the air, and this steam reacts with carbon to yield CO and hydrogen.

$$H_2O + C \rightleftharpoons CO + H_2$$

The final producer gas contains CO and $H_2$ as combustible materials, plus $N_2$ and small amounts of $CO_2$. This fuel gas has a low heating value of about 150 Btu/ft$^3$ of gas. It is used industrially as a cheap source of heat for coke ovens and other types of furnaces.

### Water Gas

Water gas is produced by passing steam through carbon at a temperature of 1200°C to give the following reaction:

$$H_2O + C \rightleftharpoons CO + H_2$$

The reaction is endothermic and causes the temperature to decrease. The undesirable reaction

$$2H_2O + C \rightleftharpoons CO_2 + 2H_2$$

occurs at temperatures below 1000°C. Therefore, when the temperature drops to 1000°C, the bed is reheated by shutting off the steam and blowing in air. The oxygen in the air reacts exothermically with the carbon to give CO and $CO_2$. When the evolved heat has increased the temperature sufficiently, the air blow is stopped and steam is again injected. Thus the "blow" and "water-gas make" periods must be alternated during the entire operation.

The final water gas contains CO and $H_2$, with small amounts of $N_2$ and $CO_2$. The heating value of this fuel gas is about 300 Btu/ft$^3$.

Enriched water gas is made by adding considerable amounts of $C_2H_4$ and $CH_4$ obtained by thermal decomposition of oil. Heating values of 500 to 600 Btu/ft$^3$ of gas may be obtained with enriched water gas.

## Coal Gas

One product obtained from the thermal decomposition (or distillation) of coal is coal gas. The coal is heated in a closed vessel to yield a variety of products including coke, coal gas, coal tar, ammonia, and sulfur. The purified coal gas contains $CH_4$ and $H_2$ with small amounts of CO, $CO_2$, $N_2$, $C_2H_4$, and $C_6H_6$. The heating value is about 550 Btu/ft$^3$ of gas.

## Coal Gasification

Coal can be converted into pipeline-quality (i.e., a gas with high enough heating value to make long-distance transfer by pipelines feasible) substitute natural gas if the right conditions of temperature, pressure, catalyst, and equipment configurations are used. The process consists of passing steam and oxygen through coal, where the following basic reactions occur in the gasifier reactor:

$$C + O_2 \longrightarrow CO_2$$

$$C + CO_2 \longrightarrow 2CO$$

$$C + H_2O \rightleftharpoons CO + H_2$$

$$C + 2H_2 \rightleftharpoons CH_4$$

$$CH_4 + 2H_2O \rightleftharpoons CO_2 + 4H_2$$

The unit is designed to operate continuously and gives a raw product gas with a heat content of only about 150 to 200 Btu/ft$^3$. Treatment of this gas to remove $CO_2$, followed by further reactions with special catalysts which serve to convert the CO to $CH_4$, can result in a final gas that has a heat content of 1000 Btu/ft$^3$ or more, which makes the gas equivalent to natural gas and therefore makes it into what is known as "pipeline-quality" gas. The two critical reactions for the final increase in the methane content take place in a catalytic unit called the methanator and are as follows:

$$CO + H_2O \underset{}{\overset{\text{catalyst}}{\rightleftharpoons}} CO_2 + H_2 \qquad \text{(shift reaction)}$$

$$CO + 3H_2 \underset{}{\overset{\text{catalyst}}{\rightleftharpoons}} CH_4 + H_2O \qquad \text{(methanation reaction, requires 3 to 1 molal ratio of } H_2 \text{ to CO)}$$

Industrial plants for coal gasification are in operation throughout the world, but they are extremely expensive and are very complicated and difficult to operate.

### Natural Gas

Beneath the earth's surface at various localities throughout the world, natural gas is found. Raw natural gas contains various by-products, such as propane, butane, and natural gasoline, which are separated and sold commercially.

Purified natural gas contains chiefly $CH_4$, with small amounts of $C_2H_6$. It may have a heating value in excess of 1000 Btu/ft$^3$. Natural gas, in addition to being used for heating, also serves as a raw material for the chemical synthesis of methanol, ethanol, formaldehyde, and other organic chemicals.

## INDUSTRIAL GASES

### Carbon Dioxide

Carbon dioxide is obtained from the combustion of coke and as a by-product from the fermentation process for producing ethyl alcohol. The carbon dioxide is purified by absorbing it in a cold aqueous solution of sodium carbonate to form sodium bicarbonate, according to the following reaction:

$$CO_2 + Na_2CO_3 + H_2O \rightleftharpoons 2NaHCO_3$$

The above reaction may be reversed by applying heat to the bicarbonate solution to yield pure $CO_2$. The main uses of $CO_2$ are in the refrigeration industry as dry ice and in beverages.

### Hydrogen

Hydrogen is produced commercially by the water-gas process.

$$H_2O + C \longrightarrow CO + H_2$$

The CO is removed by oxidation to $CO_2$ with steam at 475°C over a suitable catalyst, such as iron oxide promoted with chromium oxide. The oxidation yields additional hydrogen, as shown by the following equation:

$$H_2O + CO \longrightarrow CO_2 + H_2$$

The $CO_2$ is removed by scrubbing with water, and any residual CO is taken out of the gas by absorption in ammoniacal copper formate.

Hydrogen is also produced from natural gas by the decomposition of methane. At a temperature of 1200°C, methane breaks down into carbon black and hydrogen as follows:

$$CH_4 \longrightarrow C + 2H_2$$

The *steam-iron process* produces hydrogen by passing excess steam over iron at about 650°C.

$$3Fe + 4H_2O \longrightarrow Fe_3O_4 + 4H_2$$

The $Fe_3O_4$ is reduced back to iron by reacting it with water gas, according to the following equation:

$$Fe_3O_4 + 2CO + 2H_2 \longrightarrow 3Fe + 2CO_2 + 2H_2O$$

When a direct current is passed through a dilute aqueous solution of alkali, the water decomposes to yield hydrogen at the cathode and oxygen at the anode.

$$2H_2O \longrightarrow 2H_2 + O_2$$

Pure hydrogen is obtained with this electrolytic method by separating the two electrodes with an asbestos diaphragm.

## Oxygen

Oxygen is obtained by the fractionation of liquid air. Carbon dioxide, water vapor, and dust are first removed from the air. The purified air, consisting of nitrogen and oxygen plus small amounts of argon, neon, xenon, and krypton, is then liquified by compression and cooling.

Since liquid nitrogen boils at $-195.8$°C and liquid oxygen boils at $-183$°C (at atmospheric pressure), these two components may be separated by fractional distillation. The nitrogen, along with the contained rare gases, is withdrawn from the top part of the fractionating column, while pure oxygen is obtained at the bottom of the column.

## Acetylene

The action of water on calcium carbide produces acetylene ($C_2H_2$), as shown in the following equation:

$$2H_2O + CaC_2 \longrightarrow Ca(OH)_2 + C_2H_2$$

Since acetylene is unstable when highly compressed, it is usually sold in steel cylinders as an acetone solution. The pressure over the solution is approximately 10 atm.

The primary use of acetylene is for welding and cutting metals. It is also used as a raw material for the chemical synthesis of organic chemicals such as acetic acid and neoprene synthetic rubber.

## ELECTROLYTIC PROCESSES

Energy, in the form of electricity, is used to cause chemical reactions in the electrolytic industry. The cost of electric power is usually the deciding factor when choosing between an electrolytic method and a straight chemical method for preparing a given material. Certain substances, such as aluminum, can only be prepared economically by electrolytic methods. Therefore, a plant of this type should always be located close to a cheap power source.

### Magnesium

The electrolysis of molten magnesium chloride produces magnesium, according to the reaction

$$MgCl_2 \longrightarrow Mg + Cl_2$$

Sodium chloride is added to the magnesium chloride to lower the melting point and increase the conductivity of the mixture. The temperature of the electrolytic cell is maintained at 700°C by the use of gas-heated furnaces. The molten magnesium formed by the electrolysis rises to the surface of the bath, while dry chlorine is given off as a by-product.

### Sodium

We may obtain sodium by the electrolysis of molten sodium chloride in an electrolytic cell such as a Downs cell operated at 600°C. The reaction is

$$2\,NaCl \longrightarrow 2\,Na + Cl_2$$

The melting point of the sodium chloride is reduced by the addition of fluoride salts. The reduced melting point prevents the formation of sodium fog. Chlorine is also produced in this process, and a diaphragm is located in the cell to permit separation of the two products.

### Aluminum

One of the most important materials produced by electrolytic methods is aluminum. It is obtained by the electrolytic reduction of pure alumina ($Al_2O_3$) in a mixture of molten cryolite ($AlF_3 \cdot 3\,NaF$) and fluorspar ($CaF_2$).

Figure 13-3 illustrates an electrolytic cell for the production of aluminum. A rectangular cell with a carbon lining is used. The lining serves as the cathode, and carbon anodes dip down into the bath. The oxygen in the $Al_2O_3$ unites with carbon from the anodes to form $CO_2$ and CO, according to the reactions

$$2\,Al_2O_3 + 3\,C \longrightarrow 4\,Al + 3\,CO_2$$

$$Al_2O_3 + 3\,C \longrightarrow 2\,Al + 3\,CO$$

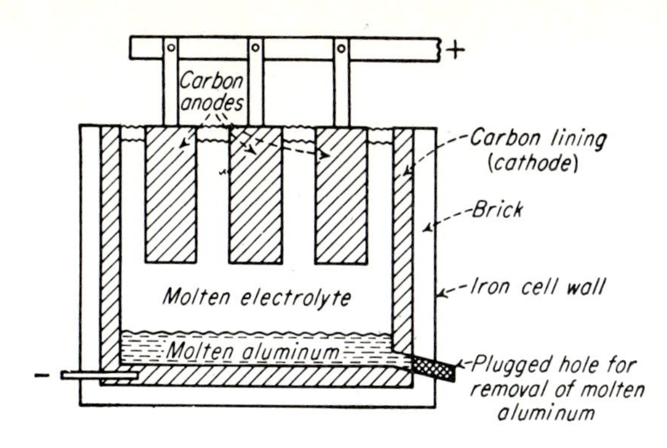


**Figure 13-3** Electrolytic cell for production of aluminum.

Aluminum settles to the bottom of the cell, where it is removed periodically. The temperature must be maintained between 900 and 1000°C to permit the aluminum to settle rapidly.

## THE PETROLEUM INDUSTRY

The petroleum industry employs a large number of chemical engineers and is one of the world's major chemical industries. Petroleum refining methods have been developed whereby the crude oil can be separated into its many constituents, and special chemical treatments have been developed to permit the refineries to produce a variety of important chemical compounds.

It is generally believed that petroleum has been formed from organic matter of past ages by a process involving fermentation and decomposition. Crude petroleum contains many different hydrocarbons, plus compounds of sulfur, nitrogen, and oxygen. Classification of the different crude grades is often based on the type of hydrocarbons predominating in the raw mixture. The three main classifications are the following:

1. Paraffin base—containing mostly open-chain compounds
2. Naphthene base—containing a large amount of naphthenic (cyclic) compounds
3. Mixed base—containing large amounts of both paraffinic and naphthenic compounds.

An oil refinery divides petroleum into various fractions by taking advantage of the boiling-point differences. The following liquid fractions are obtained in a typical refinery similar to the one outlined in Fig. 13-4:

1. Straight-run gasoline—the most volatile liquid hydrocarbons; used for blending with ordinary motor fuels
2. Naphtha—naphtha oils and waxes plus light oils; used mainly for solvents
3. Kerosene—a light oil; used for heating and solvents

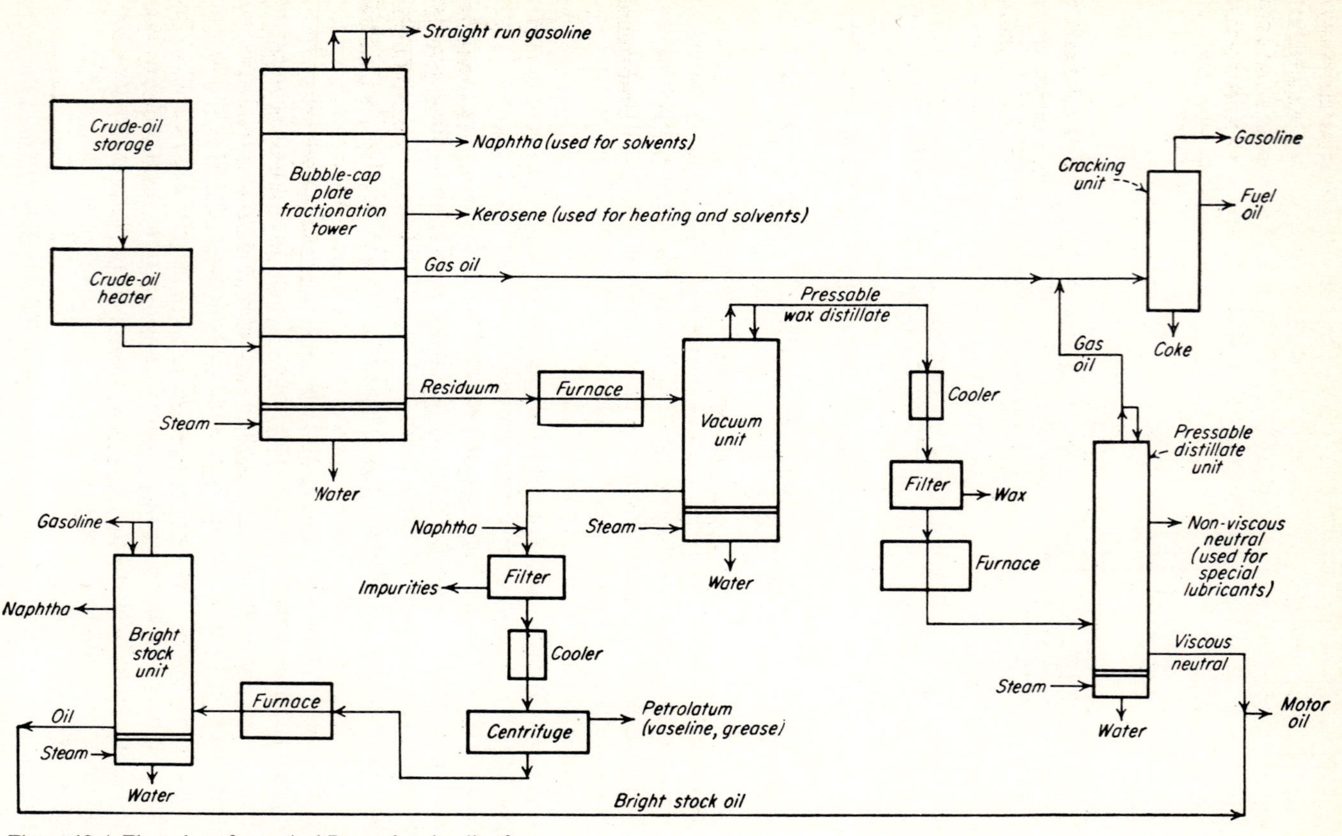


**Figure 13-4** Flow sheet for typical Pennsylvania oil refinery.

4. Gas oil—a mixture of heavy and light oils; subjected to a cracking treatment to yield gasoline and fuel oil
5. Residuum—lubricating oils, waxes, petroleum jelly, grease, motor oil, asphalt, and tar

### Cracking

A large proportion of all motor gasoline is obtained by thermal or catalytic cracking of high-molecular-weight hydrocarbons. The long-chain hydrocarbons are not sufficiently volatile for use as a gasoline, but the carbon chain can be broken by a cracking process. The short-chain hydrocarbons produced can then be separated to yield grades which are usable as gasoline.

Thermal cracking achieves cleavage of the carbon chain by the use of high temperatures and pressures. Catalytic processes use lower temperatures and pressures. Silicate-base catalysts are often used in the modern cracking plants, and proper choice of time, temperature, and catalyst form determines the type of products as well as the overall yield.

## FERMENTATION PROCESSES

Many important chemicals are produced by processes involving fermentation. A few examples are ethyl alcohol, acetic acid, acetone, butyl alcohol, and penicillin.

Certain types of microorganisms produce substances which cause a catalytic conversion of organic materials. Under controlled conditions, these catalysts (or enzymes) can be forced to produce useful chemicals by a process known as *fermentation.*

### Production of Ethyl Alcohol from Molasses

Molasses contains sugar, which can be fermented to yield ethyl alcohol. Yeast is used as the fermenting agent. The yeast produces two important enzymes known as *invertase* and *zymase.* The invertase causes the sugar (or sucrose) to hydrolyze into two molecules of monosaccharides, as shown by the following reaction:

$$\underset{\text{Sucrose}}{C_{12}H_{22}O_{11}} + H_2O \longrightarrow \underset{d\text{-glucose}}{C_6H_{12}O_6} + \underset{d\text{-fructose}}{C_6H_{12}O_6}$$

The zymase changes the monosaccharides into ethyl alcohol and carbon dioxide as follows:

$$\underset{\text{Monosaccharide}}{C_6H_{12}O_6} \longrightarrow \underset{\text{ethyl alcohol}}{2C_2H_5OH} + 2CO_2$$

Figure 13-5 presents a flow diagram for the production of ethyl alcohol from molasses. The sugar concentration in molasses is too high to support fermentation. Therefore, water is added to dilute the mixture to a concentration of about 10 wt %

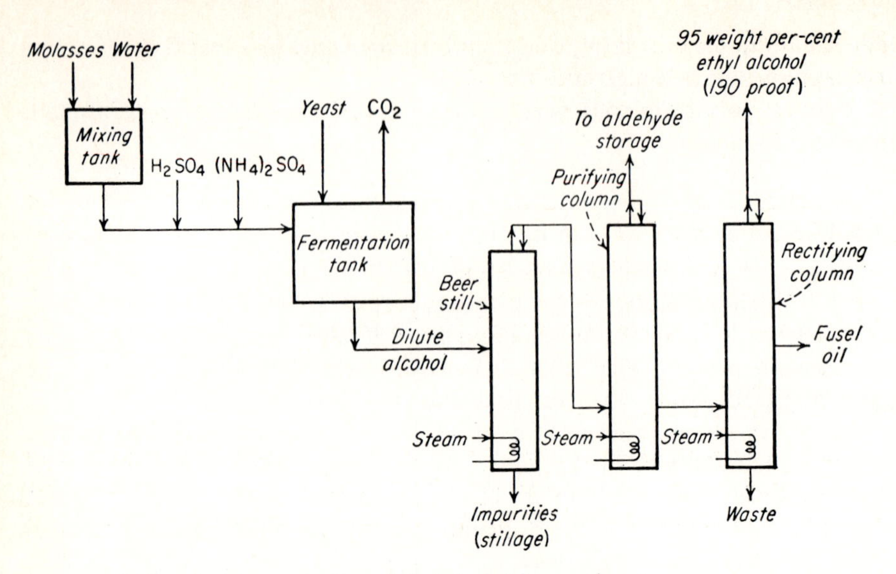


**Figure 13-5** Flow sheet for production of ethyl alcohol from molasses.

sugar. Small amounts of sulfuric acid and ammonium sulfate are added to the mixture. The sulfuric acid is added to suppress the formation of wild yeasts or bacteria and to adjust the pH of the solution. The ammonium sulfate furnishes a nutritive agent.

The ethyl alcohol is produced as a dilute mixture with water and other products, and it is necessary to purify the alcohol by distillation. Carbon dioxide is obtained as a by-product along with small amounts of aldehydes and fusel oil (an impure amyl alcohol used for solvent purposes).

Since ethyl alcohol forms an azeotrope with water at a concentration of 95.6 wt % alcohol (at 1 atm pressure), it is impossible to obtain pure alcohol by conventional methods. Higher concentrations of ethyl alcohol may be obtained by azeotropic or extractive distillation, as explained in Chap. 8 (Distillation).

## PLASTICS

Plastics and synthetic resins are organic materials which can be shaped by chemical or heat treatment to yield solid noncrystalline substances. In some cases, the final product may appear as filaments or fibers such as the common forms of nylon. Probably the best-known end products are the various solid forms, such as combs, radio cases, toothbrush handles, and electrical equipment. Plastics and synthetic resins are widely used as film-forming substances in varnishes and various types of protective coatings.

The end product is ordinarily a mixture of the plastic or synthetic resin with a filler such as wood flour and a coloring agent. Plastics and synthetic resins are produced as liquids or solid powders, and conversion to the final form is carried out by application of heat or catalysis.

There are two major classifications of these materials with reference to their behavior when heated.

1. *Thermoplastic* types are softened by the application of heat. They may be reworked and reformed with no change in chemical composition by heating until soft and then cooling to a hard product.
2. *Thermosetting* types are resistant to softening by the application of heat. Materials of this type will decompose before any softening occurs.

## Phenol Formaldehyde

An example of a thermosetting resin is phenol formaldehyde. It is prepared by a multistage polymerization of formaldehyde and phenol. As a first stage, the reaction between formaldehyde and phenol may be represented as

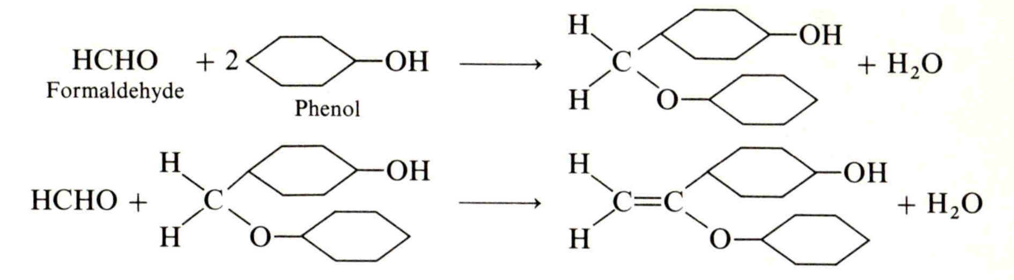


This last material may polymerize further by opening of the double bond to give

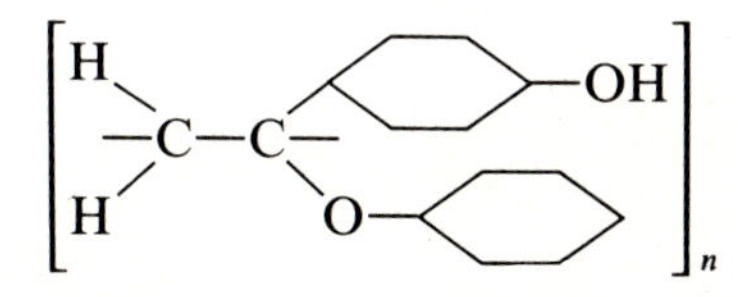


By control of temperature and concentration along with the use of special catalysts, a variety of different polymerization products may be obtained from phenol and formaldehyde. These resins are used as adhesives and as molded parts for radio cabinets, telephones, handles, and other common household items.

## Polyvinyl Chloride

One important thermoplastic resin is polyvinyl chloride. All the polyvinyl resins contain the vinyl group (—CH=CH$_2$). Vinyl chloride is produced from acetylene or ethylene according to the following reactions:

$$HC{\equiv}CH + HCl \longrightarrow Cl{-}CH{=}CH_2$$

$$CH_2{=}CH_2 + Cl_2 \longrightarrow Cl{-}CH_2{-}CH_2{-}Cl \xrightarrow{\text{alkali}} Cl{-}CH{=}CH_2 + HCl$$

The vinyl chloride may be polymerized by the use of a peroxide catalyst to give the following linear-chain type of molecule:

$$-\underset{\displaystyle Cl}{\underset{|}{CH}}-CH_2-\underset{\displaystyle Cl}{\underset{|}{CH}}-CH_2-\underset{\displaystyle Cl}{\underset{|}{CH}}-CH_2-$$

Vinyl chloride may be polymerized in the presence of vinyl acetate to give a copolymer of vinyl chloride and vinyl acetate. This copolymer has high flexibility and is used as a rubber substitute. It has also been widely used in many types of coverings such as tank linings, wire coatings, and special protective fabrics.

## NUCLEAR REACTIONS

For many years, scientists have known that large amounts of energy could be obtained from nuclear reactions, but little was known about the methods for initiating and controlling the reactions. The impetus of the Second World War caused the science of atomic energy to develop rapidly into a major field involving large numbers of physicists, chemists, and chemical engineers.

The source of atomic energy may be illustrated by considering the formation of helium from hydrogen. The tremendous heat energy originating from the sun, as well as the potential energy of the "hydrogen bomb," is a result of hydrogen changing into helium. By the application of high pressures and temperatures in the range of millions of degrees Fahrenheit or by use of energies and products from less violent nuclear reactions, the following reaction occurs:

$$2H_2 \longrightarrow \text{He} + \text{mass converted into energy} \tag{13-1}$$

The atomic mass of hydrogen may be taken as 1.00813 and that of helium as 4.00389. Therefore, when 1 g atom of helium is formed from hydrogen, the loss of mass amounts to $(4)(1.00813) - 4.00389 = 0.02863$ g. According to the Einstein relation between mass and energy, this unaccounted-for mass appears as energy. The amount of energy evolved may be expressed by the following equation:

$$\Delta E = \Delta mc^2 \tag{13-2}$$

where $\Delta E$ = the amount of energy released, ergs
$\Delta m$ = loss in mass, g
$c$ = velocity of light $= 3 \times 10^{10}$ cm/s

The energy released by the formation of 1 g atom of helium from hydrogen may be calculated by Eq. (13-2) to be $25.767 \times 10^{18}$ ergs. Since 1 erg is equivalent to $3.968 \times 10^{-7}$ J or $9.478 \times 10^{-11}$ Btu, the preceding may be expressed in conventional engineering terms as follows: When 4 g of hydrogen is converted to helium, the energy released is equivalent to 10.2 trillion J or 2.440 billion Btu.

Atomic energy may be obtained from other nuclear reactions which are easier to initiate and control. The conversion of certain uranium isotopes and plutonium to lower elements by fission constitutes the basis of the "atomic bomb." Atomic piles have been developed for producing controlled nuclear reactions based on uranium and plutonium as starting materials. Carbon and heavy water have been used as moderators to assist in the nuclear reactions. Cadmium compounds have been employed for controlling the reactions.

Controlled nuclear reactors have been constructed for the purpose of generating useful power. The heat from the reactor is transferred to a suitable medium, and the heat from this medium may be used to generate steam. Conventional methods may be employed to obtain electrical or mechanical power from the steam.

Many engineering difficulties are encountered in the construction and operation of the nuclear reactors. Construction materials having low neutron absorption must be used. The standard materials of construction, such as steel and aluminum, cannot withstand the high temperatures involved. Stainless steel and chromium have been suggested as possible materials of construction, but neither of these is ideal. Molten metals, such as lead, sodium, or potassium, have been found suitable as heat-transfer media. Gases, including helium and air, have also been considered as possible materials for removing the heat generated in the reactors. There is always the danger of radioactive contamination due to leakage of the gases or liquid, and stringent safety requirements must be met before the unit can be operated. Despite these difficulties, commercial plants are in operation for generation of power from nuclear energy, and more plants are planned.

## PROBLEMS

**13-1** A contact sulfuric acid plant converts 95 percent of the entering sulfur into sulfuric acid. How many pounds of pure sulfur will be required per day if the plant produces 100,000 lb of 95 wt % sulfuric acid per day?

**13-2** A tank contains 50,000 kg of 20 % oleum. (A 20 % oleum means that in 100 kg of material there is 20 kg of $SO_3$ dissolved in 80 kg of pure $H_2SO_4$.) How many kilograms of water must be added to the tank to give 93.2 wt % sulfuric acid (commonly called oil of vitriol)?

**13-3** An analysis of the gases entering the converter of an ammonia-oxidation nitric acid plant indicates that the gases contain 0.066 lb of ammonia and 0.934 lb of air per pound of gas. A sample of the gases leaving the converter is absorbed in hydrogen peroxide and titrated with sodium hydroxide. The results indicate that 1 lb of the gas leaving the converter contains 0.00361 lb mol of NO (assume no $NO_2$ is formed). Determine the percent of ammonia converted to NO in the converter.

**13-4** If 90 percent of the ammonia entering an ammonia-oxidation unit is converted to nitric acid, how many kilograms of ammonia are necessary to produce 60,000 kg of 62 wt % nitric acid? The 10 percent loss of ammonia is due to conversion losses, NO in the stack gases, and miscellaneous losses.

**13-5** A water gas has the following composition, expressed as mole percent:

$$CO = 42.9$$
$$CO_2 = 3.0$$
$$H_2 = 49.8$$
$$N_2 = 3.3$$
$$O_2 = 0.5$$
$$CH_4 = 0.5$$

The net heating value for carbon monoxide equals 4347 Btu/lb, for hydrogen equals 52,010 Btu/lb, and for methane equals 21,520 Btu/lb. Determine the net heating value of the water gas expressed as Btu per cubic foot of dry gas at 60°F and 1 atm pressure.

**13-6** The electrolytic reaction for the production of aluminum may be written as

$$Al_2O_3 + \tfrac{3}{2}C \longrightarrow 2Al + \tfrac{3}{2}CO_2 - 235{,}000 \text{ cal}$$

This equation shows that 235,000 cal of energy (in the form of electricity) is consumed for the production of 2 g mol of aluminum. Some CO is also formed in a reaction requiring more energy. On the basis of the reaction evolving $CO_2$, the current efficiency may be taken as 35 percent. This means that only 35 percent of the current passing through the cell is consumed in the actual reaction producing aluminum. If electricity is available at a cost of 1 cent/kWh, how much must be paid for electrical power per pound of aluminum produced?

**13-7** A water solution of molasses contains 10% by weight sucrose ($C_{12}H_{22}O_{11}$). Determine the weight percent of ethyl alcohol in the solution after 90 percent of the sucrose has been converted to ethyl alcohol by fermentation. The $CO_2$ formed in the reaction can be considered as having a negligible solubility in the solution, and the last 10 percent of the sugar can be assumed as unchanged.

**13-8** Determine the number of Btu equivalent to the amount of energy liberated when 1 g of helium is formed by the following nuclear reaction:

$$Li + \tfrac{1}{2}H_2 \longrightarrow 2He$$

In this reaction, the atomic mass of hydrogen is 1.00813, the atomic mass of lithium is 7.01820, and the atomic mass of helium is 4.00389.

**13-9** The raw gas produced from a coal-gasification plant has the following composition expressed as volume percent:

$$\begin{aligned} CO_2 &= 20 \\ CO &= 21 \\ H_2 &= 44 \\ CH_4 &= 14 \\ \text{Other as } SO_2, N_2, H_2S, \text{ etc.} &= 1 \\ \text{Total} &= 100 \end{aligned}$$

What is the net heating value for this gas expressed as Btu per cubic foot of dry gas at 60°F and 1 atm pressure? The net heating value for CO is 4347 Btu/lb, for $H_2$ is 52,010 Btu/lb, and for $CH_4$ is 21,520 Btu/lb.

**13-10** At a plant which produces sulfuric acid by the contact process, a large storage tank currently holds 100,000 kg of 20% oleum, where 20% oleum means that in 100 kg of material there is 20 kg of $SO_3$ dissolved in 80 kg of pure $H_2SO_4$. The tank has a total capacity when filled of 130,000 kg. If water is added to the tank to fill it just to its capacity, what will be the weight percent sulfuric acid in the final mixture in the tank?

**13-11** (*This problem is intended for computer program solution.*) The following flows are determined for an ammonia plant. These values are in metric ton moles per day (a metric ton is 1000 kg).

| Stream | 1 Fresh feed | 2 Mixed feed | 3 Reactor outlet | 4 Product | 5 Off gas | 6 Purge | 7 Recycle |
|---|---|---|---|---|---|---|---|
| Ar | 0.20 | 1.05 | 1.05 | 0.00 | 1.05 | 0.20 | 0.85 |
| $H_2$ | 74.20 | 225.56 | 180.45 | 0.00 | 180.45 | 29.09 | 151.36 |
| $N_2$ | 25.60 | 79.75 | 64.71 | 0.00 | 64.71 | 10.56 | 54.15 |
| $NH_3$ | 0.00 | 0.00 | 30.07 | 30.07 | 0.00 | 0.00 | 0.00 |
| Total | 100.00 | 306.36 | 276.28 | 30.07 | 246.21 | 39.85 | 206.36 |

(*a*) Write a computer program which prints out an identically laid-out table of flows in metric tons per day.

(*b*) Merge the two tables so that both moles and mass are printed for all streams

| | 1 Fresh feed | | 2 Mixed feed | | . . . |
|---|---|---|---|---|---|
| | Mols | Tons | Mols | Tons | |
| Ar | | | | | |
| $H_2$ | | | | | |
| ⋮ | | | | | |

Note: Document the basic mathematics you use.

The following is a simple process flow diagram:

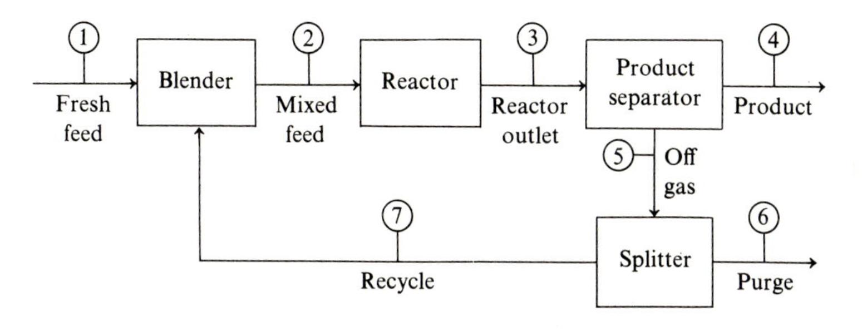

CHAPTER

# FOURTEEN

## CHEMICAL ENGINEERING ECONOMICS AND PLANT DESIGN

Chemical engineers are often called upon to design plants and equipment. In this modern age of industrial competition, the design of engineer's work must be based on economic considerations as well as on efficient and effective operation. For this reason, the chemical engineer should consider plant design and economics as one combined subject.

A good design engineer must have the ability to apply theoretical training and practical experience to new problems. It is necessary to be cost-conscious and to realize that a monetary return on any investment is of prime importance. A knowledge of chemical engineering plant design requires an understanding of the principles of stoichiometry, unit operations, and chemical technology. In addition, the design engineer must have a knowledge of many other subjects, such as plant location, plant layout, cost estimation, investments, and profits.

**Table 14-1 Nomenclature for economics and plant design**

$a$ = length of time equipment has been in use, years
$d$ = the annual depreciation, dollars/year
$D$ = total depreciation up to $a$ years, dollars
$D_i$ = optimum economic inside pipe diameter, in
$f_a$ = accrued amount if \$1 is deposited each year for $a$ years at interest rate $i$
$i$ = interest rate expressed as a fraction
$K_v$ = constant for determination of maximum allowable velocity in packed or bubble-cap towers. Value dependent on tower design, m/s or ft/s
$n$ = service life of equipment, years
$P$ = the reducing-balance factor of depreciation
$S$ = estimated scrap value of equipment when service life has expired, dollars
$S_n$ = amount of money which must be put on deposit per year at interest rate $i$ to have \$1 on deposit after $n$ years, dollars
$u$ = superficial vapor velocity (based on the total cross-sectional area), m/s or ft/s
$V$ = fixed capital investment for equipment including initial cost, cost of installation, and auxiliaries, dollars
$w$ = thousands of pounds-mass flowing per hour, 1000 lb/h

Greek symbols

$\rho$ = average density of a fluid (sub $L$ indicates liquid, sub $G$ indicates gas), kg/m$^3$ or lb/ft$^3$

## COSTS

The complete design of a chemical manufacturing plant should include information on investment and operating costs as well as an estimation of the probable profits that can be obtained by use of the proposed design. The total costs involved in a manufacturing process may be divided into seven general categories as follows: (1) fixed costs, (2) direct production costs, (3) general plant overhead, (4) general administrative and office overhead, (5) distribution costs, (6) contingencies, and (7) gross-earnings expenses.

**Fixed costs.** The expenses which remain practically constant from year to year are called *fixed costs*. They do not vary widely with increase or decrease in production rate. Interest, rent, insurance, local taxes, depreciation, research, development, and patent royalties involve expenditures included under fixed costs. Interest is normally limited to financing expenses for borrowed money needed for the capital investment, and the cost is often not included for preliminary economic analyses, with the results being presented on a basis of no interest included as cost.

**Direct production costs.** Expenses directly involved in the production operation are called *direct production costs*. Direct production costs include expenditures for raw materials; direct operating labor; supervisory and clerical labor directly connected with the manufacturing operation; plant maintenance and repairs; operating supplies such as oil, grease, and janitor supplies; payroll overhead, including pensions, vacation allowances, social security, and life insurance; power; and utilities.

**General plant overhead.** Costs for hospital and medical services, engineering and drafting, general maintenance, safety services, restaurant and recreation facilities, salvage services, and control laboratories comprise the *general plant overhead*.

**General administrative and office overhead.** Expenses for executive and clerical wages, office supplies, upkeep on office buildings, and communications are included in *general administrative and office overhead*.

**Distribution costs.** Expenses incurred in the process of distributing and selling the various products are called *distribution costs*. These costs include salaries, wages, and supplies for sales offices; salaries, commissions, and travelling expenses for salespeople; shipping and packaging and advertising.

**Contingencies.** A safety factor on the costs to take care of unforeseen developments or emergencies is provided by *contingencies*.

**Gross-earnings expenses.** Gross-earnings expenses are based on income-tax laws. These expenses are a direct function of the gross earnings made by all the various

**Table 14-2 Breakdown of annual operating costs for chemical manufacturing plants**

The percentages and costs indicated in this table are approximations applicable to ordinary chemical plants. The values will vary for different plants and localities and are included to give the reader a general idea of the relative magnitudes. All percentages are expressed on a yearly basis.

1. Fixed costs (about 20 percent of annual operating costs)
   *a.* Interest (0 to 15 percent of the total investment)
   *b.* Rent (8 percent of rented land or building value)
   *c.* Insurance (1 percent of fixed-capital investment)
   *d.* Local taxes (1 to 4 percent of fixed-capital investment)
   *e.* Depreciation (varies for different processes; for a life period of 20 years and no scrap value, depreciation amounts to approximately 5 percent of the fixed-capital investment)
   *f.* Patents and royalties (0 to 5 percent of annual operating costs)
   *g.* Research and development (5 percent of every sales dollar)
2. Direct production costs (about 60 percent of annual operating costs)
   *a.* Raw material including transportation costs (20 percent of annual operating costs)
   *b.* Direct operating labor (15 percent of annual operating costs)
   *c.* Direct supervisory and clerical labor (10 percent of direct operating labor or 1.5 percent of annual operating costs)
   *d.* Plant maintenance and repairs (7 percent of fixed-capital investment)
   *e.* Operating supplies (20 percent of cost for plant maintenance and repairs or 1 percent of fixed-capital investment)
   *f.* Payroll overhead (15 percent of costs for direct labor and supervision or 2 percent of annual operating costs)
   *g.* Power and utilities (20 percent of annual operating costs)
3. General plant overhead (about 10 percent of annual operating costs). Wages, salaries, and suppliers for:
   *a.* Hospital and medical services
   *b.* Engineering and drafting
   *c.* General maintenance
   *d.* Safety services
   *e.* Restaurant and recreation facilities
   *f.* Salvage and waste-disposal services
   *g.* Control laboratories
4. General administration and office overhead (about 2 percent of annual operating costs)
   *a.* Salaries and wages
   *b.* Office expenses
   *c.* Upkeep on office buildings
   *d.* Communications
5. Distribution costs (about 4 percent of annual operating costs)
   *a.* Salaries, wages, and suppliers for sales offices
   *b.* Salaries, commissions, and traveling expenses for sales staff
   *c.* Shipping and packaging
   *d.* Advertising
6. Contingencies (about 4 percent of annual operating costs)
7. Gross-earnings expenses (gross earnings = total annual income–annual operating costs.) Gross-earnings expenses depend on amount of gross earnings for entire company and current income-tax regulations; a general range is 30 to 50 percent of gross earnings.

interests held by a particular company. Because these costs depend on the companywide picture, they are often not included in predesign or preliminary cost estimation figures for a single plant, and the probable returns are reported on the basis of the gross earnings before income taxes obtainable with the given design.

Table 14-2 presents a breakdown of annual operating costs for chemical manufacturing plants. It also shows approximate magnitude of the various costs.

## INVESTMENTS

The capital expenditure required for a chemical plant may be divided into two general classifications: fixed-capital investment and working-capital investment.

**Fixed-capital investment.** The amount of money needed to construct and equip the entire plant is the *fixed-capital investment.* It includes costs for equipment, buildings, land, shops, laboratories, warehouses, and facilities for power, waste disposal, and shipping.

**Working-capital investment.** The total amount of money invested in (1) raw materials and supplies carried in stock, (2) finished products in stock and semifinished products in the process of being manufactured, (3) accounts receivable, and (4) cash which must be kept on hand for monthly payment of operating expenses, such as salaries, wages, raw material purchases, and miscellaneous items, is the *working-capital investment.* A basis of 1 month of operation is often used for estimating this investment.

**Total investment.** The sum of the fixed- and working-capital investments is the *total investment.* While the ratio of fixed- to working-capital investments varies for different types of plants, in general the fixed-capital investment for chemical manufacturing plants makes up 80 to 90 percent of the total investment.

## FIXED CHARGES

The five most important fixed costs (interest, rent, insurance, local taxes, and depreciation) are often grouped together and called *fixed charges.* The most convenient way to express fixed charges is on a yearly basis as a percent of the fixed-capital investment. For example, consider a piece of equipment that costs \$20,000 completely installed with all auxiliaries. If the fixed charges on this equipment are 15 percent, the sum of interest, rent, local taxes, insurance, and depreciation for the equipment would be $(0.15)(20{,}000) = \$3000$ per year.

## DEPRECIATION

Costs due to interest, rent, local taxes, and insurance are self-explanatory and need no special treatment or discussion. Depreciation costs may become rather complex. and it is necessary to understand the various ways in which depreciation may be determined. Loss of service value is measured as *depreciation*. Depreciation may be due to physical decay caused by wear and tear on the equipment, or it may be due to chemical decay caused by prolonged exposure to corrosive materials.

**Obsolescence.** One special type of depreciation, resulting in a loss of service life due to technological or economic advances, is *obsolescence*. It is not a loss in the physical sense.

**Service life.** The period during which a piece of equipment is capable of economically justifiable operation is its *service life*.

**Depletion.** Capacity loss due to materials actually consumed is measured as *depletion* (also called *production-unit depreciation*). Depletion cost equals the initial cost times the ratio of amount of material used to original amount of material purchased. This type of depreciation is particularly applicable to natural resources such as stands of timber or mineral and oil deposits.

The physical significance of depreciation cost can be best illustrated by an example. Assume that a certain evaporator has an initial value of $10,000 when completely installed and ready to operate. The service life of this piece of equipment is estimated to be 10 years. At the end of 10 years the equipment will be worth $2000 as scrap material. Therefore, the total depreciation during the 10-year period will be $8000.

### Methods for Calculating Depreciation

**The straight-line method.** The straight-line method assumes that depreciation is constant throughout the service-life period. A certain amount of money is laid away each year, and, at the end of the service life, the total amount of money must equal the total depreciation cost.

The straight-line method may be expressed in equation form as

$$d = \frac{V - S}{n} \tag{14-1}$$

where $d$ = annual depreciation, dollars per year

$V$ = fixed-capital investment for equipment, including initial cost and cost of installation and auxiliaries, dollars

$S$ = scrap value of equipment when service life has expired, dollars

$n$ = service life of equipment, years

The values of $S$ and $n$ must be estimated. Care must be taken to include the possibility of obsolescence when estimating the value of the service-life period.

The actual value (or asset value) of equipment at any time during the service-life period may be determined by use of the following straight-line-method equation:

$$\text{Asset value} = V - ad \tag{14-2}$$

where $a$ = the number of years in actual use.

**The reducing-balance method.**[1] The reducing-balance method (or fixed-percentage) method assumes that the value of the equipment falls off more rapidly during the early years of the service life than it does during the latter years. The reducing-balance factor of depreciation is designated by $P$.

At the end of the first year,

$$\text{Asset value} = V(1 - P) \tag{14-3}$$

At the end of the second year,

$$\text{Asset value} = V(1 - P)^2 \tag{14-4}$$

At the end of $n$ years,

$$\text{Asset value} = V(1 - P)^n = S \tag{14-5}$$

Therefore

$$P = 1 - \left(\frac{S}{V}\right)^{1/n} \tag{14-6}$$

and

$$\text{Asset value after } a \text{ years} = V(1 - P)^a \tag{14-7}$$

**The sinking-fund method.** The sinking-fund method assumes that a certain amount of money is set aside each year at a given interest rate. At the end of the service life, the accrued money and interest must equal the depreciation. Using this method,

$$\text{Total depreciation up to } a \text{ years} = D = (V - S)S_n f_a \tag{14-8}$$

where $S_n$ = amount of money which must be put on deposit at start of each year at an interest rate of $i$ to have \$1 on deposit after years, or

$$S_n(1 + i)^1 + S_n(1 + i)^2 + S_n(1 + i)^3 + \cdots + S_n(1 + i)^n = \$1 \tag{14-9}$$

[1] Because the reducing-balance factor given by Eq. (14-6) is so sensitive to the value of $S$, an alternate depreciation method which ignores salvage value can be used by applying the reducing-balance method, with $P$ equal to 2 times the reciprocal of the service life $n$. This is known as the *double-declining-balance method*.

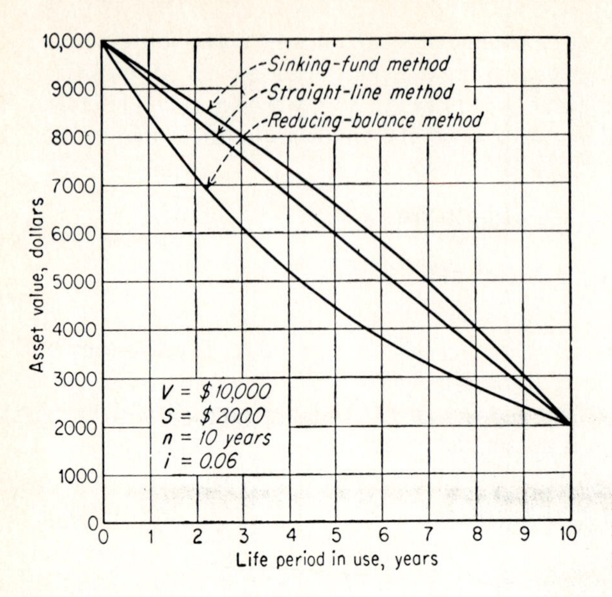


**Figure 14-1** Comparison of different methods for calculating depreciation.

$f_a$ = accrued amount if \$1 is deposited at start of each year for $a$ years at an interest rate of $i$, or

$$f_a = (1 + i)^1 + (1 + i)^2 + (1 + i)^3 + \cdots + (1 + i)^a \tag{14-10}$$

where $i$ = the annual interest rate expressed as a fraction.

With the sinking-fund method, the asset value may be expressed as

$$\text{Asset value} = V - D = V - (V - S)\frac{(1 + i)^a - 1}{(1 + i)^n - 1} \tag{14-11}$$

Figure 14-1 presents a graph showing the differences among the three methods for calculating depreciation. The reducing-balance method causes the investment to be paid off most rapidly, while the sinking-fund method permits the slowest payoff of the investment.

The choice of the particular method to use depends upon the policies of the company involved and the type of process used. In general, the straight-line method is preferable because of its simplicity. Therefore, the straight-line method should always be employed for calculating depreciation unless there is a specific reason for using one of the other methods.

**Example 14-1: Calculation of depreciation by the various methods** A piece of equipment requires a fixed capital investment of \$10,000. The service life is estimated at 10 years, and the scrap value is estimated as \$2000. Money may be invested at 6 percent interest. Determine the asset value of the equipment after 4 years by (*a*) the straight-line method, (*b*) the reducing-balance method, and (*c*) the sinking-fund method.

SOLUTION

(*a*) Straight-line method:

$$V = \$10{,}000$$

$$S = \$2000$$

$$n = 10 \text{ years}$$

$$d = \frac{V - S}{n} = \frac{8000}{10} = \$800/\text{year}$$

Asset value after 4 years is

$$V - ad = 10{,}000 - (4)(800) = \$6800$$

(*b*) Reducing-balance method:

$$P = 1 - \left(\frac{S}{V}\right)^{1/n} = 1 - \left(\frac{2000}{10{,}000}\right)^{1/10} = 0.1486$$

Asset value after 4 years is

$$V(1 - P)^a = 10{,}000(1 - 0.1486)^4 = \$5240$$

With the double-declining-balance method and ignoring salvage value,

$$P = \frac{2}{n} = \frac{2}{10} = 0.2$$

Asset value after 4 years is

$$V(1 - P)^a = 10{,}000(1 - 0.2)^4 = \$4096$$

(*c*) Sinking-fund method:

$$i = 0.06$$

$$S_n[(1.06) + (1.06)^2 + (1.06)^3 + \cdots + (1.06)^{10}] = 1.00$$

$$S_n = 0.0717$$

At 4 years,

$$f_a = (1.06) + (1.06)^2 + (1.06)^3 + (1.06)^4$$

$$f_a = 4.63$$

$$D \text{ (for 4 years)} = (10{,}000 - 2000)(0.0717)(4.63) = \$2660$$

$$\text{Asset value after 4 years} = 10{,}000 - 2660 = \$7340$$

$$= 10{,}000 - (10{,}000 - 2000)\frac{(1.06)^4 - 1}{(1.06)^{10} - 1}$$

## RETURN ON INVESTMENTS

An industrial company is not interested in making a major expenditure unless there is a chance for realizing a profit on the investment. In most cases, this return can be expressed on a tangible basis of dollars and cents. Therefore, when a proposed plant design is being considered, one of the major items of interest is the possible profit which may be obtained from the investment.

Return on investments is expressed on a yearly basis as a percentage of the total investment. Thus, if a manufacturing plant requires a total capital investment of $1 million and realizes a profit of $200,000 per year, the return on the investment is $(200{,}000/1{,}000{,}000)(100) = 20\%$.

Many companies have set policies on the percent return they require before they will consider a capital expenditure. Consider the example of a company requiring at least a 15 percent return on all investments. Two proposed plant designs have been submitted for consideration, and only one of the designs can be accepted. The first design requires an expenditure of $1 million and gives a profit of $200,000 per year. The investment for the second design is $2 million and the yearly profit is $300,000. The return on the first design is 20 percent, while that for the second design is 15 percent.

At first glance, it appears that the second design should be accepted, since it gives the greater profit and meets the required conditions. However, by comparing the two designs, it is seen that an added investment of $1 million yields a yearly return of only $100,000, or a 10 percent profit on the investment. The company policies require a 15 percent profit on any investment. The design engineer should assume this means the company can invest money in outside interests or in other projects to yield a 15 percent return. In this example, the company will make a greater profit by accepting the first design involving an investment of $1 million and putting the second $1 million in an outside interest at a 15 percent return. The process illustrated in this example is known as *the method of investment comparison*, and it should be thoroughly understood by all design engineers.

A common error made in reporting returns for manufacturing processes is to base the percent return on the fixed-capital investment rather than on the total investment. When considering an overall manufacturing process, the percent return should always be based on the total investment. When making investment comparisons for single pieces of equipment or small parts of plants, it is usually permissible to simplify the work by comparing the returns based on the fixed-capital investment.

**Example 14-2: Determination of best investment for a standard-return policy** During a 24-h day, 900,000 lb of a water–caustic soda liquor containing 5 percent by weight caustic soda must be concentrated to 40 percent by weight. A single-effect evaporator of this capacity has an initial completely installed value of $18,000. The life period is estimated as 10 years, and the scrap value of the single-effect evaporator is $6000. Fixed charges minus depreciation amount to 20 percent yearly, based on the initial completely installed value. Steam costs \$0.30 per 1000 lb, and administration, labor, and miscellaneous costs are \$40 per day, no matter how many evaporator effects are used. Where $X$ is the number of evaporator effects, $0.9X$ equals the number of pounds of water evaporated per pound of steam.

There are 300 operating days per year. The preceding data on a single-effect evaporator apply to each effect. If a company regards 20 percent as a worthwhile return, how many effects should be purchased?

## SOLUTION

**Basis**

1 operating day

$X$ = total number of evaporator effects

Depreciation per operating day (straight-line method) is

$$\frac{X(18{,}000 - 6000)}{(10)(300)} = \$4.00X \text{ per day}$$

$$\text{Fixed charges} - \text{depreciation} = \frac{X(18{,}000)(0.20)}{300}$$

$$= \$12X \text{ per day}$$

Pounds of water evaporated per day is

$$(900{,}000)(0.05)\left(\tfrac{95}{5}\right) - (900{,}000)(0.05)\left(\tfrac{60}{40}\right) = 787{,}500 \text{ lb/d}$$

$$\text{Steam costs} = \frac{(787{,}500)(0.30)}{X(0.9)(1000)} = \frac{\$262.5}{X} \text{ per day}$$

| $X$ = No. of effects | Steam costs per day | Fixed charge minus depreciation, per day | Depreciation per day | Labor, etc., per day | Total cost per day |
|---|---|---|---|---|---|
| 1 | \$262.5 | \$12 | \$ 4 | \$40 | \$318.5 |
| 2 | 131.3 | 24 | 8 | 40 | 203.3 |
| 3 | 87.5 | 36 | 12 | 40 | 175.5 |
| 4 | 65.6 | 48 | 16 | 40 | 169.6 |
| 5 | 52.5 | 60 | 20 | 40 | 172.5 |

Comparing 2 effects to 1 effect:

$$\text{Percent return} = \frac{(318.5 - 203.3)(300)(100)}{36{,}000 - 18{,}000} = 192\%$$

Therefore, 2 effects are better than 1 effect.

Comparing 3 effects to 2 effects:

$$\text{Percent return} = \frac{(203.3 - 175.5)(300)(100)}{54{,}000 - 36{,}000} = 46.3\%$$

Therefore, 3 effects are better than 2 effects.

Comparing 4 effects to 3 effects:

$$\text{Percent return} = \frac{(175.5 - 169.6)(300)(100)}{72{,}000 - 54{,}000} = 9.8\%$$

Since a return of 20 percent is required on any investment, 3 effects are better than 4 effects.

Comparing 5 effects to 3 effects:

$$\text{Percent return} = \frac{(175.5 - 172.5)(300)(100)}{90{,}000 - 54{,}000} = 2.5\%$$

Therefore, 3 effects are better than 5 effects.

Since the total daily costs increase as the number of effects increase above 5, no further comparisons need to be made.

The company should purchase an evaporator having 3 effects.

## COST ESTIMATION

The problem of cost estimation for manufacturing processes or chemical equipment is frequently encountered by chemical engineers. The accuracy of a cost estimate depends upon the extent of detailed information available on the equipment or process. Estimates can be divided into three general classifications, depending on the degree of accuracy required. The three classifications are order-of-magnitude estimates, preliminary estimates, and firm estimates.

**Order-of-magnitude estimates.** A general outline of the process or equipment is the basis for order-of-magnitude estimates. No detailed information is available, and a rough estimate is required. Little time is spent in obtaining estimates of this type, and they usually have an accuracy of ± 35 percent.

**Preliminary estimates.** The initial form of the design is the basis for the preliminary estimate. While the design may be fairly complete, not all the details are included, and changes may be required before the final form is available. Estimates of this type should have an accuracy of ± 20 percent.

**Firm estimates.** The detailed design is the basis for a firm estimate. Final drawings, and only minor changes should be necessary on these quotations. These estimates should be within 10 percent of the correct value.

It is difficult to obtain accurate cost information, since labor and material prices are changing constantly. However, by using practical judgment and contacting producers with experience in the field, it is possible to make remarkably accurate estimates. Cost indexes, special power and multiplication factors, and published information on process equipment costs are very valuable aids in making cost estimates.

## Cost Indexes for Time Changes

Most cost data which are available for immediate use in a preliminary or predesign estimate are based on conditions at some time in the past. Because prices may change considerably with time because of changes in economic conditions, some method must be used for updating cost data applicable at a past date to costs that are representative of conditions at a more recent time. This can be done by use of cost indexes.

A cost index is merely an index value for a given point in time showing the cost at that time relative to a certain base time. If the cost at some time in the past is known, the equivalent cost at the present time can be determined by multiplying the original cost by the ratio of the present index value to the index value applicable when the original cost was obtained.

$$\text{Present cost} = \text{original cost}\left(\frac{\text{index value at present time}}{\text{index value at time original cost was obtained}}\right) \tag{14-12}$$

Many different types of cost indexes are published regularly. Some of these can be used for estimating equipment costs; others apply specifically to labor, construction, materials, or other specialized fields. Two cost indexes that are useful for chemical engineering process plants and equipment are the *Marshall and Swift All-Industry Equipment Index*, which is based on the year 1926 as 100, and the *Chemical Engineering Plant Cost Index*, which is based on the years 1957 to 1959 as 100. Both of these indexes are published regularly in *Chemical Engineering*, and Table 14-3 gives a listing of annual average values of the indexes for 1964 to 1982. Table 14-3 also gives labor and material indexes for construction as adapted from the *Monthly Labor Review* for the same time period.

## Power-Factor Rule for Capacity Changes

It is often necessary to estimate the cost of a piece of equipment when no cost data are available for the particular size of operational capacity involved. Good results can be obtained by using the straight-line logarithmic relationship of cost versus capacity, known as the *six-tenths-factor rule*, if the new piece of equipment is similar to one of another capacity for which cost data are available. According to this rule, if the cost of a given unit at one capacity is known, the cost of a similar

**Table 14-3 Cost indexes as annual averages**

| Year | Marshall and Swift all-industry equipment index, 1926 = 100 | Chemical engineering plant cost index, 1957–1959 = 100 | Hourly earnings labor index for construction workers, 1967 = 100 | Construction materials producer price index, 1967 = 100 |
|---|---|---|---|---|
| 1964 | 242 | 103 | 86 | 95 |
| 1965 | 245 | 104 | 90 | 96 |
| 1966 | 253 | 107 | 95 | 99 |
| 1967 | 263 | 110 | 100 | 100 |
| 1968 | 273 | 114 | 107 | 106 |
| 1969 | 285 | 119 | 116 | 112 |
| 1970 | 303 | 126 | 128 | 113 |
| 1971 | 321 | 132 | 139 | 120 |
| 1972 | 332 | 137 | 147 | 127 |
| 1973 | 344 | 144 | 155 | 139 |
| 1974 | 398 | 165 | 164 | 161 |
| 1975 | 444 | 182 | 176 | 174 |
| 1976 | 472 | 192 | 187 | 188 |
| 1977 | 505 | 204 | 197 | 205 |
| 1978 | 545 | 219 | 210 | 228 |
| 1979 | 599 | 239 | 221 | 247 |
| 1980 | 660 | 261 | 237 | 268 |
| 1981 | 721 | 297 | 259 | 288 |
| 1982 | 746 | 314 | 277 | 294 |

unit with $X$ times the capacity of the first is approximately $(X)^{0.6}$ times the cost of the first unit.

$$\text{Cost of equip. } a = \text{cost of equip. } b\left(\frac{\text{capac. equip. } a}{\text{capac. equip. } b}\right)^{0.6} \tag{14-13}$$

## Multiplication Factors for Total Investment

Multiplication factors based on delivered equipment cost are occasionally used for rapid estimation of total plant investments including installation as well as working capital if the estimate is for a major portion of a plant. These factors have been presented in the literature for various types of processes. For the ordinary chemical industries, the total plant investment is approximately four times the delivered equipment cost.

**Example 14-3: Estimation of cost of equipment using cost indexes and power and multiplication factors** The purchased cost of a tube-and-shell one-pass heat exchanger with 18.6 m$^2$ of heat-transfer area was \$3500 in 1972. Estimate the 1980 purchased and installed cost of a similar heat exchanger with 37.2 m$^2$ of heat-transfer area. Use the *Marshall and Swift All-Industry Equipment index* for updating the cost.

SOLUTION According to Table 14-3, the *Marshall and Swift All-Industry Equipment Index* was 332 in 1972 and 660 in 1980. Therefore, by Eq. (14-12), the factor to be applied to the \$3500 to change from 1972 to 1980 is $\frac{660}{332}$.

By the six-tenths-factor rule of Eq. (14-13), the factor to correct for going from 18.6 $m^2$ to 37.2 $m^2$ capacity is $(37.2/18.6)^{0.6}$.

The multiplication factor to take care of installing the equipment is approximately 4.

Applying all three of these factors, the final installed cost of the 37.2-$m^2$-area heat exchanger in 1980 should be

$$\$3500\left(\frac{660}{332}\right)\left(\frac{37.2}{18.6}\right)^{0.6}(4) = \$42{,}200$$

## THE DESIGN PROJECT

The development of a design project always starts with an initial idea or plan. This initial idea should be stated as clearly and concisely as possible. General specifications and pertinent laboratory or chemical engineering data should be presented along with the initial idea.

Before any detailed work is done on the design, the technical and economic factors of the proposed process should be examined. The various reactions and physical processes involved must be considered, and laboratory work should be carried out if necessary information is not available from other sources. The preliminary technical and economic survey gives an immediate indication of the feasibility of the project. Following is a list of items which should be considered in making the preliminary survey:

1. Raw materials (availability, quantity, quality, cost)
2. Thermodynamics of chemical reactions involved (equilibrium, yields, rates, optimum conditions)
3. Facilities and equipment available at present
4. Facilities and equipment which must be purchased
5. Estimation of production costs and total investment
6. Profits (probable and optimum, per pound of product and per year, return on investment)
7. Materials of construction
8. Markets (present and future supply and demand, possible new uses, present buying habits)
9. Competition (overall production statistics, comparisons of various manufacturing processes, product specifications of competitors)
10. Sales and sales service
11. Effect of storage on products
12. Shipping restrictions and containers

13. Safety factors
14. Patent situation and legal restrictions
15. Plant location
16. Consideration of what is necessary for immediate development

## Procedure for the Design Project

The method for carrying out a design project may be divided into the three following classifications, depending on the accuracy and detail required:

1. Preliminary or quick-estimate designs
2. Detailed-estimate designs
3. Firm process designs

*Preliminary designs* determine order-of-magnitude costs and approximate process methods. This type of design is ordinarily used as a basis for determining the advisability of continuing with further work on the proposed process. The time spent on calculations is kept at a minimum.

*Detailed-estimate designs* may require 1 to 4 months for completion. The costs and economics of an established process are determined by detailed calculations. However, exact specifications are not given for the equipment, and drafting-room work is minimized.

*Firm process designs* require accurate calculations and cost estimations based on detailed specifications. The completed form of the design includes blueprints and sufficient information to permit immediate development of the final plans for constructing the plant. Six months to 2 years may be required for the completion of a firm process design.

The outline presented in Table 14-4 may be used as a guide for the general procedure in carrying out a design project.

## Literature Surveys

A survey of the literature will often reveal information pertinent to the development of a design project. *Chemical Abstracts*, published semimonthly by the American Chemical Society, offers the best starting point for literature surveys dealing with chemical manufacturing processes. It presents a brief outline and the original reference of published articles dealing with chemistry and related fields. Yearly and decennial indexes of subjects and authors permit rapid location of articles concerning specific topics.

Figure 14-2 shows the typical form of a card for recording information from the *Abstracts*. The notation in the upper right-hand corner indicates the *Chemical Abstracts* volume, abstract number, and year. If the article appears to be of value, further information can be obtained by examination of the original reference.

**Table 14-4 Design project procedure**

I. Preliminary survey
   A. Examine the given project carefully.
      1. Record and study all available related data and information.
      2. Understand clearly the object of the project.
   B. Make rough flow diagrams.
   C. Decide on possible types of solutions.
   D. Outline briefly the best solutions.

II. Literature search
   A. Examine the literature for additional information as to
      1. Kinetics of reactions (equilibrium, rates, heat effects)
      2. New methods of attack
      3. Physical and chemical data
      4. Special information

III. Final survey
   A. Examine all outlined information.
   B. Decide on best method of attack.
   C. Outline final method of solution.
   D. Make final rough flow diagram.

IV. Final design
   A. Make necessary calculations.
      1. Find optimum conditions.
         *a.* Consider economics and special required items.
         *b.* Consider temperature, reaction rates, equilibrium, operating difficulties, equipment required and available, and costs.
      2. Make final calculations at optimum conditions.
         *a.* Consider standard and practical sizes, complete economics, skill and demand of operators, methods for saving heat and power, and other things dependent on particular process.
   B. Make up flow diagrams.
      1. Qualitative
      2. Quantitative
      3. Drawing of individual pieces of equipment
   C. Make up tables.
      1. Costs
      2. Equipment and specifications
      3. Balances
         *a.* Material
         *b.* Energy
   D. Determine profits, return on investment, and final recommendations.
   E. Make final report integrating all the preceding information.

## Comparison of Different Processes

The development of a design project requires determination of the most suitable process for obtaining a desired product. Since several different manufacturing methods may be available for making the same material, it is necessary to compare the various processes and pick out the one best suited to the existing conditions.

C.A.: **95**, 205952 (1981)

Subject: Mixing of Fluids

Title: A Framework for Description of Mechanical Mixing of Fluids

Authors: Ottino, J. M., W. E. Ranz, and C. W. Macosko

Source: AIChE Journal, **27**, No. 4, pp 565-77 (1981)

Abstract: The mechanical mixing of fluids is described by continuum mechanics arguments. The approach provides a unified mathematical description of the mechanics of mixing.

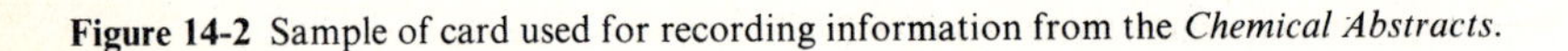

**Figure 14-2** Sample of card used for recording information from the *Chemical Abstracts*.

The following items should be considered when comparing different processes:

1. Technical factors
   - *a.* Process flexibility
   - *b.* Continuous operation
   - *c.* Special controls involved
   - *d.* Commercial yields
   - *e.* Technical difficulties involved
   - *f.* Power consumption
   - *g.* Special auxiliaries required
   - *h.* Health and safety hazards involved
2. Raw materials
   - *a.* Present and future availability
   - *b.* Processing required
   - *c.* Storage requirements
   - *d.* Handling of materials
3. Waste products and by-products
   - *a.* Amount produced
   - *b.* Value
   - *c.* Potential markets and uses
   - *d.* Manner of discard
4. Equipment
   - *a.* Availability
   - *b.* Materials of construction
   - *c.* Initial costs
   - *d.* Maintenance and installation costs

   *e.* Replacement requirements
   *f.* Special designs required
5. Plant location
   *a.* Amount of land required
   *b.* Transportation facilities
   *c.* Proximity to markets and raw-material sources
   *d.* Availability of service and power facilities
   *e.* Availability of labor
   *f.* Legal restrictions and taxes
6. Costs
   *a.* Raw materials
   *b.* Depreciation
   *c.* Other fixed charges
   *d.* Processing
   *e.* Labor
   *f.* Real estate

## Patents

A patent gives the holder exclusive rights to prevent others from using or practicing the invention for a period of 17 years following the date of granting. A new design should be examined to make certain that no patent infringements are involved. If even one legally expired patent can be found covering the details of the proposed process, the method can be used with no fear of patent difficulties.

A patent may be obtained on any new and useful process, machine, method of manufacture, or composition of matter, provided it has not been known or used for more than a year prior to the patent application. A patentable item must be new and different, and it cannot be something involving merely mechanical skill. A patent will not be granted for a change on an older item unless the change involves something entirely new.

## Plant Location

The geographical location of the final plant must be taken into consideration when a process design is developed. If the plant is located in a cold climate, costs may be increased by the necessity for constructing protective shelters around the process equipment. Cooling-water costs may be very high if the plant is not located near an ample water supply. Electrolytic processes require a cheap source of electricity, and plants of this type ordinarily cannot operate economically unless they are located near large hydroelectric installations.

The effects of the following factors on production costs are of importance in considering plant location:

1. Source of raw materials
2. Markets for finished products

3. Transportation facilities
4. Labor supply
5. Power and fuel
6. Water supply and waste disposal
7. Utilization of by-products
8. Taxation and legal restrictions
9. Water table
10. Flood and fire protection
11. Room for expansion
12. Building and land costs
13. Climatic conditions

## Plant Layout

The physical layout of a plant should be designed to permit coordination between the operation of the process equipment and the use of storage and materials-handling equipment. Scale drawings complete with elevation indications are useful for determining the best locations for the equipment and facilities.

The following items should be taken into consideration in setting up the plant layout:

1. Storage facilities readily available for use
2. Possible changes in the future
3. Process operations arranged for operators' convenience
4. Space available
5. Time interval between successive operations
6. Automatic or semiautomatic controls
7. Safety considerations
8. Availability of utilities and services
9. Waste disposal
10. Necessary control tests
11. Auxiliary equipment
12. Gravity flow

## Flow Diagrams

The visualization of an overall design for a manufacturing process is simplified by the use of flow diagrams. As discussed in Chap. 2 (Technical Introduction), these diagrams may be divided into three general types: qualitative, quantitative, and combined detail.

A qualitative flow diagram indicates the flow of materials, unit operations involved, equipment necessary, and special information as to temperatures and pressures. A quantitative flow diagram shows the quantities of materials required for the process operation. Figure 2-1 shows a qualitative flow diagram for the production of nitric acid, while Fig. 2-2 presents a quantitative flow diagram for the same process. A typical combined-detail flow diagram showing how the information from qualitative and quantitative flow diagrams can be combined into one diagram is illustrated in Fig. 2-3.

In the development of a design project, it is frequently necessary to prepare two or three preliminary diagrams before the final qualitative, quantitative, and combined-detail diagrams can be drawn. By the use of tables cross-referenced to the drawings, a large amount of pertinent data on the process can be presented in a condensed form. In this manner, the drawings do not lose their effectiveness by the inclusion of too much information, and the necessary data are readily available by direct reference to the tables. This method is often used for indicating equipment specifications such as the necessary capacity, materials of construction, number required, and availability.

**Table 14-5 Organization of design reports**

1. Letter of transmittal
   *a.* Indicates why report has been prepared
   *b.* gives essential results that have been *specifically* requested
   *c.* Transmits the report
2. Title page
   *a.* Includes title of report, date, writer's name, and organization
3. Table of contents
   *a.* Includes titles of figures and tables and all major sections
4. Summary
   *a.* Presents essential results, recommendations, and conclusions in a clear and precise manner
5. Introduction
   *a.* Presents a brief discussion to explain what the report is about and the reason for the report No results are included
6. Discussion
   *a.* Outlines method of attack on project
   *b.* Discusses technical matters of importance
   *c.* Indicates assumptions made and the reasons
   *d.* May include literature-survey results of importance
   *e.* Indicates possible sources of error
   *f.* Gives a general discussion of results and proposed design
7. Final recommended design with appropriate data
   *a.* Drawings of proposed design
      (1) Qualitative flow sheets
      (2) Quantitative flow sheets
      (3) Combined-detail flow sheets
   *b.* Tables listing equipment and specifications
   *c.* Tables indicating quantities of materials involved
   *d.* Economics including costs, profits, and return on investment
8. Sample calculations
   *a.* Presents and explains clearly one example for each type of calculation
9. Table of nomenclature
   *a.* Indicates sample units
10. Bibliography
11. Tables of data employed with reference to sources

### The Design Report

A design engineer must be able to write up his or her work in a clear and understandable fashion. The report should explain what has been done and how the work has been carried out. The design report usually goes to company executives who have no time and patience for sloppy or unorganized material. All the pertinent information should be included, but care must be taken to avoid inclusion of trivial details. This is particularly important in the report summary. The summary includes the important results, recommendations, and conclusions and, in a few concise pages, should give the reader the essential contents of the report.

Table 14-5 presents a proposed outline for design reports. This plan of organization can be changed to fit the requirements of individual designs. The use of subheadings throughout the report adds to the effectiveness of the presentation.

## DESIGN CALCULATIONS

Design calculations involve the application of theoretical and empirical relationships. For example, the design of a heat exchanger requires use of theoretical principles concerning the total heat-transfer resistance offered by the combined individual resistances. Empirical relationships are used to determine the values of the individual film coefficients. Thus, the final design of the heat exchanger is based on both theoretical and empirical principles.

Calculations are based on the maximum capacity or requirement for the proposed process or individual piece of equipment. Since the calculations cannot be assumed to have 100 percent accuracy, it is necessary to apply practical safety factors when determining the final design. The calculated values should be increased by 10 to 100 percent to give an adequate safety factor. This size of the safety factor is dictated by the accuracy of the data and methods employed in carrying out the calculations.

The design engineer must show ingenuity in using available information for new process designs. He or she must be willing to make assumptions; however, any assumption should be justified and taken into consideration when the final safety factor is estimated.

### Transfer of Materials

Pump sizes and power requirements must often be determined in the course of a plant design. The amount of power required for a given pumping duty can be calculated from the total mechanical-energy balance by methods illustrated in Chap. 5 (Flow of Fluids).

Pipe size is an important factor in the determination of pumping-power costs. With a constant flow rate, the pressure drop due to friction increases as the diameter of the containing pipe decreases. More pumping power must then be supplied to overcome the larger pressure drop. Considering only pumping power,

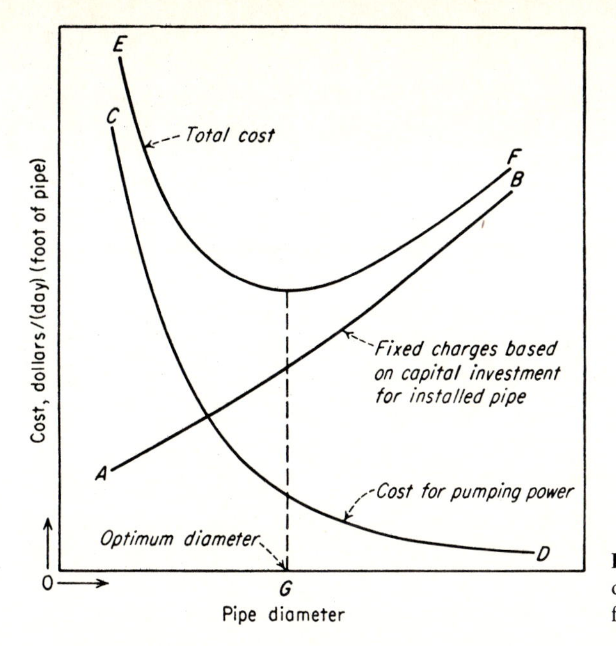


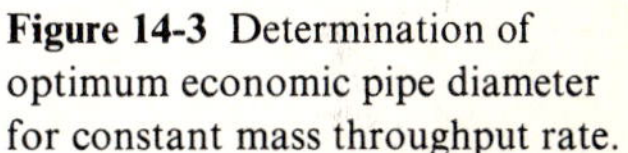
**Figure 14-3** Determination of optimum economic pipe diameter for constant mass throughput rate.

it would appear that a pipe of infinite diameter should be used. However, the investment required for the pipe increases as the size increases. There must be some optimum pipe size where the total of power cost and fixed changes based on investment is a minimum.

Figure 14-3 shows one method for determining optimum pipe diameter. Curve *AB* represents the fixed cost of the installed pipe expressed on a convenient basis of dollars per day per foot of pipe length. Curve *CD* shows the cost of pumping power as dollars per day per foot of pipe length. The total cost is represented by curve *EF*. This curve has a minimum value at a pipe diameter of *G*. Therefore *G* represents the optimum pipe diameter. A practical design engineer does not immediately recommend the use of the optimum size. He or she finds the nearest standard pipe size and then compares to smaller pipe sizes until the return on the added investment is worthwhile.

An approximate value for the optimum diameter of steel pipe may be obtained from the following dimensional equation:[2]

$$D_i = \frac{2.2w^{0.45}}{\rho^{0.32}} \tag{14-14}$$

where $D_i$ = optimum inside pipe diameter, in
$w$ = mass flow, 1000 lb/h
$\rho$ = density of the flowing fluid, lb/ft$^3$

[2] See App. B for derivation of Eq. (14-14).

Equation (14-14) is applicable only when the fluid flow is turbulent. It may be used for estimating the optimum diameter of steel pipe under ordinary plant conditions if the viscosity of the fluid is between 0.02 and 30 cP. It does not apply when the flowing fluid is steam. Diameters determined by the use of this equation are conservative in that they are usually on the high side.

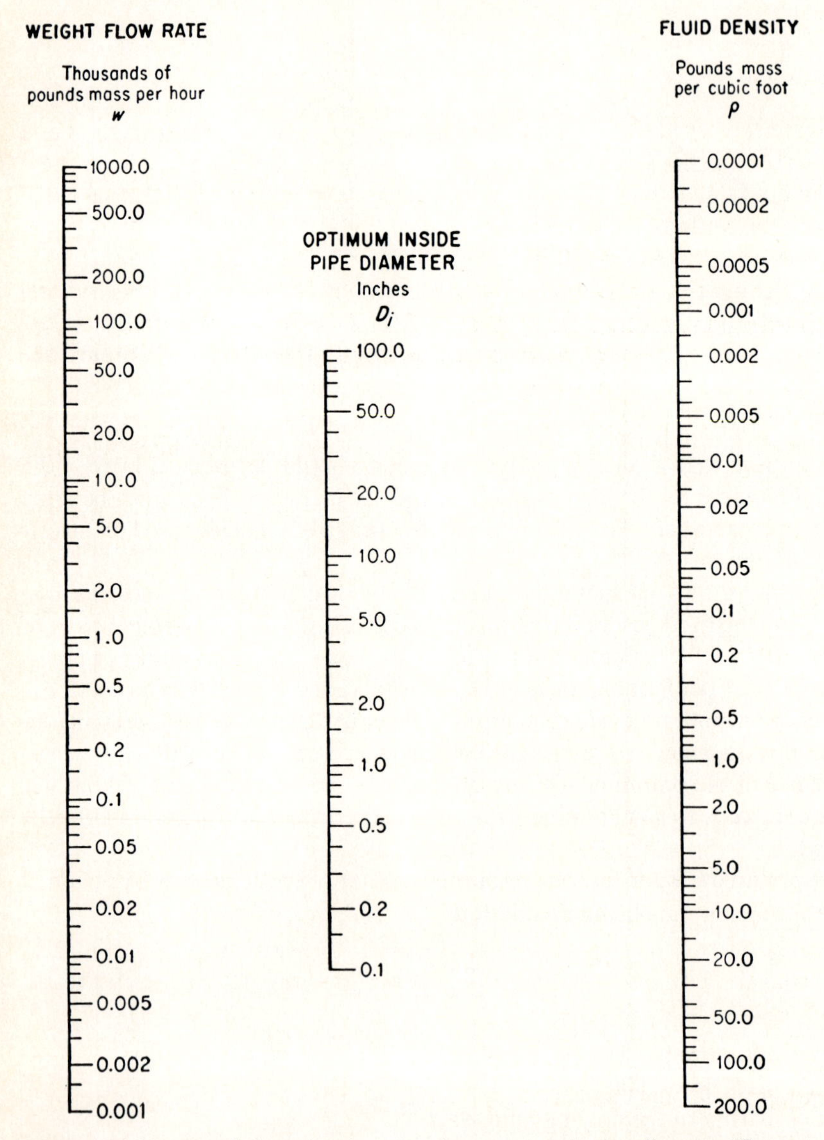


**Figure 14-4** Nomograph for estimation of optimum economic pipe diameter. (Connect values of $w$ and $\rho$ by a straight line to obtain optimum pipe diameter.) (Kg/h $\times$ 0.000454 = 1000 lbm/h; kg/m$^3$ $\times$ 0.0624 = lb/ft$^3$; in $\times$ 0.0254 = m.)

Figure 14-4 presents a nomograph for estimating economic (or optimum) pipe diameters. This nomograph is based on Eq. (14-14).

A general rule of thumb for estimating the optimum diameters of pipes containing water or steam may be stated as follows: The velocity of water flowing in a pipe should not exceed 1.2 to 1.5 m/s (4 to 5 ft/s), and the velocity of steam flowing in a pipe at pressures between 240 and 450 kPa (20 and 50 lb/in$^2$, gage) should not exceed 15 to 18 m/s (50 to 60 ft/s).

## Heat Transfer

The design of heat-transfer equipment requires consideration of power costs and fixed-capital-investment costs. Consider the example of the design for a heat exchanger containing a given number of tubes where liquid flows through the exchanger at a constant mass rate. As the tube diameter in the exchanger is decreased, the heat-transfer coefficients usually increase because of the increased velocity of the fluid flowing inside the tubes, resulting in a drop of necessary total area and fixed-capital investment. However, as the tube diameter becomes smaller, the pressure drop and power costs increase. Thus, there is an optimum tube diameter where the sum of power and capital-investment costs is a minimum.

Many heat-transfer equations have been developed which are directly applicable for design calculations. These relationships are generally obtained by setting up equations for the sum of capital-investment costs and operating costs. By differentiating the total cost with respect to the crucial variable (such as diameter, amount of cooling water, or temperature difference) and setting the result equal to zero, a design equation for minimum cost can be obtained.

The following simplified example illustrates a common type of design calculation involving heat transfer.

**Example 14-4: Determination of economic insulation thickness** Insulation is to be purchased for use on a heat exchanger. The insulation can be obtained in thickness of 1, 2, 3, or 4 in. The following data have been determined for the different insulation thicknesses:

| | 1 in | 2 in | 3 in | 4 in |
|---|---|---|---|---|
| Btu/h saved | 300,000 | 350,000 | 370,000 | 380,000 |
| Cost for installed insulation | \$1200 | \$1600 | \$1800 | \$1870 |
| Annual fixed charges as % of installed cost | 10% | 10% | 10% | 10% |

What thickness of insulation should be used? The value of heat is 30 cents per 1 million Btu. Fifteen percent of the fixed capital investment is considered a worthwhile annual return for this type of investment. The exchanger operates 300 days per year.

SOLUTION

**Basis**

1 year

For 1-in insulation:

$$\text{Money saved on heat} = \frac{(300{,}000)(24)(300)(0.30)}{1{,}000{,}000} = \$648$$

$$\text{Fixed charges} = (0.1)(1200) = \$120$$

$$\text{Total saving} = 648 - 120 = \$528$$

For 2-in insulation:

$$\text{Money saved on heat} = \frac{(350{,}000)(24)(300)(0.30)}{1{,}000{,}000} = \$756$$

$$\text{Fixed charges} = (0.1)(1600) = \$160$$

$$\text{Total saving} = 756 - 160 = \$596$$

In a similar manner:

For 3-in insulation, total saving = \$620

For 4-in insulation, total saving = \$634

Comparing 1-in insulation to 2-in insulation:

$$\text{Return on investment} = \frac{596 - 528}{1600 - 1200}(100) = 17\%$$

Since at least a 15 percent return is required, the 2-in insulation is preferable to the 1-in insulation.

Comparing 2-in insulation to 3-in insulation:

$$\text{Return on investment} = \tfrac{24}{200}(100) = 12\%$$

The 2-in insulation is preferable to the 3-in insulation.

Comparing 2-in insulation to 4-in insulation:

$$\text{Return on investment} = \tfrac{38}{270}(100) = 14\%$$

The 2-in insulation is preferable to the 4-in insulation.

Therefore, the 2-in insulation thickness should be recommended.

## Mass Transfer

Absorption and distillation columns must be designed to operate below flooding conditions. When the maximum allowable vapor velocity is exceeded in packed or plate towers, liquid cannot descend through the unit in a normal manner, and a condition known as *flooding* occurs. Liquid builds up in the top of the tower and may eventually flood the vapor-condensing system. There may also be an excessive entrainment or carry-over of liquid droplets in the rising vapors.

Towers for use in absorption and distillation operations are ordinarily designed to handle a given throughput, expressed as mass or volume of product per day. The amount of separation and reflux ratio required determine the total number of trays or packed height necessary. The maximum allowable vapor velocity under the given operating conditions determines the diameter of the tower. In general, design practice calls for towers to be operated at 75 to 85 percent of the maximum allowable velocity.

**Plate columns.** The maximum allowable vapor velocity for bubble-cap plate columns may be predicted from the following empirical equation:

$$u = K_v \sqrt{\frac{\rho_L - \rho_G}{\rho_G}} \tag{14-15}$$

where $u$ = superficial vapor velocity in column based on total inside cross-sectional area of column, m/s or ft/s

$\rho_L$ = average density of downflow liquid in column, $kg/m^3$ or $lb/ft^3$

$\rho_G$ = average density of rising vapors in column, $kg/m^3$ or $lb/ft^3$

$K_v$ = a constant dependent on tower design, m/s or ft/s

The value of $K_v$ may be obtained by using data from other similar plate columns having the same liquid-seal depth and plate spacing as the proposed column. If these data are not available, Fig. 14-5, giving approximations of the value for $K_v$ in feet per second at various plate spacings, can be used.

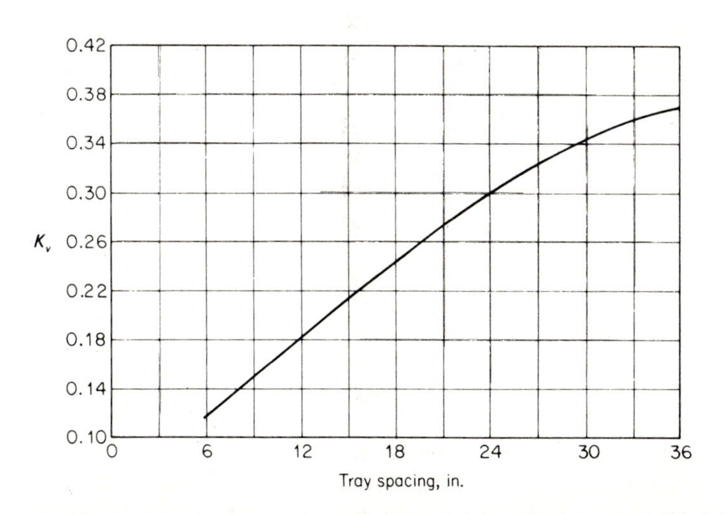


**Figure 14-5** Rough estimate values (±25%) for $K_v$ in units of feet per second for Eq. (14-15) for maximum allowable velocities in plate towers (weir height less than 15% of plate spacing).

The maximum allowable vapor velocity in plate columns is usually about 0.9 m/s or 3 ft/s when operation is at atmospheric pressure. At reduced pressures, the maximum velocity may be estimated by the following equation:

$$u_R = u_A \sqrt{\frac{(\rho_{G_A})(\rho_{L_R})}{(\rho_{G_R})(\rho_{L_A})}} \tag{14-16}$$

where subscript $R$ = at reduced pressure
subscript $A$ = at atmospheric (or any) pressure

**Packed columns.** Equations (14-15) and (14-16) can be used for determining maximum allowable vapor velocities in packed columns. If one set of accurate flooding data can be found for a given packing, the value of $K_v$ can be determined for that packing and applied to other pressures and separating mixtures. If no experimental data are available, various forms of empirical methods can be found in the literature and used for approximating the maximum allowable velocity.

**Pressure drop.** Empirical methods are available in the literature for estimating pressure drops over plate columns or packed columns. In some types of installations, the pressure drop is an important factor because of the power required for forcing material into the system. In most vacuum units, a large pressure drop relative to the operating pressure is not permissible. The pressure drop may be decreased by lowering the vapor velocity in the tower. This may be accomplished in the design by increasing the diameter of the column.

## CHEMICAL KINETICS IN DESIGN

Rates of chemical reactions and equilibrium limitations must be taken into consideration when designing chemical process plants. If a reaction proceeds at a slow rate, it may be necessary to design the reactor with a large volume in order to permit a practical degree of conversion. In some cases, reaction rates can be increased by the use of higher temperatures. The effect of temperature on equilibrium must also be considered, since a change in temperature may tend to reverse the chemical reaction and cause reduced yields.

The design of the conventional contact process for the production of sulfuric acid includes an excellent example of kinetics application. At a temperature of 450°C, the equilibrium relationships between $SO_2$, $O_2$, and $SO_3$ indicate that 97 percent of the $SO_2$ may be oxidized to $SO_3$. However, at this temperature, the reaction rate is slow. By examination of reaction rates and equilibrium conditions at various temperatures, design engineers might recommend the use of two reactors. The first reactor operates at 550°C where the reaction rate is sufficiently high to permit rapid conversion of 80 percent of the $SO_2$ to $SO_3$. The second reactor operates at the optimum equilibrium temperature of 450°C and permits an overall conversion of almost 97 percent.

# PROBLEMS

**14-1** The initial installed cost of a heat exchager is $4000. If the yearly depreciation is 10 percent, what is the asset value of the heat exchanger after it has been in use for 7 years? The straight-line method may be used, and the service-life period is greater than 7 years.

**14-2** The sum of interest, rent, insurance, and local taxes for a certain piece of equipment may be taken as 15 percent per year, based on an initial installed cost of $20,000. The service life is estimated as 8 years, and the scrap value is estimated at $4000. Determine the yearly percent fixed charges.

**14-3** A company must purchase one new evaporator. Four evaporators have been designed, all of which are equally capable of carrying out the required evaporation. The following data apply to the four designs:

| | Design 1 | Design 2 | Design 3 | Design 4 |
|---|---|---|---|---|
| Fixed capital investment | $10,000 | $12,000 | $14,000 | $16,000 |
| Sum of operating and fixed costs per year | 3000 | 2800 | 2350 | 2100 |

If the company demands a 15 percent return on any unnecessary investment, which of the four designs should be accepted?

**14-4** A cost estimate is being prepared as part of a process design. A reactor requiring special construction specifications is included in the design, and it is known that a similar reactor cost $40,000 in 1974. Using the data given in Table 14-3, estimate the cost of the reactor in 1980 by (*a*) the *Marshall and Swift All-Industry Equipment Index*, (*b*) the *Chemical Engineering Plant Cost Index*, and (*c*) construction labor and materials indexes, assuming that 50 percent of the cost is for labor and 50 percent of the cost is for materials.

**14-5** An article dealing with advances in distillation is abstracted in the *Chemical Abstracts*; **95**:205934 (1981). List the title, author, and source for this article.

**14-6** Estimate the optimum economic inside diameter of steel pipe when 10,000 gal of a liquid having a density of 58 $lb/ft^3$ flow through the pipe per hour. The flow is turbulent.

**14-7** Air at 90°F and 1 atm pressure flows through a steel pipe having an inside diameter of 6 in. By the use of Eq. (14-14) or Fig. 14-4, estimate how many cubic feet of the air per minute should flow through the pipe for optimum economic operation.

**14-8** A bubble-cap plate absorption tower is designed for a liquid depth of 2 in on each plate. The plates are spaced 24 in apart. The average density of the liquid descending the tower is 40.5 $lb/ft^3$. The density of the gas rising through the tower is approximately constant at 0.193 $lb/ft^3$. The tower is to operate at 85 percent of the maximum allowable vapor velocity. If the gas rises through the tower at a rate of 180 lb/min, estimate the necessary inside diameter of the tower.

**14-9** When a packed distillation column operates at atmospheric pressure and 85 percent of the maximum allowable velocity, the gas rate for a given mixture is 100 kg/hr. Estimate the boil-up rate for the same mixture and column as kilograms per hour at 85 percent of the maximum velocity if the operating pressure is changed to 100 mmHg. Atmospheric pressure equals 760 mmHg. The average liquid density at the atmospheric-pressure boiling point is 801 $kg/m^3$. The average liquid density at the 100 mmHg boiling point is 833 $kg/m^3$. The average vapor density (by the perfect-gas law) at the atmospheric-pressure boiling point is 3.2 $kg/m^3$. The average vapor density (by the perfect-gas law) at the 100 mmHg boiling point is 0.497 $kg/m^3$.

**14-10** At 80°F, the equilibrium constant for the reaction $2NO_2 \rightleftharpoons N_2O_4$ is $K = p_{N_2O_4}/(p_{NO_2})^2 = 5.77\ atm^{-1}$. Determine the equilibrium density (as pounds per cubic foot) of a gas containing only $NO_2$ and $N_2O_4$ at 80°F and 760 mmHg pressure. The perfect-gas law is applicable under these conditions.

**14-11** An $SO_2$ recovery unit for stack gases is proposed for addition to an existing plant, and four designs for the unit have been completed by your design team. Your company management demands at least a 10 percent return before federal taxes on any investment by $SO_2$-recovery dollar value before they will proceed with the installation. Following are the results of your calculations for the four designs:

| Design number | Total installed cost for recovery unit, $ | Added annual operating costs due to unit, $ | Value of $SO_2$ saved per year, $ | Fixed costs (depreciation, rent, local taxes, insurance, interest on borrowed money), % of total installed cost per year |
|---|---|---|---|---|
| 1 | 100,000 | 1000 | 41,000 | 20 |
| 2 | 160,000 | 1000 | 56,000 | 20 |
| 3 | 200,000 | 1000 | 73,000 | 20 |
| 4 | 260,000 | 1000 | 90,000 | 20 |

Which, if any, of the four designs would you recommend to your management? Neglect effects of working capital.

**14-12** In the design of a plant, we are trying to save money by recovery of heat. We have made four heat-exchanger designs with various special features, and we have completed all price, costs, and savings calculations for each of the four designs. These results are as follows:

| | Design 1 | Design 2 | Design 3 | Design 4 |
|---|---|---|---|---|
| Total investment (total installed cost with auxiliaries) | $10,000 | $16,000 | $20,000 | $26,000 |
| All operating costs, constant, $/yr | 100 | 100 | 100 | 100 |
| Value of heat saved, $/yr | 4100 | 6000 | 6900 | 8850 |
| Fixed costs per year, % of total investment | 20% ($2000) | 20% ($3200) | 20% ($4000) | 20% ($5200) |

Only one of the designs can be put to use, and you have unlimited money available for use. Which of the four designs (if any) would you recommend if your company requires a 10 percent return on any unnecessary investment (i.e., can always invest at an annual return of 10 percent of the investment)? Ignore working capital and effect of taxes.

APPENDIX

# A

# THE INTERNATIONAL SYSTEM OF UNITS (SI)[1]

As the International System of Units, or the so-called "SI units," become accepted in the U.S., there will be a long transition period when both the American engineering system and SI will be in use simultaneously. The engineer, in particular, will need to be able to think and work in both systems because of the wide variety of persons involved in engineering considerations. Accordingly, this text has used a mixture of the two systems.

The purpose of this appendix is to provide a description of SI units along with rules for conversion and rules for usage in the written form. Conversion factors are given with one table presenting a full and detailed list with extensive footnote explanations and another table giving conversion factors in a simplified form for units commonly encountered by chemical engineers.

The name SI is derived from Système International d'Unités and has evolved from an original basis of a given length (meter) and mass (kilogram) established by members of the Paris Academy of Science in the late eighteenth century. The original system was known as the metric system, but there are differences in the modern SI system and the old metric system based primarily on new names being added for derived terms.

The current International System of Units (SI) is a metric system of measurement which has been adopted internationally by the General Conference of Weights and Measures and is described in an *International Standard* (ISO 1000)[2]

[1] Adapted from M. S. Peters and K. D. Timmerhaus *Plant Design and Economics for Chemical Engineers*, 3d ed., McGraw-Hill, New York, 1980. (*With permission.*)

[2] International Standard, "SI Units and Recommendations for the Use of Their Multiples and of Certain Other Units," ISO 1000-1973(E), American National Standards Institute, 1430 Broadway, New York City, NY 10018.

## SI base units on which the entire system is founded

| Name, symbol | Definition |
|---|---|
| meter,* m (length) | The meter is the length equal to 1,650,763.73 wavelengths in vacuum of the radiation corresponding to the transition between the levels $2p_{10}$ and $5d_5$ of the krypton-86 atom. |
| kilogram, kg (mass) | The kilogram is a unit of mass (not force). A prototype of the kilogram made of platinum-iridium is kept at the International Bureau of Weights and Measures, Sèvres, France. (The kilogram is the only base unit having a prefix and defined by an artifact.) |
| second, s (time) | The second is the duration of 9,192,631,770 periods of the radiation corresponding to the transition between the two hyperfine levels of the ground state of the cesium-133 atom. |
| ampere, A (electric current) | The ampere is that constant current which, if maintained in two straight parallel conductors of infinite length and of negligible circular cross section, and placed 1 meter apart in a vacuum, would produce between these conductors a force equal to $2 \times 10^{-7}$ $m \cdot kg/s^2$ (newton) per meter of length. |
| kelvin, K (temperature) | The kelvin is the fraction 1/273.16 of the thermodynamic temperature of the triple point of water.<br>The kelvin is a unit of thermodynamic temperature (T). The word (or symbol) "degree" is not used with kelvin.<br>The Celsius (formerly centigrade) temperature is also used. Celsius temperature (symbol $t$) is defined by the equation $t = T - T_0$, where $T_0$ equals 273.15 K. A degree Celsius (°C) is thus equal to 1 kelvin.<br>The term centigrade should not be used because of possible confusion with the French unit of angular measurement, the grade. |
| mole, mol (amount of substance) | The mole is the amount of substance of a system which contains as many elementary entities as there are atoms in 0.012 kilograms of carbon 12.<br>When the mole is used, the elementary entities must be specified and may be atoms, molecules, ions, electrons, other particles, or specified groups of such particles. |
| candela, cd (luminous intensity) | The candela is the luminous intensity, in the perpendicular direction, of a surface of 1/600,000 square meter of a blackbody at the temperature of freezing platinum (2045 K) under a pressure of 101,325 $m^{-1} \cdot kg \cdot s^{-2}$. |

Supplementary units

| Name, symbol | Definition |
|---|---|
| radian, rad (plane angle) | The radian is the plane angle between two radii of a circle which cuts off, on the circumference, an arc equal in length to the radius. |
| steradian, sr (solid angle) | The steradian is the solid angle, which, having its vertex in the center of a sphere, cuts off an area of the surface of the sphere equal to that of a square with sides of length equal to the radius of the sphere. |

* The spelling of *metre* (and *litre*) is commonly accepted internationally and is recommended by the ASTM. However, the spelling as *meter* (and *liter*) is widely used in the United States and is the spelling used in this book.

**Table A-1 Common derived units with special names and symbols acceptable in SI**

| Name | Symbol | Quantity | Expression in terms of SI base units | Expression in terms of other units |
|---|---|---|---|---|
| becquerel | Bq | radioactivity | $s^{-1}$ | |
| coulomb | C | quantity of electricity or electric charge | $A \cdot s$ | |
| farad | F | electric capacitance | $m^{-2} \cdot kg^{-1} \cdot s^4 \cdot A^2$ | C/V |
| gray | Gy | absorbed radiation | $m^2 \cdot s^{-2}$ | J/kg |
| henry | H | electric inductance | $m^2 \cdot kg \cdot s^{-2} \cdot A^{-2}$ | Wb/A |
| hertz | Hz | frequency | $s^{-1}$ | |
| joule | J | energy, work, or quantity of heat | $m^2 \cdot kg \cdot s^{-2}$ | $N \cdot m$ |
| lumen | lm | luminous flux | $cd \cdot sr$ | |
| lux | lx | illuminance | $m^{-2} \cdot cd \cdot sr$ | $lm/m^2$ |
| newton | N | force | $m \cdot kg \cdot s^{-2}$ | $J \cdot m^{-1}$ |
| ohm | Ω | electric resistance | $m^2 \cdot kg \cdot s^{-3} \cdot A^{-2}$ | V/A |
| pascal | Pa | pressure or stress | $m^{-1} \cdot kg \cdot s^{-2}$ | $N/m^2$ |
| siemens | S | electric conductance | $m^{-2} \cdot kg^{-1} \cdot s^3 \cdot A^2$ | A/V |
| tesla | T | magnetic flux density | $kg \cdot s^{-2} \cdot A^{-1}$ | $Wb/m^2$ |
| volt | V | electric potential, potential difference, or electromotive force | $m^2 \cdot kg \cdot s^{-3} \cdot A^{-1}$ | W/A |
| watt | W | power or radiant flux | $m^2 \cdot kg \cdot s^{-3}$ | J/s |
| weber | Wb | magnetic flux | $m^2 \cdot kg \cdot s^{-2} \cdot A^{-1}$ | $V \cdot s$ |

and in numerous other publications.[3] Usage differences among countries have been resolved by a series of international conferences resulting in a set of seven base units, two supplementary units, and derived units as given in the table on page 310.

## Derived Units

Derived units are algebraic combinations of the seven base units or two supplementary units with some of the combinations being assigned special names and symbols. Examples are shown in Table A-1.

For the chemical engineer, the seven base units and two supplementary units are no problem because they have been used regularly in technical work of a chemical nature. However, the SI units for some of the derived terms, such as for

[3] The basic English document for SI is the National Bureau of Standards Special Publication 330 which can be obtained from the Superintendent of Documents, U.S. Government Printing Office, Washington, D.C. 20402 as document SD Catalog No. C13.10:330/3. This is the authorized English translation of the official document of the international body. For guidance in U.S. usage, the most widely recognized document in use is the ASTM *Standard for Metric Practice E 380* available from the American Society for Testing and Materials, 1916 Race St., Philadelphia, PA 19103.

pressure, are not familiar or in common usage in the American engineering system of units. The SI pressure unit is the pascal (Pa) (rhymes with *rascal*) which is a newton (N) per square meter, or $N \cdot m^{-2}$. Since a newton is an SI derived unit of force as mass (kg) times acceleration ($m/s^2$), the net expression of the pascal in terms of SI base units is

$$Pa = N \cdot m^{-2} = kg \cdot m \cdot s^{-2} \cdot m^{-2} = m^{-1} \cdot kg \cdot s^{-2}$$

Chemical engineers have commonly used atmospheres as a unit for pressure. Although the unit of atmosphere (1 atm = 101.325 kPa) was internationally authorized as an SI derived unit, this authorization was granted for a limited time only and its use should be minimized.

Another common set of units used by chemical engineers is the calorie (or British thermal unit) for energy. The units of calorie [1 cal = 4.1868 J, where J is the symbol of joule (rhymes with *pool*) which is a newton meter with base units of $m^2 \cdot kg \cdot s^{-2}$], and British thermal unit (1 Btu = $1.055056 \times 10^3$ J) are not acceptable with SI units.

In the SI system, the kilogram is restricted to the unit of mass so that it is not acceptable to use a unit of force as kilogram-force which would be analogous to the American engineering unit of pound-force. The newton is the unit of force in the SI system and should be used in place of kilogram-force. Confusion can occur because the term *weight* is used to mean either *force* or *mass*. In common everyday use, the term *weight* normally means mass, but in physics weight usually means the force exerted by gravity. Because of the ambiguity involved in the dual use of the term *weight*, the term should be avoided in technical practice unless the conditions are such that the meaning is totally clear.

Table A-1 lists common derived SI units with special names. The table also gives the approved SI symbol and the expression for the term in base units and in terms of other units. Table A-2 gives examples of other derived units which are commonly used in chemical engineering, including a description and SI units. Table A-3 shows units which are not officially recognized as usable with SI but which are authorized for use to a certain extent, while Table A-4 gives units which are not acceptable for use with SI.

## Advantages and Guidelines for the SI System

An advantage of the SI system is its total coherence in that all of the units are related by unity. Thus, as can be seen from Table A-1, a force of one newton exerted over a length of one meter gives an energy of one joule, while one joule occurring over a time period of one second results in a power of one watt. Mass is always measured in kilograms and force in newtons when dealing with the SI system, so that the confusion often found in the American engineering system of using both pounds-force and pounds-mass is eliminated.

A fundamental characteristic of the SI system is the fact that each defined quantity has only one unit. Thus, the fundamental SI unit of energy is the joule, and the fundamental SI unit of power is the watt. While a joule is defined as a

**Table A-2 Other derived units commonly used in chemical engineering with description in terms of acceptable SI units**

| Quantity | Description | Symbol | Expression in terms of SI base units |
|---|---|---|---|
| acceleration | meter per second squared | $m/s^2$ | $m \cdot s^{-2}$ |
| area | square meter | $m^2$ | $m^2$ |
| coefficient of heat transfer (U.S. symbol of $h$ or $U$) | watt per square meter kelvin | $W/(m^2 \cdot K)$ $J/(m^2 \cdot K \cdot s)$ | $kg \cdot s^{-3} \cdot K^{-1}$ |
| concentration (of amount of substance) | mole per cubic meter | $mol/m^3$ | $mol \cdot m^{-3}$ |
| current density | ampere per square meter | $A/m^2$ | $A \cdot m^{-2}$ |
| density (mass density) (U.S. symbol of $\rho$) | kilogram per cubic meter | $kg/m^3$ | $kg \cdot m^{-3}$ |
| electric charge density | coulomb per cubic meter | $C/m^3$ | $m^{-3} \cdot s \cdot A$ |
| electric field strength | volt per meter | $V/m$ | $m \cdot kg \cdot s^{-3} \cdot A^{-1}$ |
| electric flux density | coulomb per square meter | $C/m^2$ | $m^{-2} \cdot s \cdot A$ |
| energy density | joule per cubic meter | $J/m^3$ | $m^{-1} \cdot kg \cdot s^{-2}$ |
| force | newton | N or J/m | $m \cdot kg \cdot s^{-2}$ |
| heat capacity or entropy | joule per kelvin | J/K | $m^2 \cdot kg \cdot s^{-2} \cdot K^{-1}$ |
| heat flow rate (U.S. symbol of $Q$ or $q$) | watt | W or J/s | $m^2 \cdot kg \cdot s^{-3}$ |
| heat flux density or irradiance | watt per square meter | $W/m^2$ | $kg \cdot s^{-3}$ |
| luminance | candella per square meter | $cd/m^2$ | $cd \cdot m^{-2}$ |
| magnetic field strength | ampere per meter | A/m | $A \cdot m^{-1}$ |
| modulus of elasticity or Young's modulus | gigapascal | GPa | $10^{-9} \cdot m^{-1} \cdot kg \cdot s^{-1}$ |
| molar energy | joule per mole | J/mol | $m^{-2} \cdot kg \cdot s^{-2} \cdot mol^{-1}$ |
| molar entropy or molar heat capacity | joule per mole kelvin | $J/(mol \cdot K)$ | $m^2 \cdot kg \cdot s^{-2} \cdot K^{-1} \cdot mol^{-1}$ |
| moment of force or torque | newton meter | $N \cdot m$ | $m^2 \cdot kg \cdot s^{-2}$ |
| moment of inertia | kilogram meter squared | $kg \cdot m^2$ | $kg \cdot m^2$ |
| momentum | kilogram meter per second | $kg \cdot m/s$ | $kg \cdot m \cdot s^{-1}$ |
| permeability | henry per meter | H/m | $m \cdot kg \cdot s^{-2} \cdot A^{-2}$ |
| permittivity | farad per meter | F/m | $m^{-3} \cdot kg^{-1} \cdot s^4 \cdot A^2$ |
| power | kilowatt | kW | $10^{-3} \cdot m^2 \cdot kg \cdot s^{-3}$ |
| pressure (U.S. symbol of $P$ or $p$) | kilopascal | kPa | $10^{-3} \cdot m^{-1} \cdot kg \cdot s^{-2}$ |
| specific energy | joule per kilogram | J/kg | $m^2 \cdot s^{-2}$ |
| specific heat capacity or specific entropy (U.S. symbol of $c_p$, $c_v$, or $s$) | joule per kilogram kelvin | $J/(kg \cdot K)$ | $m^2 \cdot s^{-2} \cdot K^{-1}$ |
| specific volume | cubic meter per kilogram | $m^3/kg$ | $m^3 \cdot kg^{-1}$ |
| stress | megapascal | MPa | $10^{-6} \cdot m^{-1} \cdot kg \cdot s^{-2}$ |
| surface tension | newton per meter | N/m | $kg \cdot s^{-2}$ |
| thermal conductivity (U.S. symbol of $k$) | watt per meter kelvin | $W/(m \cdot K)$ | $m \cdot kg \cdot s^{-3} \cdot K^{-1}$ |
| torque | newton meter | $N \cdot m$ | $m^2 \cdot kg \cdot s^{-2}$ |
| velocity or speed | meter per second | m/s | $m \cdot s^{-1}$ |
| viscosity–absolute or dynamic (U.S. symbol of $\mu$) | pascal second | $Pa \cdot s$ | $m^{-1} \cdot kg \cdot s^{-1}$ |
| viscosity–kinematic (U.S. symbol of $v$) | square meter per second | $m^2/s$ | $m^2 \cdot s^{-1}$ |
| volume | cubic meter | $m^3$ | $m^3$ |
| wave number | 1 per meter | 1/m | $m^{-1}$ |
| work energy (U.S. symbol of $W$ in foot-pounds force) | joule | J or $N \cdot m$ | $m^2 \cdot kg \cdot s^{-2}$ |

## Table A-3 Non-SI units which are acceptable for use

The following common units have been authorized for use with SI to a certain extent and continue in use on an unofficially accepted basis.

| Name | Symbol | Value in SI units |
|---|---|---|
| time–minute, hour, day, year | min, h, d, yr | 60s, 3 600s, 86 400s, $\approx$ 365d |
| angle–degree, minute, second | °, ′, ″ | $(\pi/180)$ rad, $(1/60)°$, $(1/60)'$ |
| liter* | l (or L) | 1 $dm^3$ |
| nautical mile | nautical mile | 1 852 m |
| knot (1 nautical mile per hour) | knot | 0.513 9 m/s |
| hectare | ha | $10^4$ $m^2$ |
| ångström | Å | 0.1 nm = $10^{-10}$ m |
| are | a | $10^2$ $m^2$ |
| atmosphere pressure | atm | 101.325 kPa |
| bar pressure | bar | $10^5$ Pa |
| galileo or gal | Gal | $10^{-2}$ $m/s^2$ |
| metric ton | t | $10^3$ kg |

* The SI unit of volume is the cubic meter, and this unit or one of its regular multiples is preferred for all cases. However, the special name *liter* has been approved for the cubic decimeter, but the use of this unit is restricted to the measurements of liquids and gases. No prefix other than *milli* should be used with liter.

## Table A-4 Common units which are not acceptable with SI

Despite the fact that the following units have been used commonly in the past, they are not acceptable with SI

| Name | Symbol | Value in SI units |
|---|---|---|
| British thermal unit | Btu | 1.055 056 × $10^3$ J |
| calorie | cal | 4.186 8 J |
| dyne | dyn | $10^{-5}$ N |
| erg | erg | $10^{-7}$ J |
| fermi | Fm | $10^{-15}$ m |
| gamma | $\gamma$ | $10^{-9}$ T |
| gauss | Gs, G | $10^{-4}$ T |
| kilogram-force | kgf | 9.806 65 N |
| lambda | $\lambda$ | $10^{-6}$ liter |
| maxwell | Mx | $10^{-8}$ Wb |
| metric carat | | 200 mg |
| micron | $\mu$ | 1 micrometer |
| oersted | Oe | $(1000/4\pi)A \cdot m^{-1}$ |
| phot | ph | $10^4$ lx |
| poise | P | 0.1 Pa · s |
| stere | st | 1 $m^3$ |
| stilb | sb | 1 $cd/cm^2$ |
| stokes | St | 1 $cm^2/s$ |
| torr | | 101 325/760 Pa |
| X unit | | 1.002 × $10^{-4}$ nm (approximately) |

newton meter, it refers to a unit force moving through a unit distance. The expression "newton meter" is used in the SI system to refer to torque in which there is no indication of motion or movement. Thus, the SI system is very explicit that joule and newton meter are different units.

The SI system has a series of approved prefixes and symbols for decimal multiples as shown in Table A-5.

The common usage of "psi" and "atmosphere" for units of pressure will be replaced by the pascal in the SI system. Because a pascal, as a force of one newton against an area of one square meter, is a very small unit, it is convenient to deal with kilopascals (kPa) rather than pascals in many cases.[4] The following conversion factors are useful for making the transition from the American engineering system to the SI system for pressure designations:

| To convert to kPa from | Multiply by |
|---|---|
| psi, $lbf/in^2$ | 6.895 |
| atmosphere | 101.325 |
| torr | 0.1333 |
| bar | 100.000 |

## Rules for Use of SI Units

**1. Periods.** A period is never used after a symbol of an SI unit unless it is used to designate the end of a sentence.

**2. Capitalization.** Capitals are not used to start units that are written out except at the beginning of a sentence. However, when the units are expressed as symbols, the first letter of the symbol is capitalized when the name of the unit was derived from the name of a person. For example, it is correct to write

5 pascals or 5 Pa
5 newtons or 5 N
5 meters or 5 m
300 kelvins or 300 K

But note that the following temperature forms are correct:

200 degrees Celsius or 300°C
100 degrees Fahrenheit or 100°F

[4] To give an idea as to the approximate magnitude of a pressure of one pascal, it would be equivalent to the extra pressure exerted on the palm of an open hand when a person blows a sharp breath on the hand.

## Table A-5 SI unit prefixes

| Multiplication factor | Prefix | Symbol | Pronunciation (USA) (1) | Meaning (in USA) | Meaning (in other countries) |
|---|---|---|---|---|---|
| 1 000 000 000 000 000 000 = $10^{18}$ | exa (2) | E | ex'a (a as in about) | One quintillion times (3) | trillion |
| 1 000 000 000 000 000 = $10^{15}$ | peta (2) | P | as in petal | One quadrillion times (3) | thousand billion |
| 1 000 000 000 000 = $10^{12}$ | tera | T | as in terrace | One trillion times (3) | billion |
| 1 000 000 000 = $10^{9}$ | giga | G | jig'a (a as in about) | One billion times (3) | milliard |
| 1 000 000 = $10^{6}$ | mega | M (4) | as in megaphone | One million times | |
| 1 000 = $10^{3}$ | kilo | k | as in kilowatt | One thousand times | |
| 100 = $10^{2}$ | hecto | h (5) | heck'toe | One hundred times | |
| 10 = 10 | deka | da (5) | deck'a (a as in about) | Ten times | |
| 0.1 = $10^{-1}$ | deci | d (5) | as in decimal | One tenth of | |
| 0.01 = $10^{-2}$ | centi | c (5) | as in sentiment | One hundredth of | |
| 0.001 = $10^{-3}$ | milli | m | as in military | One thousandth of | |
| 0.000 001 = $10^{-6}$ | micro | $\mu$ (6) | as in microphone | One millionth of | |
| 0.000 000 001 = $10^{-9}$ | nano | n | nan'oh (an as in ant) | One billionth of (3) | milliardth |
| 0.000 000 000 001 = $10^{-12}$ | pico | p | peek'oh | One trillionth of (3) | billionth |
| 0.000 000 000 000 001 = $10^{-15}$ | femto | f | fem'toe (fem as in feminine) | One quadrillionth of (3) | thousand billionth |
| 0.000 000 000 000 000 001 = $10^{-18}$ | atto | a | as in anatomy | One quintillionth of (3) | trillionth |

1. The first syllable of every prefix is accented to assure that the prefix will retain its identity. Therefore, the preferred pronunciation of kilometer places the accent on the first syllable, not the second.
2. Approved by the 15th General Conference of Weights and Measures (CGPM), May–June 1975.
3. These terms should be avoided in technical writing because the denominators above one million are different in most other countries, as indicated in the last column.
4. The symbol M often means 1 000 when used with American engineering units.
5. While hecto, deka, deci, and centi are SI prefixes, their use should generally be avoided except for the SI unit-multiples for area and volume and nontechnical use of centimeter, as for body and clothing measurement. The prefix hecto should be avoided also because the longhand symbol h may be confused with k.
6. Although SI rules prescribe vertical (roman) type, the sloping (*italics*) form is usually acceptable in the USA for the Greek letter $\mu$ because of the scarcity of the upright style.

In the SI system, it is very important to follow the precise, agreed-upon use of uppercase and lowercase letters. This importance is shown by the following examples taken from Tables A-3 and A-5 and base-unit definitions:

G for giga; g for gram
K for kelvin; k for kilo
M for mega; m for milli
N for newton; n for nano
T for tera; t for metric ton

**3. Plurals.** As indicated in some of the preceding examples, the plural is used in the normal grammatical sense when the units are written out as words, but plurals are never used with the unit symbols. For numerical values greater than 1, equal to 0, or less than —1, the names of units are plural. All other values take the singular form for the unit names. For example, the following forms are correct:

| | | | |
|---|---|---|---|
| 200 | kilograms | or | 200 kg |
| 1.05 | meters | or | 1.05 m |
| 0 | degrees Celsius | or | 0°C |
| −2 | degrees Celsius | or | −2°C |
| 3 | kelvins | or | 3 K |
| 0.9 | meter | or | 0.9 m |
| −0.5 | degree Celsius | or | −0.5°C |
| 1 | kelvin | or | 1 K |
| −1 | degree Celsius | or | −1°C |

An "s" is added to form the plurals of unit names as illustrated in the preceding except that hertz, lux, and siemens remain unchanged and henry becomes henries.

**4. Groupings of numbers and decimal points.** The common U.S. practice of using commas to separate multiples of 1000 is not followed with SI which uses a space instead of a comma to separate the multiples of 1000. For decimals, the space is filled on both sides of the decimal point. The decimal point is placed on the line as a regular period for U.S. usage rather than at mid-line height or use of a comma as is frequent European practice. When writing numbers with values less than one, a zero should be placed ahead of the decimal.

Numbers with many digits should be set off in groups of three digits away from the decimal point on both the left and the right. For example, the following forms are correct:

57 321 684.521 69
0.431 684 2

If there are only four digits to the left or right of the decimal point, the use of the space is optional unless there is a column of figures which is aligned on the decimal point with one or more numbers having more than four digits to the left or right of the decimal point. Thus, the following forms are correct:

3200 or 3 200
0.6854 or 0.685 4

$$\begin{array}{r} 13.6 \\ +15\,957 \\ +\ 3\,200 \\ \hline 18\,270.6 \end{array}$$

**5. Spacing, hyphens, and italics.** When a unit symbol is given after a number, a space is always left between the number and the symbol with the exception of cases where the symbol appears in the superscript position, such as degree, minute, and second of plane angles. The symbol for degree Celsius may be written either with or without a space before the degree symbol. For example,

68 kHz
60 mm
$10^6$ N

plane angle of 20° 24′ 26″
20°C or 20 °C (20°C is preferred and 20° C is *not* acceptable)

For both symbols and names of units having prefixes, no space is left between letters making up the symbol or name. For example,

kA, kiloampere; mg, milligram

The symbols when printed are always given as roman (vertical) type. Sloping letters (or *italics*) are reserved for quantity symbols such as *m* for mass, *l* for length, or general algebraic quantities such as *a*, *b*, or *c*. When the algebraic quantity is used, there is no space used between the algebraic quantity and the numerical coefficient. For example,

5 m means a distance of 5 meters,
but 5*m* means 5 times the algebraic quantity *m*

When a quantity is used in an adjectival sense, a hyphen should be used between the number and the symbol except for symbols appearing in the superscript position. For example,

He bought a 35-mm film; *but*, the width of the film is 35 mm.
He bought a 5-kg ham; *but*, the mass of the ham is 5 kg.
However, it is correct to write:
He bought a 100°C thermometer which covers a temperature range of 100°C.

A space should be left on each side of signs for multiplication, division, addition, and subtraction except within a compound symbol. The product dot (as in N · m)

is used for the derived unit symbol with no space on either side. The product dot should not be used as a multiplier symbol for calculations. For example,

Write $6\ \text{m} \times 8\ \text{m}$ (not $6\ \text{m}{\times}8\ \text{m}$ or $6\ \text{m} \cdot 8\ \text{m}$)
$\text{kg/m}^3$ or $\text{kg} \cdot \text{m}^{-3}$
$\text{m}^2 \cdot \text{kg} \cdot \text{s}^{-2}$

**6. Prefixes.** In general, it is desirable to keep numerical values between 0.1 and 1000 by the use of appropriate prefixes shown prior to the unit symbol. Prefixes and symbols along with pronunciations and meanings as acceptable in the SI system are given in table A-5. Some typical examples are

$$5\,527\ \text{Pa} = 5.527 \times 10^3\ \text{Pa} = 5.527\ \text{kPa}$$
$$0.051\ \text{m} = 51 \times 10^{-3}\ \text{m} = 51\ \text{mm}$$
$$0.235 \times 10^{-6}\ \text{s} = 0.235\ \mu\text{s} = 235 \times 10^{-3}\ \mu\text{s} = 235\ \text{ns}$$

Two or more SI prefixes should not be used simultaneously for the designation of a unit. For example,

write 1 pF instead of 1 $\mu\mu$F

For cases that fall outside the range covered by single prefixes, the situation should be handled by expressing the value with powers of ten as applied to the base unit.

With reference to the spelling with prefixes, there are three cases where the final vowel in a prefix is omitted. These are megohm, kilohm, and hectare. In all other cases, both vowels are retained and both are pronounced. No space or hyphen should be used.

**7. Combination of units.** It is desirable to avoid the use of prefixes in the denominator of compound units with the one exception of the base unit kg. For example,

use kN/m instead of N/mm
use kg/s instead of g/ms

The single exception is to use J/kg instead of mJ/g.

Use a solidus(/) to indicate a division factor. Avoid the use of a double solidus. For example,

write $\text{J}/(\text{s} \cdot \text{m})^2$ or $\text{J} \cdot \text{s}^{-2} \cdot \text{m}^{-2}$ instead of $\text{J/s}^2/\text{m}^2$

When the denominator of a unit expression is a product, it should normally be shown in parentheses. For example,

$$\text{W}/(\text{m}^2 \cdot \text{K})$$

If an expression is given for units raised to a power, such as square millimeters, the power number refers to the entire unit and not just to the last symbol. For example,

$\text{mm}^2$ means $(\text{mm})^2$ instead of milli(square meters) or $\text{m}(\text{m}^2)$

Symbols and unit names should not be used together in the same expression. For example,

write joules per kilogram or J/kg instead of
joules/kilogram or joules/kg or joules · $kg^{-1}$

**8. Guidelines for calculations.** It is generally desirable to carry out calculations in base units and then convert the final answers to appropriate-size numbers by use of correct prefixes.

**9. Confusion of meaning of billion in U.S.A. and other countries.** In the United States, *billion* means a thousand million (prefix *giga*), but, in most other countries, it means a million million (prefix *tera*). Because of possible confusion as to the meaning, the term *billion* should be avoided in technical writing. As shown in Table A-5, the same possible confusion exists with *quintillion* (prefix *exa* in U.S.A.), *quadrillion* (prefix *peta* in U.S.A.), and *trillion* (prefix *tera* in U.S.A.).

**10. Round-offs in conversions.** In making a conversion of a number to new units, the number of significant digits should not be increased or decreased. It is, therefore, necessary to use sufficient precision in the conversion factor to preserve the precision of the quantity converted.

**11. Conversions between SI and American engineering units.** SI and American engineering units can be presented with the American engineering units first followed by SI units in parentheses, as 2.45 in (62.2 mm) or as the preferred SI units first with the American engineering units in parentheses, such as 170 kPa (24.7 $lb/in^2$).

Table A-6 presents a detailed list of conversion factors that can be used to convert between U.S.-British units and SI units, while Table A-7 gives a simplified and abbreviated list of equivalences for converting unacceptable units commonly used by chemical engineers into acceptable SI units.

**Table A-6 Conversion factors for converting from American engineering units to SI units—alphabetical listing in detail†**

| To convert from | To | Multiply by |
|---|---|---|
| abampere | ampere (A) | 1.000 000*E + 01 |
| abcoulomb | coulomb (C) | 1.000 000*E + 01 |
| abfarad | farad (F) | 1.000 000*E + 09 |
| abhenry | henry (H) | 1.000 000*E − 09 |
| abmho | siemens (S) | 1.000 000*E + 09 |
| abohm | ohm (Ω) | 1.000 000*E − 09 |
| abvolt | volt (V) | 1.000 000*E − 08 |
| acre foot (U.S. survey)‡ | $meter^3$ ($m^3$) | 1.233 489 E + 03 |
| acre (U.S. survey)‡ | $meter^2$ ($m^2$) | 4.046 873 E + 03 |

† See end of Table.
‡ See end of Table.

**Table A-6 Conversion factors for converting from American engineering units to SI units—alphabetical listing in detail** (*Continued*)

| To convert from | To | Multiply by |
|---|---|---|
| ampere hour | coulomb (C) | 3.600 000*E + 03 |
| are | meter$^2$ (m$^2$) | 1.000 000*E + 02 |
| ångstrom | meter (m) | 1.000 000*E − 10 |
| astronomical unit | meter (m) | 1.495 979 E + 11 |
| atmosphere (standard) | pascal (Pa) | 1.013 250*E + 05 |
| atmosphere (technical = 1kgf/cm$^2$) | pascal (Pa) | 9.806 650*E + 04 |
| bar | pascal (Pa) | 1.000 000*E + 05 |
| barn | meter$^2$ (m$^2$) | 1.000 000*E − 28 |
| barrel (for petroleum, 42 gal) | meter$^3$ (m$^3$) | 1.589 873 E − 01 |
| board foot | meter$^3$ (m$^3$) | 2.359 737 E − 03 |
| British thermal unit (International Table)§ | joule (J) | 1.055 056 E + 03 |
| British thermal unit (mean)§ | joule (J) | 1.055 87 E + 03 |
| British thermal unit (thermochemical)§ | joule (J) | 1.054 350 E + 03 |
| British thermal unit (39°F) | joule (J) | 1.059 67 E + 03 |
| British thermal unit (59°F) | joule (J) | 1.054 80 E + 03 |
| British thermal unit (60°F) | joule (J) | 1.054 68 E + 03 |
| Btu (International Table) · ft/h · ft$^2$ · °F (*k*, thermal conductivity) | watt per meter kelvin (W/m · K) | 1.730 735 E + 00 |
| Btu (thermochemical) · ft/h · ft$^2$ · °F (*k*, thermal conductivity) | watt per meter kelvin (W/m · K) | 1.729 577 E + 00 |
| Btu (International Table) · in/h · ft$^2$ · °F (*k*, thermal conductivity) | watt per meter kelvin (W/m · K) | 1.442 279 E − 01 |
| Btu (thermochemical) · in/h · ft$^2$ · °F (*k*, thermal conductivity) | watt per meter kelvin (W/m · K) | 1.441 314 E − 01 |
| Btu (International Table) · in/s · ft$^2$ · °F (*k*, thermal conductivity) | watt per meter kelvin (W/m · K) | 5.192 204 E + 02 |
| Btu (thermochemical) · in/s · ft$^2$ · °F (*k*, thermal conductivity) | watt per meter kelvin (W/m · K) | 5.188 732 E + 02 |
| Btu (International Table)/h | watt (W) | 2.930 711 E − 01 |
| Btu (thermochemical)/h | watt (W) | 2.928 751 E − 01 |
| Btu (thermochemical)/min | watt (W) | 1.757 250 E + 01 |
| Btu (thermochemical)/s | watt (W) | 1.054 350 E + 03 |
| Btu (International Table)/ft$^2$ | joule per meter$^2$ (J/m$^2$) | 1.135 653 E + 04 |
| Btu (thermochemical)/ft$^2$ | joule per meter$^2$ (J/m$^2$) | 1.134 893 E + 04 |
| Btu (International Table)/ft$^2$ · h | watt per meter$^2$ (W/m$^2$) | 3.154 591 E + 00 |
| Btu (thermochemical)/ft$^2$ · h | watt per meter$^2$ (W/m$^2$) | 3.152 481 E + 00 |
| Btu (themochemical/ft$^2$ · min | watt per meter$^2$ (W/m$^2$) | 1.891 489 E + 02 |
| Btu (thermochemical)/ft$^2$ · s | watt per meter$^2$ (W/m$^2$) | 1.134 893 E + 04 |
| Btu (thermochemical)/in$^2$ · s | watt per meter$^2$ (W/m$^2$) | 1.634 246 E + 06 |
| Btu (International Table)/h · ft$^2$ · °F (*C*, thermal conductance) | watt per meter$^2$ kelvin (W/m$^2$ · K) | 5.678 263 E + 00 |
| Btu (thermochemical)/h · ft$^2$ · °F (*C*, thermal conductance) | watt per meter$^2$ kelvin (W/m$^2$ · K) | 5.674 466 E + 00 |
| Btu (International Table)/s · ft$^2$ · °F | watt per meter$^2$ kelvin (W/m$^2$ · K) | 2.044 175 E + 04 |
| Btu (thermochemical)/s · ft$^2$ · °F | watt per meter$^2$ kelvin (W/m$^2$ · K) | 2.042 808 E + 04 |
| Btu (International Table)/lb | joule per kilogram (J/kg) | 2.326 000*E + 03 |

§ See end of Table.

*(Continued)*

**Table A-6 Conversion factors for converting from American engineering units to SI units—alphabetical listing in detail** (*Continued*)

| To convert from | To | Multiply by |
|---|---|---|
| Btu (thermochemical)/lb | joule per kilogram (J/kg) | 2.324 444 E + 03 |
| Btu (International Table)/lb · °F (*c*, heat capacity) | joule per kilogram kelvin (J/kg · K) | 4.186 800*E + 03 |
| Btu (thermochemical)/lb · °F (*c*, heat capacity) | joule per kilogram kelvin (J/kg · K) | 4.184 000 E + 03 |
| bushel (U.S.) | meter$^3$ (m$^3$) | 3.523 907 E − 02 |
| caliber (inch) | meter (m) | 2.540 000*E − 02 |
| calorie (International Table) | joule (J) | 4.186 800*E + 00 |
| calorie (mean) | joule (J) | 4.190 02 E + 00 |
| calorie (thermochemical) | joule (J) | 4.184 000*E + 00 |
| calorie (15°C) | joule (J) | 4.185 80 E + 00 |
| calorie (20°C) | joule (J) | 4.181 90 E + 00 |
| calorie (kilogram, International Table)§ | joule (J) | 4.186 800*D + 03 |
| calorie (kilogram, mean)§ | joule (J) | 4.190 02 E + 03 |
| calorie (kilogram, thermochemical)§ | joule (J) | 4.184 000*E + 03 |
| cal (thermochemical)/cm$^2$ | joule per meter$^2$ (J/m$^2$) | 4.184 000*E + 04 |
| cal (International Table)/g | joule per kilogram (J/kg) | 4.186 800*E + 03 |
| cal (thermochemical)/g | joule per kilogram (J/kg) | 4.184 000*E + 03 |
| cal (International Table)/g · °C | joule per kilogram kelvin (J/kg · K) | 4.186 800*E + 03 |
| cal (thermochemical)/g · °C | joule per kilogram kelvin (J/kg · K) | 4.184 000*E + 03 |
| cal (thermochemical)/min | watt (W) | 6.973 333 E − 02 |
| cal (thermochemical)/s | watt (W) | 4.184 000*E + 00 |
| cal (thermochemical)/cm$^2$ · min | watt per meter$^2$ (W/m$^2$) | 6.973 333 E + 02 |
| cal (thermochemical)/cm$^2$ · s | watt per meter$^2$ (W/m$^2$) | 4.184 000*E + 04 |
| cal (thermochemical)/cm · s · °C | watt per meter kelvin (W/m · K) | 4.184 000*E + 02 |
| carat (metric) | kilogram (kg) | 2.000 000*E − 04 |
| centimeter of mercury (0°C) | pascal (Pa) | 1.333 22 E + 03 |
| centimeter of water (4°C) | pascal (Pa) | 9.806 38 E + 01 |
| centipoise | pascal second (Pa · s) | 1.000 000*E − 03 |
| centistokes | meter$^2$ per second (m$^2$/s) | 1.000 000*E − 06 |
| circular mil | meter$^2$ (m$^2$) | 5.067 075 E − 10 |
| clo | kelvin meter$^2$ per watt (K · m$^2$/W) | 2.003 712 E − 01 |
| cup | meter$^3$ (m$^3$) | 2.365 882 E − 04 |
| curie | becquerel (Bq) | 3.700 000*E + 10 |
| day (mean solar) | second (s) | 8.640 000 E + 04 |
| day (sidereal) | second (s) | 8.616 409 E + 04 |
| degree (angle) | radian (rad) | 1.745 329 E − 02 |
| degree Celsius | kelvin (K) | $t_K = t_{°C} + 273.15$ |
| degree Fahrenheit | degree Celsius | $t_{°C} = (t_{°F} - 32)/1.8$ |
| degree Fahrenheit | kelvin (K) | $t_K = (t_{°F} + 459.67)/1.8$ |
| degree Rankine | kelvin (K) | $t_K = t_{°R}/1.8$ |
| °F · h · ft$^2$/Btu (International Table) (*R*, thermal resistance) | kelvin meter$^2$ per watt (K · m$^2$/W) | 1.761 102 E − 01 |
| °F · h · ft$^2$/Btu (thermochemical) (*R*, thermal resistance) | kelvin meter$^2$ per watt (K · m$^2$/W) | 1.762 280 E − 01 |
| denier | kilogram per meter (kg/m) | 1.111 111 E − 07 |
| dyne | newton (N) | 1.000 000*E − 05 |

§ See end of Table.

**Table A-6 Conversion factors for converting from American engineering units to SI units—alphabetical listing in detail** (*Continued*)

| To convert from | To | Multiply by |
|---|---|---|
| dyne · cm | newton meter (N · m) | 1.000 000*E − 07 |
| dyne/cm$^2$ | pascal (Pa) | 1.000 000*E − 01 |
| electronvolt | joule (J) | 1.602 19 E − 19 |
| EMU of capacitance | farad (F) | 1.000 000*E + 09 |
| EMU of current | ampere (A) | 1.000 000*E + 01 |
| EMU of electric potential | volt (V) | 1.000 000*E − 08 |
| EMU of inductance | henry (H) | 1.000 000*E − 09 |
| EMU of resistance | ohm (Ω) | 1.000 000*E − 09 |
| ESU of capacitance | farad (F) | 1.112 650 E − 12 |
| ESU of current | ampere (A) | 3.335 6 E − 10 |
| ESU of electric potential | volt (V) | 2.997 9 E + 02 |
| ESU of inductance | henry (H) | 8.987 554 E + 11 |
| ESU of resistance | ohm (Ω) | 8.987 554 E + 11 |
| erg | joule (J) | 1.000 000*E − 07 |
| erg/cm$^2$ · s | watt per meter$^2$ (W/m$^2$) | 1.000 000*E − 03 |
| erg/s | watt (W) | 1.000 000*E − 07 |
| faraday (based on carbon-12) | coulomb (C) | 9.648 70 E + 04 |
| faraday (chemical) | coulomb (C) | 9.649 57 E + 04 |
| faraday (physical) | coulomb (C) | 9.652 19 E + 04 |
| fathom | meter (m) | 1.828 8 E + 00 |
| fermi (femtometer) | meter (m) | 1.000 000*E − 15 |
| fluid ounce (U.S.) | meter$^3$ (m$^3$) | 2.957 353 E − 05 |
| foot | meter (m) | 3.048 000*E − 01 |
| foot (U.S. survey)‡ | meter (m) | 3.048 006 E − 01 |
| foot of water (39.2°F) | pascal (Pa) | 2.988 98 E + 03 |
| ft$^2$ | meter$^2$ (m$^2$) | 9.290 304*E − 02 |
| ft$^2$/h (thermal diffusivity) | meter$^2$ per second (m$^2$/s) | 2.580 640*E − 05 |
| ft$^2$/s | meter$^2$ per second (m$^2$/s) | 9.290 304*E − 02 |
| ft$^3$ (volume; section modulus) | meter$^3$ (m$^3$) | 2.831 685 E − 02 |
| ft$^3$/min | meter$^3$ per second (m$^3$/s) | 4.719 474 E − 04 |
| ft$^3$/s | meter$^3$ per second (m$^3$/s) | 2.831 685 E − 02 |
| ft$^4$ (moment of section) | meter$^4$ (m$^4$) | 8.630 975 E − 03 |
| ft/h | meter per second (m/s) | 8.466 667 E − 05 |
| ft/min | meter per second (m/s) | 5.080 000*E − 03 |
| ft/s | meter per second (m/s) | 3.048 000*E − 01 |
| ft/s$^2$ | meter per second$^2$ (m/s$^2$) | 3.048 000*E − 01 |
| footcandle | lux (lx) | 1.076 391 E + 01 |
| footlambert | candela per meter$^2$ (cd/m$^2$) | 3.426 259 E + 00 |
| ft · lbf | joule (J) | 1.355 818 E + 00 |
| ft · lbf/h | watt (W) | 3.766 161 E − 04 |
| ft · lbf/min | watt (W) | 2.259 697 E − 02 |
| ft · lbf/s | watt (W) | 1.355 818 E + 00 |
| ft · poundal | joule (J) | 4.214 011 E − 02 |
| free fall, standard ($g$) | meter per second$^2$ (m/s$^2$) | 9.806 650*E + 00 |
| gal | meter per second$^2$ (m/s$^2$) | 1.000 000*E − 02 |
| gallon (Canadian liquid) | meter$^3$ (m$^3$) | 4.546 090 E − 03 |
| gallon (U.K. liquid) | meter$^3$ (m$^3$) | 4.546 092 E − 03 |

‡ See end of Table.

(*Continued*)

**Table A-6 Conversion factors for converting from American engineering units to SI units—alphabetical listing in detail** (*Continued*)

| To convert from | To | Multiply by |
|---|---|---|
| gallon (U.S. dry) | meter$^3$ (m$^3$) | 4.404 884 E − 03 |
| gallon (U.S. liquid) | meter$^3$ (m$^3$) | 3.785 412 E − 03 |
| gal (U.S. liquid)/day | meter$^3$ per second (m$^3$/s) | 4.381 264 E − 08 |
| gal (U.S. liquid)/min | meter$^3$ per second (m$^3$/s) | 6.309 020 E − 05 |
| gal (U.S. liquid)/hp · h(SFC, specific fuel consumption) | kilogram per joule (kg/J) | 1.410 089 E − 09 |
| gamma | tesla (T) | 1.000 000*E − 09 |
| gauss | tesla (T) | 1.000 000*E − 04 |
| gilbert | ampere | 7.957 747 E − 01 |
| gill (U.K.) | meter$^3$ (m$^3$) | 1.420 654 E − 04 |
| gill (U.S.) | meter$^3$ (m$^3$) | 1.182 941 E − 04 |
| grad | degree (angular) | 9.000 000*E − 01 |
| grad | radian (rad) | 1.570 796 E − 02 |
| grain (1/7000 lb avoirdupois) | kilogram (kg) | 6.479 891*E − 05 |
| grain (lb avoirdupois/7000)/gal (U.S. liquid) | kilogram per meter$^3$ (kg/m$^3$) | 1.711 806 E − 02 |
| gram | kilogram (kg) | 1.000 000*E − 03 |
| g/cm$^3$ | kilogram per meter$^3$ (kg/m$^3$) | 1.000 000*E + 03 |
| gram-force/cm$^2$ | pascal (Pa) | 9.806 650*E + 01 |
| hectare | meter$^2$ (m$^2$) | 1.000 000*E + 04 |
| horsepower (550 ft · lbf/s) | watt (W) | 7.456 999 E + 02 |
| horsepower (boiler) | watt (W) | 9.809 50 E + 03 |
| horsepower (electric) | watt (W) | 7.460 000*E + 02 |
| horsepower (metric) | watt (W) | 7.354 99 E + 02 |
| horsepower (water) | watt (W) | 7.460 43 E + 02 |
| horsepower (U.K.) | watt (W) | 7.457 0 E + 02 |
| hour (mean solar) | second (s) | 3.600 000 E + 03 |
| hour (sidereal) | second (s) | 3.590 170 E + 03 |
| hundredweight (long) | kilogram (kg) | 5.080 235 E + 01 |
| hundredweight (short) | kilogram (kg) | 4.535 924 E + 01 |
| inch | meter (m) | 2.540 000*E − 02 |
| inch of mercury (32°F) | pascal (Pa) | 3.386 38 E + 03 |
| inch of mercury (60°F) | pascal (Pa) | 3.376 85 E + 03 |
| inch of water (39.2°F) | pascal (Pa) | 2.490 82 E + 02 |
| inch of water (60°F) | pascal (Pa) | 2.488 4 E + 02 |
| in$^2$ | meter$^2$ (m$^2$) | 6.541 600*E − 04 |
| in$^3$ (volume; section modulus) | meter$^3$ (m$^3$) | 1.638 706 E − 05 |
| in$^3$/min | meter$^3$ per second (m$^3$/s) | 2.731 177 E − 07 |
| in$^4$ (moment of section) | meter$^4$ (m$^4$) | 4.162 314 E − 07 |
| in/s | meter per second (m/s) | 2.540 000*E − 02 |
| in/s$^2$ | meter per second$^2$ (m/s$^2$) | 2.540 000*E − 02 |
| kayser | 1 per meter (1/m) | 1.000 000*E + 02 |
| kelvin | degree Celsius | $t_{°C} = t_K - 273.15$ |
| kilocalorie (International Table) | joule (J) | 4.186 800*E + 03 |
| kilocalorie (mean) | joule (J) | 4.190 02 E + 03 |
| kilocalorie (thermochemical) | joule (J) | 4.184 000*E + 03 |
| kilocalorie (thermochemical)/min | watt (W) | 6.973 333 E + 01 |
| kilocalorie (thermochemical)/s | watt (W) | 4.184 000*E + 03 |
| kilogram-force (kgf) | newton (N) | 9.806 650*E + 00 |

**Table A-6 Conversion factors for converting from American engineering units to SI units—alphabetical listing in detail** (*Continued*)

| To convert from | To | Multiply by |
|---|---|---|
| kgf · m | newton meter (N · m) | 9.806 650*E + 00 |
| kgf · $s^2$/m (mass) | kilogram (kg) | 9.806 650*E + 00 |
| kgf/$cm^2$ | pascal (Pa) | 9.806 650*E + 04 |
| kgf/$m^2$ | pascal (Pa) | 9.806 650*E + 00 |
| kgf/$mm^2$ | pascal (Pa) | 9.806 650*E + 06 |
| km/h | meter per second (m/s) | 2.777 778 E − 01 |
| kilopond | newton (N) | 9.806 650*E + 00 |
| kW · h | joule (J) | 3.600 000*E + 06 |
| kip (1000 lbf) | newton (N) | 4.448 222 E + 03 |
| kip/$in^2$ (ksi) | pascal (Pa) | 6.894 757 E + 06 |
| knot (international) | meter per second (m/s) | 5.144 444 E − 01 |
| lambert | candela per $meter^2$ (cd/$m^2$) | $1/\pi$ *E + 04 |
| lambert | candela per $meter^2$ (cd/$m^2$) | 3.183 099 E + 03 |
| langley | joule per $meter^2$ (J/$m^2$) | 4.184 000*E + 04 |
| league | meter (m) | [see footnote ‡] |
| light year | meter (m) | 9.460 55 E + 15 |
| liter | $meter^3$ ($m^3$) | 1.000 000*E − 03 |
| maxwell | weber (Wb) | 1.000 000*E − 08 |
| mho | siemens (S) | 1.000 000*E + 00 |
| microinch | meter (m) | 2.540 000*E − 08 |
| micron | meter (m) | 1.000 000*E − 06 |
| mil | meter (m) | 2.540 000*E − 05 |
| mile (international) | meter (m) | 1.609 344*E + 03 |
| mile (statute) | meter (m) | 1.609 3 E + 03 |
| mile (U.S. survey)‡ | meter (m) | 1.609 347 E + 03 |
| mile (international nautical) | meter (m) | 1.852 000*E + 03 |
| mile (U.K. nautical) | meter (m) | 1.853 184*E + 03 |
| mile (U.S. nautical) | meter (m) | 1.852 000*E + 03 |
| $mi^2$ (international) | $meter^2$ ($m^2$) | 2.589 988 E + 06 |
| $mi^2$ (U.S. survey)‡ | $meter^2$ ($m^2$) | 2.589 998 E + 06 |
| mi/h (international) | meter per second (m/s) | 4.470 400*E − 01 |
| mi/h (international) | kilometer per hour (km/h) | 1.609 344*E + 00 |
| mi/min (international) | meter per second (m/s) | 2.682 240*E + 01 |
| mi/s (international) | meter per second (m/s) | 1.609 344*E + 03 |
| millibar | pascal (Pa) | 1.000 000*E + 02 |
| millimeter of mercury (0°C) | pascal (Pa) | 1.333 22 E + 02 |
| minute (angle) | radian (rad) | 2.908 882 E − 04 |
| minute (mean solar) | second (s) | 6.000 000 E + 01 |
| minute (sidereal) | second (s) | 5.983 617 E + 01 |
| month (mean calendar) | second (s) | 2.628 000 E + 06 |
| oersted | ampere per meter (A/m) | 7.957 747 E + 01 |
| ohm centimeter | ohm meter (Ω · m) | 1.000 000*E − 02 |
| ohm circular-mil per foot | ohm $millimeter^2$ per meter (Ω · $mm^2$/m) | 1.662 426 E − 03 |
| ounce (avoirdupois) | kilogram (kg) | 2.834 952 E − 02 |
| ounce (troy or apothecary) | kilogram (kg) | 3.110 348 E − 02 |
| ounce (U.K. fluid) | $meter^3$ ($m^3$) | 2.841 307 E − 05 |

‡ See end of Table.

(*Continued*)

**Table A-6 Conversion factors for converting from American engineering units to SI units—alphabetical listing in detail** (*Continued*)

| To convert from | To | Multiply by |
|---|---|---|
| ounce (U.S. fluid) | meter$^3$ (m$^3$) | 2.957 353 E − 05 |
| ounce-force | newton (N) | 2.780 139 E − 01 |
| ozf · in | newton meter (N · m) | 7.061 552 E − 03 |
| oz (avoirdupois)/gal (U.K. liquid) | kilogram per meter$^3$ (kg/m$^3$) | 6.236 021 E + 00 |
| oz (avoirdupois)/gal (U.S. liquid) | kilogram per meter$^3$ (kg/m$^3$) | 7.489 152 E + 00 |
| oz (avoirdupois)/in$^3$ | kilogram per meter$^3$ (kg/m$^3$) | 1.729 994 E + 03 |
| oz (avoirdupois)/ft$^2$ | kilogram per meter$^2$ (kg/m$^2$) | 3.051 517 E − 01 |
| oz (avoirdupois)/yd$^2$ | kilogram per meter$^2$ (kg/m$^2$) | 3.390 575 E − 02 |
| parsec | meter (m) | 3.085 678 E + 16 |
| peck (U.S.) | meter$^3$ (m$^3$) | 8.809 768 E − 03 |
| pennyweight | kilogram (kg) | 1.555 174 E − 03 |
| perm (0°C) | kilogram per pascal second meter$^2$ (kg/Pa · s · m$^2$) | 5.721 35 E − 11 |
| perm (23°C) | kilogram per pascal second meter$^2$ (kg/Pa · s · m$^2$) | 5.745 25 E − 11 |
| perm · in (0°C) | kilogram per pascal second meter (kg/Pa · s · m) | 1.453 22 E − 12 |
| perm · in (23°C) | kilogram per pascal second meter (kg/Pa · s · m) | 1.459 29 E − 12 |
| phot | lumen per meter$^2$ (lm/m$^2$) | 1.000 000*E + 04 |
| pica (printer's) | meter (m) | 4.217 518 E − 03 |
| pint (U.S. dry) | meter$^3$ (m$^3$) | 5.506 105 E − 04 |
| pint (U.S. liquid) | meter$^3$ (m$^3$) | 4.731 765 E − 04 |
| point (printer's) | meter (m) | 3.514 598*E − 04 |
| poise (absolute viscosity) | pascal second (Pa · s) | 1.000 000*E − 01 |
| pound (lb avoirdupois) | kilogram (kg) | 4.535 924 E − 01 |
| pound (troy or apothecary) | kilogram (kg) | 3.732 417 E − 01 |
| lb · ft$^2$ (moment of inertia) | kilogram meter$^2$ (kg · m$^2$) | 4.214 011 E − 02 |
| lb · in$^2$ (moment of inertia) | kilogram meter$^2$ (kg · m$^2$) | 2.926 397 E − 04 |
| lb/ft · h | pascal second (Pa · s) | 4.133 789 E − 04 |
| lb/ft · s | pascal second (Pa · s) | 1.488 164 E + 00 |
| lb/ft$^2$ | kilogram per meter$^2$ (kg/m$^2$) | 4.882 428 E + 00 |
| lb/ft$^3$ | kilogram per meter$^3$ (kg/m$^3$) | 1.601 846 E + 01 |
| lb/gal (U.K. liquid) | kilogram per meter$^3$ (kg/m$^3$) | 9.977 633 E + 01 |
| lb/gal (U.S. liquid) | kilogram per meter$^3$ (kg/m$^3$) | 1.198 264 E + 02 |
| lb/h | kilogram per second (kg/s) | 1.259 979 E − 04 |
| lb/hp · h (SFC, specific fuel consumption) | kilogram per joule (kg/J) | 1.689 659 E − 07 |
| lb/in$^3$ | kilogram per meter$^3$ (kg/m$^3$) | 2.767 990 E + 04 |
| lb/min | kilogram per second (kg/s) | 7.559 873 E − 03 |
| lb/s | kilogram per second (kg/s) | 4.535 924 E − 01 |
| lb/yd$^3$ | kilogram per meter$^3$ (kg/m$^3$) | 5.932 764 E − 01 |
| poundal | newton (N) | 1.382 550 E − 01 |
| poundal/ft$^2$ | pascal (Pa) | 1.488 164 E + 00 |
| poundal · s/ft$^2$ | pascal second (Pa · s) | 1.488 164 E + 00 |
| pound-force (lbf) | newton (N) | 4.448 222 E + 00 |
| lbf · ft | newton meter (N · m) | 1.355 818 E + 00 |
| lbf · ft/in | newton meter per meter (N · m/m) | 5.337 866 E + 01 |
| lbf · in | newton meter (N · m) | 1.129 848 E − 01 |
| lbf · in/in | newton meter per meter (N · m/m) | 4.448 222 E + 00 |

**Table A-6 Conversion factors for converting from American engineering units to SI units—alphabetical listing in detail** (*Continued*)

| To convert from | To | Multiply by |
|---|---|---|
| lbf · s/ft$^2$ | pascal second (Pa · s) | 4.788 026 E + 01 |
| lbf/ft | newton per meter (N/m) | 1.459 390 E + 01 |
| lbf/ft$^2$ | pascal (Pa) | 4.788 026 E + 01 |
| lbf/in | newton per meter (N/m) | 1.751 268 E + 02 |
| lbf/in$^2$ (psi) | pascal (Pa) | 6.894 757 E + 03 |
| lbf/lb (thrust/weight [mass] ratio) | newton per kilogram (N/kg) | 9.806 650 E + 00 |
| quart (U.S. dry) | meter$^3$ (m$^3$) | 1.101 221 E − 03 |
| quart (U.S. liquid) | meter$^3$ (m$^3$) | 9.463 529 E − 04 |
| rad (radiation dose absorbed) | gray (Gy) | 1.000 000*E − 02 |
| rhe | 1 per pascal second (1/Pa · s) | 1.000 000*E + 01 |
| rod | meter (m) | [see footnote ‡] |
| roentgen | coulomb per kilogram (C/kg) | 2.58 E − 04 |
| second (angle) | radian (rad) | 4.848 137 E − 06 |
| second (sidereal) | second (s) | 9.972 696 E − 01 |
| section | meter$^2$ (m$^2$) | [see footnote ‡] |
| shake | second (s) | 1.000 000*E − 08 |
| slug | kilogram (kg) | 1.459 390 E + 01 |
| slug/ft · s | pascal second (Pa · s) | 4.788 026 E + 01 |
| slug/ft$^3$ | kilogram per meter$^3$ (kg/m$^3$) | 5.153 788 E + 02 |
| statampere | ampere (A) | 3.335 640 E − 10 |
| statcoulomb | coulomb (C) | 3.335 640 E − 10 |
| statfarad | farad (F) | 1.112 650 E − 12 |
| stathenry | henry (H) | 8.987 554 E + 11 |
| statmho | siemens (S) | 1.112 650 E − 12 |
| statohm | ohm (Ω) | 8.987 554 E + 11 |
| statvolt | volt (V) | 2.997 925 E + 02 |
| stere | meter$^3$ (m$^3$) | 1.000 000*E + 00 |
| stilb | candela per meter$^2$ (cd/m$^2$) | 1.000 000*E + 04 |
| stokes (kinematic viscosity) | meter$^2$ per second (m$^2$/s) | 1.000 000*E − 04 |
| tablespoon | meter$^3$ (m$^3$) | 1.478 676 E − 05 |
| teaspoon | meter$^3$ (m$^3$) | 4.928 922 E − 06 |
| tex | kilogram per meter (kg/m) | 1.000 000*E − 06 |
| therm | joule (J) | 1.055 056 E + 08 |
| ton (assay) | kilogram (kg) | 2.916 667 E − 02 |
| ton (long, 2240 lb) | kilogram (kg) | 1.016 047 E + 03 |
| ton (metric) | kilogram (kg) | 1.000 000*E + 03 |
| ton (nuclear equivalent of TNT) | joule (J) | 4.184 E + 09 |
| ton (refrigeration) | watt (W) | 3.516 800 E + 03 |
| ton (register) | meter$^3$ (m$^3$) | 2.831 685 E + 00 |
| ton (short, 2000 lb) | kilogram (kg) | 9.071 847 E + 02 |
| ton (long)/yd$^3$ | kilogram per meter$^3$ (kg/m$^3$) | 1.328 939 E + 03 |
| ton (short)/h | kilogram per second (kg/s) | 2.519 958 E − 01 |
| ton-force (2000 lbf) | newton (N) | 8.896 444 E + 03 |
| tonne | kilogram (kg) | 1.000 000*E + 03 |
| torr (mmHg, 0°C) | pascal (Pa) | 1.333 22 E + 02 |
| township | meter$^2$ (m$^2$) | [see footnote ‡] |
| unit pole | weber (Wb) | 1.256 637 E − 07 |
| W · h | joule (J) | 3.600 000*E + 03 |

(*Continued*)

**Table A-6 Conversion factors for converting from American engineering units to SI units—alphabetical listing in detail** (*Continued*)

| To convert from | To | Multiply by |
|---|---|---|
| W · s | joule (J) | 1.000 000*E + 00 |
| W/Cm$^2$ | watt per meter$^2$ (W/m$^2$) | 1.000 000*E + 04 |
| W/in$^2$ | watt per meter$^2$ (W/m$^2$) | 1.550 003 E + 03 |
| yard | meter (m) | 9.144 000*E − 01 |
| yd$^2$ | meter$^2$ (m$^2$) | 8.361 274 E − 01 |
| yd$^3$ | meter$^3$ (m$^3$) | 7.645 549 E − 01 |
| yd$^3$/min | meter$^3$ per second (m$^3$/s) | 1.274 258 E − 02 |
| year (calendar) | second (s) | 3.153 600 E + 07 |
| year (sidereal) | second (s) | 3.155 815 E + 07 |
| year (tropical) | second (s) | 3.155 693 E + 07 |

* Exact equivalence.

† Adapted from ASTM *Standard for Metric Practice E 380–76*. The conversion factors are listed in standard form for computer readout as a number greater than one or less than ten with six or less decimal points. The number is followed by the letter E (for exponent), a plus or minus symbol, and two digits which indicate the power of 10 by which the number must be multiplied. An asterisk (*) after the sixth decimal place indicates that the conversion factor is exact and that all subsequent digits are zero. All other conversion factors have been rounded to the figures given. Where less than six decimal places are shown, more precision is not warranted.

For example, 1.013 250*E + 05 is exactly $1.013\ 250 \times 10^5$ or 101 325.0.
1.589 873 E − 01 has the last digit rounded off to 3 and is
$1.589\ 873 \times 10^{-1}$ or 0.158 987 3.

‡ Since 1893, the U.S. basis of length measurement has been derived from metric standards. In 1959, a small refinement was made in the definition of the yard to resolve discrepancies both in this country and abroad which changed its length from 3600/3937 m to 0.9144 m exactly. This resulted in the new value being shorter by two parts in a million. At the same time, it was decided that any data in feet derived from and published as a result of geodesic surveys within the U.S. would remain with the old standard (1 ft = 1200/3937 m) until further decision. This foot is named the U.S. survey foot. As a result, all U.S. land measurements in American engineering units will relate to the meter by the old standard. All the conversion factors in this table for units referenced to this footnote are based on the U.S. survey foot rather than on the international foot.

Conversion factors for the land measures given below may be determined from the following relationships:

1 league = 3 miles (exactly)
1 rod = $16\frac{1}{2}$ feet (exactly)
1 section = 1 square mile (exactly)
1 township = 36 square miles (exactly)

§ By definition, one calorie (International Table) is exactly 4.186 8 absolute joules which converts to $1.055\ 056 \times 10^3$ joules for one Btu (International Table). Also, by definition, one calorie (thermochemical) is exactly 4.184 absolute joules which converts to $1.054\ 350 \times 10^3$ joules for one Btu (thermochemical). A *mean* calorie is $\frac{1}{100}$th of the heat required to raise the temperature of one gram of water at one atmosphere pressure from 0°C to 100°C and equals 4.190 02 absolute joules. In all cases, the relationship between calorie and British thermal unit is established by 1 cal/(g · °C) = 1 Btu/(lb · °F). A *mean* Btu, therefore, is $\frac{1}{180}$th of the heat required to raise the temperature of one pound of water at one atmosphere pressure from 32°F to 212°F and equals $1.055\ 87 \times 10^3$ joules. When values are given as Btu or calories, the type of unit (International Table, thermochemical, mean, or temperature of determination) should be given. In all cases for this table, conversions involving joules are based on the absolute joule.

**Table A-7 Abbreviated list of equivalences for converting units commonly used by chemical engineers to acceptable SI units**

| Unacceptable unit | Acceptable SI unit with unit conversion factor |
|---|---|
| ångström | 0.1 nm* |
| atmosphere (standard) | 101.325 kPa |
| Btu† | 1.055 056 kJ |
| Btu/(lbm · °F) (heat capacity) | 4.186 8 kJ/(kg · K)* |
| Btu/h | 0.293 971 1 W |
| Btu/ft$^2$ | 11.356 53 kJ/m$^2$ |
| Btu/(ft$^2$ · h · °F) (heat transfer coefficient) | 5.678 263 J/(m$^2$ · s · K) |
| Btu/(ft$^2$ · h) (heat flux) | 3.154 591 J/(m$^2$ · s) |
| Btu/(ft · h · °F) (thermal conductivity) | 1.730 735 J/(m · s · K) |
| calorie† | 4.186 8 J* |
| cal/(g · °C) (heat capacity) | 4.186 8 kJ/(kg · K)* |
| centipoise (absolute viscosity) | 1.0 mPa · s* |
| centistoke (kinematic viscosity) | 1.0 × $10^{-6}$ m$^2$/s* |
| $t$(°F) | ($t$ + 459.67)/(1.8) K |
| $t$(°R) | $t$/(1.8) K* |
| dyne | 10.0 μN* |
| erg | 100 pJ* |
| foot‡ | 0.3048 m* |
| ft$^2$ | 9.290 304 × $10^{-2}$ m$^2$* |
| ft$^3$ | 2.831 685 × $10^{-2}$ m$^3$ |
| gallon (U.S. liquid) | 3.785 412 × $10^{-3}$ m$^3$ |
| horsepower (550 ft · lbf/s) | 745.699 9 W |
| inch | 2.54 × $10^{-2}$ m* |
| in Hg (60°F) (inches mercury pressure) | 3.376 85 kPa |
| in $H_2O$ (60°F) (inches water pressure) | 0.248 84 kPa |
| kgf (kilogram force) | 9.806 65 N* |
| mile | 1 609.344 m* |
| mmHg (0°C) (millimeters mercury pressure) | 0.133 322 kPa |
| poise (absolute viscosity) | 0.1 Pa · s* |
| lbf (pounds force) | 4.448 222 N |
| lbm (pounds mass–avoirdupois) | 0.453 592 4 kg |
| lb/in$^2$ (pounds per square inch pressure) | 6.894 757 kPa |
| stoke (kinematic viscosity) | 1.0 × $10^{-4}$ m$^2$/s* |
| yard | 0.9144 m* |

* Exact equivalence.

† British thermal unit and calorie are reported as the International Table values as adopted in 1956 for all cases in this table. The exact conversion factor for Btu (International Table) to kJ is 1.055 055 852 62. The Btu (thermochemical) is 1.054 350 kJ and the calorie (thermochemical) is exactly 4.184 J. (See footnote § for Table 6.)

‡ The foot is reported as the International Table value and holds for all cases of length in this table.

APPENDIX

# B

# DERIVATION OF SPECIAL EQUATIONS



## DIMENSIONAL ANALYSIS

The methods of dimensional analysis are widely used in the field of fluid dynamics as well as in many other fields. Before considering the actual methods of dimensional analysis, it is necessary to indicate exactly what is meant by the term "dimensions." This term is used here to designate the type of quantity which is measured. For example, a distance may be measured in meters, feet, miles, or any other *length* dimension. However, a time interval cannot be measured in length dimensions. A different type of dimension with units such as seconds, minutes, or hours must be used for measuring time.

Any quantity is either dimensional or dimensionless. A distance measured in feet is dimensional. It has a length dimension with the units of feet. The number represented by $\pi$ in the equation for the area of a circle ($A = \pi r^2$) is dimensionless. It is a pure number having no dimensions.

There are three fundamental dimensions encountered in fluid dynamics. These are length ($L$), mass ($M$), and time ($\theta$). Force ($F$) is sometimes considered as a fundamental dimension along with the two other dimensions of length and time. According to Newton's law of motion, force is proportional to mass times acceleration, or

$$\text{Force} = \alpha ma$$

**Table B-1 Dimensions of common mechanical quantities**

| Quantity | Net dimensions | |
|---|---|---|
| | $M, L, \theta$ | $F, L, \theta$ |
| Length | $L$ | $L$ |
| Diameter | $L$ | $L$ |
| Equivalent roughness of pipe surface | $L$ | $L$ |
| Area | $L^2$ | $L^2$ |
| Volume | $L^3$ | $L^3$ |
| Velocity | $L/\theta$ | $L/\theta$ |
| Acceleration | $L/\theta^2$ | $L/\theta^2$ |
| Mass | $M$ | $F\theta^2/L$ |
| Force | $ML/\theta^2$ | $F$ |
| Energy | $ML^2/\theta^2$ | $FL$ |
| Mass rate of flow | $M/\theta$ | $F\theta/L$ |
| Density | $M/L^3$ | $F\theta^2/L^4$ |
| Pressure | $M/L\theta^2$ | $F/L^2$ |
| Viscosity | $M/L\theta$ | $F\theta/L^2$ |

where $\alpha$ is the proportionality constant with dimensions and a numerical value dependent on the units involved. The dimensions of acceleration are length per unit time per unit time, or $L/\theta^2$. Thus, the units of force may be expressed in terms of the fundamental length, time, and mass dimensions as

$$\text{Force} = \alpha ML/\theta^2$$

For some purposes, force, mass, length, and time are considered as fundamental, but they are definitely interrelated by Newton's law of motion.

Table B-1 lists the dimensions for some of the common mechanical quantities ordinarily encountered in chemical engineering work. The dimensions are indicated on the basis of the mass-length-time and force-length-time systems.

Additional fundamental dimensions may be involved in fields other than fluid mechanics. For example, the dimensions of temperature ($T$) or temperature difference as well as a fundamental heat dimension may appear in heat-transfer considerations.

**Dimensionless Groups.** True equations in their simplest form must be dimensionally homogeneous. Thus, if an equation indicates that velocity equals acceleration times time ($V = at$), the dimensions of velocity ($L/\theta$) must be consistent with the dimensions of acceleration ($L/\theta^2$) multiplied by the dimensions of time ($\theta$). Since $L/\theta = (L/\theta^2)(\theta)$, this equation is dimensionally homogeneous.

Since any true equation must be dimensionally homogeneous, it is possible to express all the variables in terms of dimensionless groups. If each side of the equation $V = at$ is divided by $at$, the result is $V/at = 1$. In this example, $V/at$ is a dimensionless group as can be seen by checking the dimensions.

$$(L/\theta)(\theta^2/L)(1/\theta) = 1$$

## Dimensional Analysis

**Introduction.** The principle of dimensional homogeneity is made use of in determining the relationship among physical variables in certain types of processes. Buckingham[1] in conjunction with the development of his "pi theorem," deduced that the number of dimensionless groups involved in a mathematical representation of a physical process is equal to the number of physical variables involved, minus the number of fundamental dimensions used to express them. If there were no exceptions to this rule, it would be possible for the novice to carry out a dimensional analysis by merely inspecting the dimensions of the different variables and then arranging the variables into pertinent dimensionless groups. This is essentially the method used in the pi theorem.

The simple rule given in the preceding paragraph is not always true, and some of the exceptions are not immediately obvious. Owing to the possible violations of the general rule, the analytical algebraic method is recommended for carrying out a dimensional analysis.

The results obtained by the methods of dimensional analysis can serve only as a guide to show the possible relationships among the variables. The results cannot be used until they have been checked by experimental tests. If too many variables are included or if all the pertinent variables are not included, the results of a dimensional analysis will be meaningless and cannot be checked experimentally.

The basic principle underlying dimensional analysis may be expressed as follows: A general relationship between the essential physical quantities involved in a process may be expressed in a form involving only the dimensionless products of the physical variables and the necessary relating constants.

**Algebraic method of dimensional analysis (the Fanning equation).** The algebraic method for carrying out a dimensional analysis is best shown by an example. The Fanning equation, as presented in Chap. 5 (Fluid Flow), may be derived by dimensional analysis, and its derivation is given here.

When a fluid flows through a long pipe of constant diameter at a constant mass flow rate, the energy dissipated because of friction ($E_F$) is a function of the properties of the fluid and the confining system. The essential properties of the fluid are its linear velocity ($V$), its density ($\rho$), and its viscosity ($\mu$). The essential properties of the confining system are the pipe diameter ($D$), pipe length ($L$), and the equivalent surface roughness ($\epsilon$).

Assuming a basis of unit mass of fluid flowing, and designating the energy dissipated due to friction per unit mass of fluid as $E_F/m$, the overall function may be expressed as follows:

$$\text{Energy dissipated due to friction} = \frac{E_F}{m} = \phi_1[V, \rho, \mu, D, L, \epsilon] \tag{1}$$

[1] E. Buckingham, *Phys. Rev*, **4**: 345–376 (1914).

This expression may be written in the following form:

$$\frac{E_F}{m} = z(V)^a(\rho)^b(\mu)^c(D)^d(L)^e(\epsilon)^h \tag{2}$$

where $z$ is a dimensionless factor and the dimensionless exponents may have any value required by the situation. The values of $z$ and all the exponents are constant for point conditions or for any conditions when the physical quantities remain constant. Equation (2) is merely an alternate way of expressing Eq. (1), and it has no true *physical* significance.

The following dimensional equation is obtained by substituting the mass-length-time dimensions from Table B-1 for the variables in Eq. (2):

$$\frac{ML^2}{M\theta^2} = z(L/\theta)^a(M/L^3)^b(M/L\theta)^c(L)^d(L)^e(L)^h \tag{3}$$

Since this expression must be dimensionally homogeneous, the exponents are subject to the following limitations:

$$\text{For } M: \quad 0 = b + c \tag{4}$$

$$\text{For } L: \quad 2 = a - 3b - c + d + e + h \tag{5}$$

$$\text{For } \theta: \quad -2 = -a - c \tag{6}$$

These three separate equations involve six unknowns. Therefore, they may be solved in terms of three arbitrarily chosen unknowns, such as $c$, $e$, and $h$, to give

$$a = 2 - c$$

$$b = -c$$

$$d = -e - h - c$$

Substituting these values into Eq. (2) gives

$$\frac{E_F}{m} = z(V)^{2-c}(\rho)^{-c}(\mu)^c(D)^{-e-h-c}(L)^e(\epsilon)^h \tag{7}$$

Equation (7) may be rearranged into dimensionless groups raised to powers involving the constants $c$, $e$, and $h$, as follows:

$$\frac{E_F}{mV^2} = z(L/D)^e(DV\rho/\mu)^{-c}(\epsilon/D)^h \tag{8}$$

The constants $c$, $e$, and $h$ are arbitrary and may have any value. Experience has shown that frictional energy is directly proportional to the length of the pipe. Therefore, $e$ must have a value of unity, and

$$\frac{E_F}{m} = V^2\frac{L}{D}\phi_2\left[\left(\frac{DV\rho}{\mu}\right), \left(\frac{\epsilon}{D}\right)\right] \tag{9}$$

The expression $E_F/m$ represents energy dissipated because of friction per unit mass of flowing fluid. According to Newton's law of motion, this term must be divided by the conversion factor $g_c$ to convert it to American engineering units of foot-pounds force per pound-mass, or merely by 1.0 if SI units are used to give joules per kilogram. Designating the symbol $F$ as the amount of frictional energy dissipated in units of foot-pounds force per pound-mass or joules per kilogram,

$$F = \frac{E_F}{mg_c} = \frac{V^2L}{g_cD}\phi_2\left[\left(\frac{DV\rho}{\mu}\right), \left(\frac{\epsilon}{D}\right)\right] \text{ (with American engineering units)} \quad (10a)$$

or

$$F = \frac{E_F}{m} = \frac{V^2L}{D}\phi_2\left[\left(\frac{DV\rho}{\mu}\right), \left(\frac{\epsilon}{D}\right)\right] \text{ (with SI units)} \quad (10b)$$

Letting

$$2f = \phi_2\left[\left(\frac{DV\rho}{\mu}\right), \left(\frac{\epsilon}{D}\right)\right],$$

the following common form of the Fanning equation is obtained:

$$F = 2f\frac{V^2L}{g_cD} \text{ (with American engineering units)} \quad (11a)$$

or

$$F = 2f\frac{V^2L}{D} \text{ (with SI units)} \quad (11b)$$

This equation applies strictly only to point conditions or to conditions where the velocity, density, and viscosity of the fluid are essentially constant. Experimental tests have indicated that it is a correct representation of the actual relationship among the different variables. Therefore, the Fanning equation has been derived by dimensional analysis, and empirical data have indicated that the dimensional analysis took all the essential variables into consideration.

The frictional energy term $F$ is equal to the pressure drop over the system due to friction, divided by the density of the flowing fluid. Consequently, the Fanning equation could have been derived by considering pressure drop due to friction as a variable in place of the term $E_F/m$. The same result would also have been obtained by using the force-length-time or force-mass-length-time system for the fundamental dimensions.

## EXPERIMENTAL METHOD FOR DETERMINING VALUES OF THE FANNING FRICTION FACTOR

Values for the Fanning friction factor ($f$) are based on experimental data. The general procedure for determining these data will be described here. A fluid, such as water, is passed at a constant rate through a long horizontal pipe of known length, diameter, and effective roughness. The pressure drop over the pipe is determined by means of a manometer connected between the two ends of the pipe. The manometer legs are connected to the pipe by means of flush wall taps, and the manometer reading indicates the difference in static pressures between the entrance and exit of the system. The manometer taps should be located at a distance of about 50 pipe diameters from any bends or fittings in order to obtain constant flow conditions.

The water-flow rate may be measured by means of a calibrated measuring device such as a rotameter or by collecting the water in a tared container over a measured time interval. The temperature of the water indicates its density and viscosity. The pressure drop over the known length of pipe divided by the water density is equivalent to the mechanical-energy loss due to friction ($F$ in the Fanning equation). The linear flow velocity can be determined from the mass rate of flow, the density of the liquid, and the pipe diameter.

The fluid density, fluid viscosity, linear velocity, and pipe diameter set the value of the Reynolds number. Knowing the pipe length, the linear velocity, the pipe diameter, and the value of $F$, it is possible to determine the value of $f$ from the Fanning equation.

By making runs at different flow rates, sufficient data can be obtained to make a plot of Reynolds number versus friction factor. This curve should correspond to the curve for the particular equivalent roughness of the test pipe as shown in Fig. 5-1 (Chap. 5, Fluid Flow).

## DERIVATION OF THE TOTAL MECHANICAL-ENERGY BALANCE FROM THE TOTAL ENERGY BALANCE

The following integrated form of the total energy balance for conditions of steady flow was developed in Chap. 5 (Fluid Flow):

$$Z_1 \frac{g}{g_c} + p_1 v_1 + \frac{V_{i_1}^2}{2g_c} + u_1 + Q + W_o = Z_2 \frac{g}{g_c} + p_2 v_2 + \frac{V_{i_2}^2}{2g_c} + u_2 \qquad (1)$$

The differential form of the total energy balance is

$$\frac{g}{g_c}\,dZ + d(pv) + \frac{V_i\,dV_i}{g_c} + du = \delta Q + \delta W_o \qquad (2)$$

where $Z$ = vertical distance above an arbitrarily chosen datum plane, m or ft
$p$ = absolute pressure, Pa, $N/m^2$, or $lbf/ft^2$
$v$ = specific volume of the fluid, $m^3/kg$ or $ft^3/lbm$
$V_i$ = instantaneous or point linear velocity, m/s or ft/s
$u$ = internal energy, J/kg or ft · lbf/lbm
$Q$ = net heat energy imparted to the fluid from an outside source, J/kg or ft · lbf/lbm
$W_o$ = mechanical work imparted to the fluid from an outside source, J/kg or ft · lbf/lbm

The following assumptions are made in the development of the total mechanical-energy balance:

1. The fluid flows through the system at a constant mass rate.
2. No chemical changes occur in the system.
3. Minor energy changes, such as those caused by surface effects, are negligible.

When a fluid flows through a system, some mechanical energy is lost as such as a result of irreversible changes caused by friction. From a consideration of the first and second laws of thermodynamics, the change in internal energy during a reversible process may be expressed as

$$du = T\,dS - p\,dv = \delta Q_{\text{rev}} - p\,dv \qquad (3)$$

where $T$ = absolute temperature
$S$ = entropy
$Q_{\text{rev}}$ = reversible heat energy added to the system

Any real flow process involves a certain amount of irreversibility because of frictional effects. Because of this irreversibility, mechanical energy is lost by frictional conversion into heat. The actual net amount of heat added to the system plus the mechanical-energy losses caused by irreversibilities may, by definition, be equated to the heat which would have been added to the system if the process were reversible. Therefore,

$$\delta Q + \delta F = \delta Q_{\text{rev}} \tag{4}$$

where $\delta F$ represents the difference between the reversible heat and the actual net heat added to the differential system.

Combining Eqs. (4) and (3) with Eq. (2) gives

$$\frac{g}{g_c}\,dZ + d(pv) + \frac{V_i dV_i}{g_c} - p\,dv = \delta W_o - \delta F \tag{5}$$

Since $d(pv) = p\,dv + v\,dp$, Eq. (5) may be expressed as

$$\frac{g}{g_c}\,dZ + v\,dp + \frac{V_i\,dV_i}{g_c} = \delta W_o - \delta F \tag{6}$$

Integration of Eq. (5) or (6) between points 1 and 2 gives

$$Z_1\frac{g}{g_c} + p_1 v_1 + \frac{V_{i_1}^2}{2g_c} + \int_1^2 p\,dv + W_o = Z_2\frac{g}{g_c} + p_2 v_2 + \frac{V_{i_2}^2}{2g_c} + \Sigma F \tag{7}$$

where $\Sigma F$ represents the frictional effects involved and is, in reality, best defined as the term necessary to balance Eq. (7).

The instantaneous linear velocity ($V_i$) may be replaced by the average linear velocity ($V$) if a correction factor is included. This substitution yields the following integrated form of the *total mechanical-energy balance*.

$$Z_1\frac{g}{g_c} + p_1 v_1 + \int_1^2 p\,dv + \frac{V_1^2}{2\alpha_1 g_c} + W_o = Z_2\frac{g}{g_c} + p_2 v_2 + \frac{V_2^2}{2\alpha_2 g_c} + \Sigma F \tag{8}$$

where $\alpha = 1.0$ if the flow is turbulent
$\alpha = 0.5$ if the flow is streamline

## DERIVATION OF THE GENERAL RAYLEIGH EQUATION FOR SIMPLE BATCH (DIFFERENTIAL) DISTILLATION

In a simple batch distillation, the composition of the evolved vapors and the composition of the liquid in the still pot are continuously changing. Therefore, the mathematical analysis of the operation must be based on differential changes.

If the total number of moles of liquid in the still pot at any time is designated as $L$, the moles of vapor formed over a differential unit of time must be $-dL$. Let the mole fraction of one component in the liquid be represented by $x$. Let the mole fraction of the same component in the vapors be represented by $y$, where $y$ is the vapor composition in equilibrium with a liquid having a composition of $x$.

The moles of the component under consideration appearing in the vapors over the differential unit of time must be $-y\,dL$. The number of moles of the component in the liquid at any time is $xL$; therefore, over the differential unit of time, the moles of the given component leaving the liquid must be $-d(xL)$. By a material balance, the moles of the component leaving the liquid must equal the moles of the component appearing in the vapor, or

$$-d(xL) = -y\,dL \tag{1}$$

and

$$L\,dx + x\,dL = y\,dL \tag{2}$$

Rearranging Eq. (2) gives the following differential form of the Rayleigh equation:

$$\frac{dL}{L} = \frac{dx}{y - x} \tag{3}$$

Integrating Eq. (3) between the limits of time 1 and time 2 gives the following integrated form of the general Rayleigh equation:

$$\ln \frac{L_1}{L_2} = \int_{x_2}^{x_1} \frac{dx}{y - x} \tag{4}$$

where $L_1$ and $L_2$ represent the moles of liquid in the still pot at time 1 and time 2. Similarly, $x_1$ and $x_2$ represent the mole fraction of the component in the liquid contained in the still pot at time 1 and time 2.

If a mathematical expression relating $x$ and $y$ is known, it may be possible to evaluate the integral term in Eq. (4) analytically. However, it is usually necessary to resort to a graphical-integration method requiring the evaluation of the area under a curve of $1/(y - x)$ vs. $x$ between the limits of $x_2$ and $x_1$.

## DERIVATION OF EQUATION FOR DRYING TIME DURING FALLING-RATE PERIOD

In a drying operation, the rate of drying during the falling-rate period is a function of the free moisture content of the material involved. As the free moisture content decreases, the rate of drying decreases.

Over any differential unit of time, the drying rate per unit mass of dry solid may be expressed as

$$\text{Rate of drying per unit mass of dry solid} = -\frac{dW}{d\theta} \tag{1}$$

where rate of drying per unit mass of dry solid = rate at which water is removed from the drying material, kg water/(kg dry solid)(h) or lb water/(lb dry solid)(h)

$W$ = free moisture content, kg water/kg dry solid or lb water/lb dry solid

$\theta$ = time, h

If $Q$ is the total mass of dry solid as kilograms or pounds, the net instantaneous rate of drying must be

$$\text{Rate of drying (as kg or lb of water removed/h)} = -Q\frac{dW}{d\theta} \tag{2}$$

Since the rate of drying during the falling-rate period is a function of the free moisture content,

$$\text{Rate of drying (as kg or lb of water removed/h)} = f(W) \tag{3}$$

Combining Eqs. (2) and (3) and rearranging,

$$d\theta = -Q\frac{dW}{f(W)} \tag{4}$$

Equation (4) may be integrated between the limits imposed by the falling-rate period as follows:

$$\int_0^{\theta_f} d\theta = -Q\int_{W_C}^{W_F} \frac{dW}{f(W)} = \theta_f \tag{5}$$

where $\theta_f$ = time of drying in falling-rate period, h

$W_C$ = critical free moisture content = free moisture content at start of falling-rate period, kg water/kg dry solid or lb water/lb dry solid

$W_F$ = free moisture content at end of drying, kg water/kg dry solid or lb water/lb dry solid

If experimental data are available giving the relationship between the drying rate and the free moisture content, the integral in Eq. (5) can be evaluated by plotting $1/f(W)$ vs. $W$ and determining the area included between the limits of $W_C$ and $W_F$. If these data are not available, an approximate solution may be obtained by assuming that the drying rate decreases linearly with the free moisture content between the critical free moisture content and zero free moisture content (see Fig. 11-1). With this assumption, the net instantaneous rate of drying may be expressed as a function of the free moisture content as follows:

$$\text{Rate of drying (as kg or lb of water removed/h)} = \frac{R_c}{W_C} W \tag{6}$$

where $R_c$ is the rate of drying during the constant-rate period expressed as kilograms or pounds of water removed per hour.

Equations (6) and (2) may be combined to give

$$d\theta = -Q \frac{W_C}{R_c} \frac{dW}{W} \tag{7}$$

Integration of Eq. (7) between the limits imposed by the falling-rate period gives

$$\int_0^{\theta_f} d\theta = -Q \frac{W_C}{R_c} \int_{W_C}^{W_F} \frac{dW}{W} \tag{8}$$

Therefore

$$\theta_f = \frac{W_C Q}{R_c} \ln \frac{W_C}{W_F} = \frac{2.3 W_C Q}{R_c} \log \frac{W_C}{W_F} \tag{9}$$

where $\theta_f$ = time of drying in falling-rate period, h
$Q$ = total mass of dry solid, kg or lb
$R_c$ = rate of drying during the constant-rate period, kg water removed/h or lb water removed/h
$W_C$ = critical free moisture content, kg water/kg dry solid or lb water/lb dry solid
$W_F$ = free moisture content at end of drying, kg water/kg dry solid or lb water/lb dry solid

Equation (9) applies when the drying-rate curve over the falling-rate period is a straight line connecting the critical point with the origin.

## OPTIMUM ECONOMIC PIPE DIAMETER

The cost for piping installations represents one of the major investments in industrial processes. Consequently, it is important to consider the economic aspects when designing piping systems.

The pipe diameter is a critical factor in the determination of pumping power costs. With a constant flow rate, the pressure drop due to friction increases as the pipe diameter decreases. More pumping power must then be supplied to overcome the larger pressure drop. However, as the pipe diameter decreases, the initial investment for the pipe also decreases. This means the fixed charges on the piping equipment are less for smaller-diameter pipes, but this beneficial economic effect is counterbalanced by increased pumping power costs.

There must be some optimum pipe diameter at each fluid flow rate where the total of annual pumping power costs and fixed charges based on investment costs is a minimum. This particular diameter is designated as the *optimum economic pipe diameter*. A general expression for the optimum economic pipe diameter may be derived on the basis of the total mechanical-energy balance and pipe-cost relationships.[1]

For any given operating conditions involving the flow of a noncompressible fluid through a relatively long pipe of constant diameter, the total mechanical-energy balance can be reduced to the following form:

$$W_o = \frac{2fV^2L(1 + J)}{g_c D} + B \tag{1}$$

where $W_o$ = mechanical work added to system from an external mechanical source, J/kg or ft · lbf/lbm
$f$ = Fanning friction factor, dimensionless
$V$ = average linear velocity, m/s or ft/s
$L$ = length of pipe, m or ft
$g_c$ = gravitational conversion factor, 32.17 ft · lbm/(s$^2$)(lbf) with American engineering units, $g_c$ = 1.0 with SI units
$D$ = inside diameter of pipe, m or ft
$J$ = frictional loss due to fittings, expressed as equivalent fractional loss in a straight pipe
$B$ = constant taking all other factors of the mechanical energy balance into consideration

In the region of turbulent flow, $f$ may be approximated for new steel pipes by the following equation:

$$f = \frac{0.04}{(N_{Re})^{0.16}} \tag{2}$$

where $N_{Re}$ is the Reynolds number or $DV\rho/\mu$.

[1] Based on R. P. Genereaux, *Ind Eng Chem*, **29**:385 (1937).

The term $W_o$ represents the amount of mechanical energy added to the system from an external source such as a pump. By combining Eqs. (1) and (2) and applying the necessary conversion factors, the following dimensional equation may be obtained representing the annual pumping cost as dollars per year per foot of pipe length:

$$C_{\text{pumping}} = \frac{0.0072 w^{2.84} \mu_c^{0.16} Y K (1 + J)}{D_i^{4.84} \rho^2 E} + B' \tag{3}$$

where $C_{\text{pumping}}$ = annual pumping costs as dollars per year foot of pipe length
$w$ = thousands of pounds mass flowing per hour, 1000 lb/h
$\mu_c$ = fluid viscosity, cP
$Y$ = hours of operation per year
$K$ = cost of electrical energy, \$/kWh
$D_i$ = inside diameter of pipe, in
$\rho$ = fluid density, lb/ft$^3$
$E$ = efficiency of motor and pump expressed as a fraction
$B'$ = constant independent of $D_i$

Because Eq. (3) is dimensional, it only holds when the units indicated here are used for the variables.

The annual costs for the piping equipment depend on the initial pipe cost, fixed charges (i.e., interest, rent, local taxes, insurance, and depreciation), installation costs including fittings and valves, and maintenance costs. A plot of the logarithm of the pipe diameter versus the logarithm of the purchase cost per foot of pipe is essentially a straight line for most types of pipe. Therefore, the purchase cost for pipe may be represented by the following dimensional equation:

$$C'_{\text{pipe}} = X D_i^n \tag{4}$$

where $C'_{\text{pipe}}$ = purchase cost of new pipe as dollars per foot of pipe length if pipe diameter is $D_i$ in
$X$ = purchase cost of new pipe as dollars per foot of pipe length if the pipe diameter is 1 in
$D_i$ = diameter of pipe, in
$n$ = a constant for each pipe material equals 1.5 for steel pipe

The annual cost for the installed piping equipment per foot of pipe length may be expressed as follows:

$$C_{\text{pipe}} = (a + b)(F + 1) X D_i^n \tag{5}$$

where $C_{\text{pipe}}$ = annual cost for installed piping equipment as dollars per year per foot of pipe length
$a$ = fixed charges expressed as a fraction (based on cost of installed equipment)
$b$ = annual maintenance costs expressed as a fraction (based on cost of installed equipment)

$F$ = ratio of total costs for fittings and erection to total cost of new pipe. (If the new pipe costs \$1000 and the cost for the equipment installation and the necessary fittings is \$1400, the value of $F$ would be 1.4.)

An expression for the total annual cost per foot of pipe length for the pumping and the piping equipment may be obtained by adding Eqs. (3) and (5). The only variable in the resulting cost expression is the pipe diameter. The optimum economic pipe diameter is that diameter which makes the total cost a minimum. Therefore, the optimum economic pipe diameter can be found by taking the derivative of the total cost with respect to the diameter, setting the resulting expression equal to zero, and solving for $D_i$. The following dimensional equation may be obtained by carrying out this mathematical operation:

$$D_{i(\text{optimum})}^{4.84+n} = \frac{0.0348w^{2.84}\mu_c^{0.16}YK(1 + J)}{n(a + b)(F + 1)XE\rho^2} \tag{6}$$

The value of $n$ for steel pipes is 1.5. Substituting this value in Eq. (6) gives

$$D_{i(\text{optimum})} = \frac{\omega^{0.448}\mu_c^{0.025}}{\rho^{0.315}}\left[\frac{0.0232YK(1 + J)}{(a + b)(F + 1)XE}\right]^{0.158} \tag{7}$$

Equation (7) is applicable to any steady-state flow of liquids through steel pipes when the flow is in the turbulent range. It also applies to the flow of compressible fluids if the total pressure drop over the system is less than 10 percent of the downstream pressure.

The exponent involved in the square-bracketed term in Eq. (7) indicates that the optimum diameter is relatively insensitive to most of the terms involved. Inflation effects tend to cancel out because the two cost variables in the bracketed term appear with 1 in the numerator and 1 in the denominator. Therefore, it is possible to simplify the equation by substituting average numerical values for some of the less critical terms. The following values are applicable under ordinary industrial conditions:

$K$ = \$0.055/kWh
$Y$ = 8760 h/yr
$J$ = 0.35 or 35%
$E$ = 0.50 or 50%
$a$ = 0.15 or 15%
$b$ = 0.05 or 5%
$F$ = 1.4
$X$ = \$0.45/ft for standard 1-in diameter steel pipe

Substituting these values into Eq. (7) gives

$$D_{i(\text{optimum})} = \frac{2.2w^{0.448}\mu_c^{0.025}}{\rho^{0.315}} \tag{8}$$

Since the exponent on the viscosity term is very small, the value of $\mu_c^{0.025}$ may be taken as unity. Equation (8) may then be reduced to the following simplified dimensional equation.

$$D_{i(\text{optimum})} = \frac{2.2w^{0.45}}{\rho^{0.32}} \tag{9}$$

where $D_{i(\text{optimum})}$ = estimated optimum economic pipe diameter, in
$w$ = thousands of pounds-mass flowing per hour, 1000 lb/h
$\rho$ = density of the flowing fluid, lb/ft$^3$

Depending on the accuracy desired, Eq. (9), (8), (7), or (6) may be used to determine optimum economic pipe diameters if the flow is turbulent. Equation (9) is sufficiently accurate for design estimates under ordinary plant conditions if the viscosity of the fluid is between 0.02 and 30 cP. The equation gives conservative results in that it gives diameters which may be larger than necessary. Consequently, it is permissible to select the closest standard pipe diameter below the actual diameter calculated by Eq. (9).

The results of Eq. (9) should not be applied directly if the flowing fluid is steam because the pressure drop in steam lines involves a loss in available energy. The amount of available energy lost depends on the pressure and temperature level of the steam.

Other methods have also been developed for calculating optimum economic pipe diameters; these methods generally use the same approach as has been used here but take other factors into account, such as cost of the pump, federal income tax rate, required rate of return, or cost of capital. Almost all of these methods give final results for the optimum economic diameter that are smaller than would be given by the equations in this section. Thus, the results from Eq. (9), (8), (7), or (6) are conservative and give safe results, since they tend to give a diameter somewhat larger than the true optimum economic diameter.

APPENDIX

# C

# CONVERSIONS AND PHYSICAL PROPERTIES

**Table C-1 General engineering conversion factors and constants***

| Length | | Mass | |
|---|---|---|---|
| 1 inch | 2.54 centimeters | 1 pound† | 16.0 ounces |
| 1 foot | 30.48 centimeters | 1 pound† | 453.6 grams |
| 1 yard | 91.44 centimeters | 1 pound† | 7000 grains |
| 1 meter | 100.000 centimeters | 1 ton (short) | 2000 pounds† |
| 1 meter | 39.37 inches | 1 kilogram | 1000 grams |
| 1 micron | $10^{-6}$ meter | 1 kilogram | 2.205 pounds† |
| 1 mile | 5280 feet | | |
| 1 kilometer | 0.6214 mile | | |

† Avoirdupois.

| Volume | |
|---|---|
| 1 cubic inch | 16.39 cubic centimeters |
| 1 liter | 61.03 cubic inches |
| 1 liter | 1.057 quarts |
| 1 cubic foot | 28.32 liters |
| 1 cubic foot | 1728 cubic inches |
| 1 cubic foot | 7.481 U.S. gallons |
| 1 U.S. gallon | 4.0 quarts |
| 1 U.S. gallon | 3.785 liters |
| 1 U.S. bushel | 1.244 cubic feet |

Density

1 gram per cubic centimeter 62.43 pounds per cubic foot
1 gram per cubic centimeter 8.345 pounds per U.S. gallon
1 gram mole of an ideal gas at 0°C and 760 mmHg is equivalent to 22.414 liters
1 pound mole of an ideal gas at 0°C and 760 mmHg is equivalent to 359.0 cubic feet
Density of dry air at 0°C and 760 mmHg 1.293 grams per liter = 0.0807 pound per cubic foot
Density of mercury 13.6 grams per cubic centimeter (at −2°C)

| Pressure | |
|---|---|
| 1 pound per square inch | 2.04 inches of mercury |
| 1 pound per square inch | 51.71 millimeters of mercury |
| 1 pound per square inch | 2.31 feet of water |
| 1 atmosphere | 760 millimeters of mercury |
| 1 atmosphere | 2116.2 pounds per square foot |
| 1 atmosphere | 33.93 feet of water |
| 1 atmosphere | 29.92 inches of mercury |
| 1 atmosphere | 14.7 pounds per square inch |
| 1 atmosphere | 101.325 kilopascals |

| Temperature scales | |
|---|---|
| Degrees Fahrenheit (F) | 1.8 (degrees C) + 32 |
| Degrees Celsius (C) | (degrees F − 32)/1.8 |
| Kelvin (K) | degrees C + 273.15 |
| Degrees Rankine (R) | degrees F + 459.7 |

| Power | |
|---|---|
| 1 kilowatt | 737.56 foot-pounds force per second |
| 1 kilowatt | 56.87 Btu per minute |
| 1 kilowatt | 1.341 horsepower |
| 1 horsepower | 550 foot-pounds force per second |
| 1 horsepower | 0.707 Btu per second |
| 1 horsepower | 745.7 watts |

**Table C-1 General engineering conversion factors and constants** *(Continued)*

Heat, energy, and work equivalents

| | cal | Btu | ft · lb | kWh |
|---|---|---|---|---|
| cal | 1 | $3.97 \times 10^{-3}$ | 3.086 | $1.162 \times 10^{-6}$ |
| Btu | 252 | 1 | 778.16 | $2.930 \times 10^{-4}$ |
| ft · lb | 0.3241 | $1.285 \times 10^{-3}$ | 1 | $3.766 \times 10^{-7}$ |
| kWh | 860,565 | 3412.8 | $2.655 \times 10^{6}$ | 1 |
| hp-h | 641,615 | 2545.0 | $1.980 \times 10^{6}$ | 0.7455 |
| joules | 0.239 | $9.478 \times 10^{-4}$ | 0.7376 | $2.773 \times 10^{-7}$ |
| liter-atm | 24.218 | $9.604 \times 10^{-2}$ | 74.73 | $2.815 \times 10^{-5}$ |

| | hp-h | joules | liter-atm |
|---|---|---|---|
| cal | $1.558 \times 10^{-6}$ | 4.1840 | $4.129 \times 10^{-2}$ |
| Btu | $3.930 \times 10^{-4}$ | 1055 | 10.41 |
| ft-lb | $5.0505 \times 10^{-7}$ | 1.356 | $1.338 \times 10^{-2}$ |
| kWh | 1.341 | $3.60 \times 10^{6}$ | 35,534.3 |
| hp-h | 1 | $2.685 \times 10^{6}$ | 26,494 |
| joules | $3.725 \times 10^{-7}$ | 1 | $9.869 \times 10^{-3}$ |
| liter-atm | $3.774 \times 10^{-5}$ | 101.33 | 1 |

Constants

| | |
|---|---|
| $e$ | 2.7183 |
| $\pi$ | 3.1416 |

Gas-law constants:

| | |
|---|---|
| $R$ | 1.987 (cal)/(g mol) (K) or (Btu)/(lb mol) (°R) |
| $R$ | 82.06 ($cm^3$) (atm)/(g mol) (K) |
| $R$ | 10.73 (lb/$in^2$) ($ft^3$)/(lb mol) (°R) |
| $R$ | 0.730 (atm) ($ft^3$)/(lb mol) (°R) |
| $R$ | 1545.0 (lb/$ft^2$) ($ft^3$)/(lb mol) (°R) |
| $R$ | 8.314(kPa) ($m^3$)/(kg mol) (K) or (J)/(g mol) (K) |
| $R$ | 21.9(inHg) ($ft^3$)/(lb mol) (°R) |
| $g_c$ | 32.17 (ft) (lbm)/($s^2$) (lbf) |

Analysis of air

By weight: oxygen, 23.2%; nitrogen, 76.8%
By volume: oxygen, 21.0%; nitrogen, 79.0%
Average molecular weight of air on above basis = 28.84 (usually rounded off to 29)
True molecular weight of dry air (including argon) = 28.96

Viscosity

| | |
|---|---|
| 1 centipoise | 0.001 kg/(m) (s) |
| 1 centipose | 0.000672 lb(s) (ft) |
| 1 centipose | 2.42 lb/(h) (ft) |

* See also Tables A-6 and A-7 in App. A for SI conversion factors and more exact conversion factors.

**Table C-2 Viscosities of gases: Coordinates for use with Fig. C-1**

| No. | Gas | *X* | *Y* | No. | Gas | *X* | *Y* |
|---|---|---|---|---|---|---|---|
| 1 | Acetic acid | 7.7 | 14.3 | 29 | Freon 113 | 11.3 | 14.0 |
| 2 | Acetone | 8.9 | 13.0 | 30 | Helium | 10.9 | 20.5 |
| 3 | Acetylene | 9.8 | 14.9 | 31 | Hexane | 8.6 | 11.8 |
| 4 | Air | 11.0 | 20.0 | 32 | Hydrogen | 11.2 | 12.4 |
| 5 | Ammonia | 8.4 | 16.0 | 33 | $3H_2 + 1N_2$ | 11.2 | 17.2 |
| 6 | Argon | 10.5 | 22.4 | 34 | Hydrogen bromide | 8.8 | 20.9 |
| 7 | Benzene | 8.5 | 13.2 | 35 | Hydrogen chloride | 8.8 | 18.7 |
| 8 | Bromine | 8.9 | 19.2 | 36 | Hydrogen cyanide | 9.8 | 14.9 |
| 9 | Butene | 9.2 | 13.7 | 37 | Hydrogen iodide | 9.0 | 21.3 |
| 10 | Butylene | 8.9 | 13.0 | 38 | Hydrogen sulfide | 8.6 | 18.0 |
| 11 | Carbon dioxide | 9.5 | 18.7 | 39 | Iodine | 9.0 | 18.4 |
| 12 | Carbon disulfide | 8.0 | 16.0 | 40 | Mercury | 5.3 | 22.9 |
| 13 | Carbon monoxide | 11.0 | 20.0 | 41 | Methane | 9.9 | 15.5 |
| 14 | Chlorine | 9.0 | 18.4 | 42 | Methyl alcohol | 8.5 | 15.6 |
| 15 | Chloroform | 8.9 | 15.7 | 43 | Nitric oxide | 10.9 | 20.5 |
| 16 | Cyanogen | 9.2 | 15.2 | 44 | Nitrogen | 10.6 | 20.0 |
| 17 | Cyclohexane | 9.2 | 12.0 | 45 | Nitrosyl chloride | 8.0 | 17.6 |
| 18 | Ethane | 9.1 | 14.5 | 46 | Nitrous oxide | 8.8 | 19.0 |
| 19 | Ethyl acetate | 8.5 | 13.2 | 47 | Oxygen | 11.0 | 21.3 |
| 20 | Ethyl alcohol | 9.2 | 14.2 | 48 | Pentane | 7.0 | 12.8 |
| 21 | Ethyl chloride | 8.5 | 15.6 | 49 | Propane | 9.7 | 12.9 |
| 22 | Ethyl ether | 8.9 | 13.0 | 50 | Propyl alcohol | 8.4 | 13.4 |
| 23 | Ethylene | 9.5 | 15.1 | 51 | Propylene | 9.0 | 13.8 |
| 24 | Fluorine | 7.3 | 23.8 | 52 | Sulfur dioxide | 9.6 | 17.0 |
| 25 | Freon 11 | 10.6 | 15.1 | 53 | Toluene | 8.6 | 12.4 |
| 26 | Freon 12 | 11.1 | 16.0 | 54 | 2,3,3-Trimethylbutane | 9.5 | 10.5 |
| 27 | Freon 21 | 10.8 | 15.3 | 55 | Water | 8.0 | 16.0 |
| 28 | Freon 22 | 10.1 | 17.0 | 56 | Xenon | 9.3 | 23.0 |

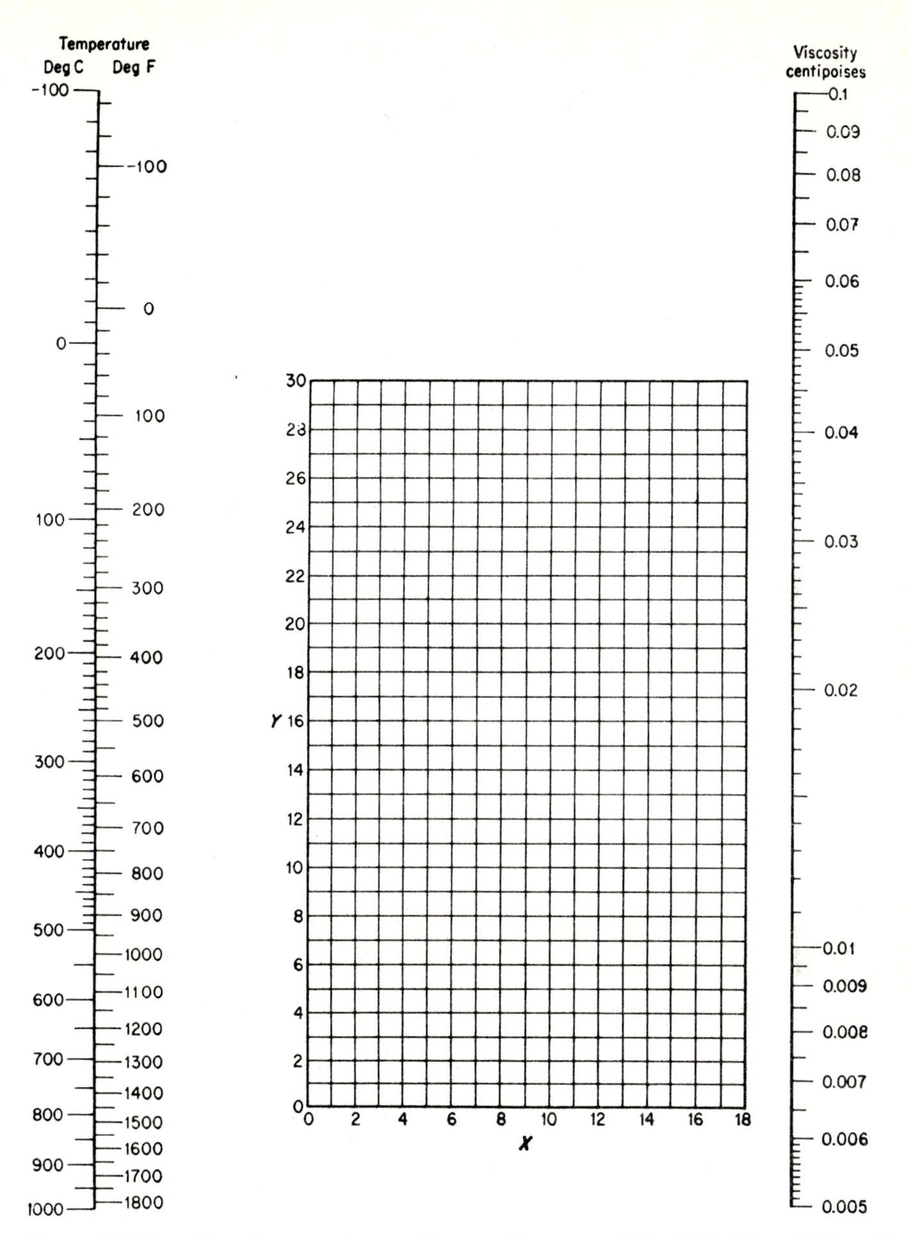


**Figure C-1** Viscosities of gases at 1 atm. (For coordinates see Table C-2.) [°F = 1.8 K − 459.7; cP × 0.001 = kg/(m)(s).]

**Table C-3 Viscosities of liquids: Coordinates for use with Fig. C-2**

| No. | Liquid | *X* | *Y* | No. | Liquid | *X* | *Y* |
|---|---|---|---|---|---|---|---|
| 1 | Acetaldehyde | 15.2 | 4.8 | 56 | Freon 22 | 17.2 | 4.7 |
| 2 | Acetic acid, 100% | 12.1 | 14.2 | 57 | Freon 113 | 12.5 | 11.4 |
| 3 | Acetic acid, 70% | 9.5 | 17.0 | 58 | Glycerol, 100% | 2.0 | 30.0 |
| 4 | Acetic anhydride | 12.7 | 12.8 | 59 | Glycerol, 50% | 6.9 | 19.6 |
| 5 | Acetone, 100% | 14.5 | 7.2 | 60 | Heptene | 14.1 | 8.4 |
| 6 | Acetone, 35% | 7.9 | 15.0 | 61 | Hexane | 14.7 | 7.0 |
| 7 | Allyl alcohol | 10.2 | 14.3 | 62 | Hydrochloric acid, 31.5% | 13.0 | 16.6 |
| 8 | Ammonia, 100% | 12.6 | 2.0 | 63 | Isobutyl alcohol | 7.1 | 18.0 |
| 9 | Ammonia, 26% | 10.1 | 13.9 | 64 | Isobutyric acid | 12.2 | 14.4 |
| 10 | Amyl acetate | 11.8 | 12.5 | 65 | Isopropyl alcohol | 8.2 | 16.0 |
| 11 | Amyl alcohol | 7.5 | 18.4 | 66 | Kerosene | 10.2 | 16.9 |
| 12 | Aniline | 8.1 | 18.7 | 67 | Linseed oil, raw | 7.5 | 27.2 |
| 13 | Anisole | 12.3 | 13.5 | 68 | Mercury | 18.4 | 16.4 |
| 14 | Arsenic trichloride | 13.9 | 14.5 | 69 | Methanol, 100% | 12.4 | 10.5 |
| 15 | Benzene | 12.5 | 10.9 | 70 | Methanol, 90% | 12.3 | 11.8 |
| 16 | Brine, $CaCl_2$, 25% | 6.6 | 15.9 | 71 | Methanol, 40% | 7.8 | 15.5 |
| 17 | Brine, NaCl, 25% | 10.2 | 16.6 | 72 | Methyl acetate | 14.2 | 8.2 |
| 18 | Bromine | 14.2 | 13.2 | 73 | Methyl chloride | 15.0 | 3.8 |
| 19 | Bromotoluene | 20.0 | 15.9 | 74 | Methyl ethyl ketone | 13.9 | 8.6 |
| 20 | Butyl acetate | 12.3 | 11.0 | 75 | Naphthalene | 7.9 | 18.1 |
| 21 | Butyl alcohol | 8.6 | 17.2 | 76 | Nitric acid, 95% | 12.8 | 13.8 |
| 22 | Butyric acid | 12.1 | 15.3 | 77 | Nitric acid, 60% | 10.8 | 17.0 |
| 23 | Carbon dioxide | 11.6 | 0.3 | 78 | Nitrobenzene | 10.6 | 16.2 |
| 24 | Carbon disulfide | 16.1 | 7.5 | 79 | Nitrotoluene | 11.0 | 17.0 |
| 25 | Carbon tetrachloride | 12.7 | 13.1 | 80 | Octane | 13.7 | 10.0 |
| 26 | Chlorobenzene | 12.3 | 12.4 | 81 | Octyl alcohol | 6.6 | 21.1 |
| 27 | Chloroform | 14.4 | 10.2 | 82 | Pentachloroethane | 10.9 | 17.3 |
| 28 | Chlorosulfonic acid | 11.2 | 18.1 | 83 | Pentane | 14.9 | 5.2 |
| 29 | Chlorotoluene, ortho | 13.0 | 13.3 | 84 | Phenol | 6.9 | 20.8 |
| 30 | Chlorotoluene, meta | 13.3 | 12.5 | 85 | Phosphorus tribromide | 13.8 | 16.7 |
| 31 | Chlorotoluene, para | 13.3 | 12.5 | 86 | Phosphorus trichloride | 16.2 | 10.9 |
| 32 | Cresol, meta | 2.5 | 20.8 | 87 | Proponic acid | 12.8 | 13.8 |
| 33 | Cyclohexanol | 2.9 | 24.3 | 88 | Propyl alcohol | 9.1 | 16.5 |
| 34 | Dibromoethane | 12.7 | 15.8 | 89 | Propyl bromide | 14.5 | 9.6 |
| 35 | Dichloroethane | 13.2 | 12.2 | 90 | Propyl chloride | 14.4 | 7.5 |
| 36 | Dichloromethane | 14.6 | 8.9 | 91 | Propyl iodide | 14.1 | 11.6 |
| 37 | Diethyl oxalate | 11.0 | 16.4 | 92 | Sodium | 16.4 | 13.9 |
| 38 | Dimethyl oxalate | 12.3 | 15.8 | 93 | Sodium hydoxide, 50% | 3.2 | 25.8 |
| 39 | Diphenyl | 12.0 | 18.3 | 94 | Stannic chloride | 13.5 | 12.8 |
| 40 | Dipropyl oxalate | 10.3 | 17.7 | 95 | Sulfur dioxide | 15.2 | 7.1 |
| 41 | Ethyl acetate | 13.7 | 9.1 | 96 | Sulfuric acid, 110% | 7.2 | 27.4 |
| 42 | Ethyl alcohol, 100% | 10.5 | 13.8 | 97 | Sulfuric acid, 98% | 7.0 | 24.8 |
| 43 | Ethyl alcohol, 95% | 9.8 | 14.3 | 98 | Sulfuric acid, 60% | 10.2 | 21.3 |
| 44 | Ethyl alcohol, 40% | 6.5 | 16.6 | 99 | Sulfuryl chloride | 15.2 | 12.4 |
| 45 | Ethyl benzene | 13.2 | 11.5 | 100 | Tetrachloroethane | 11.9 | 15.7 |
| 46 | Ethyl bromide | 14.5 | 8.1 | 101 | Tetrachloroethylene | 14.2 | 12.7 |
| 47 | Ethyl chloride | 14.8 | 6.0 | 102 | Titanium tetrachloride | 14.4 | 12.3 |
| 48 | Ethyl ether | 14.5 | 5.3 | 103 | Toluene | 13.7 | 10.4 |
| 49 | Ethyl formate | 14.2 | 8.4 | 104 | Trichloroethylene | 14.8 | 10.5 |
| 50 | Ethyl iodide | 14.7 | 10.3 | 105 | Turpentine | 11.5 | 14.9 |
| 51 | Ethylene glycol | 6.0 | 23.6 | 106 | Vinyl acetate | 14.0 | 8.8 |
| 52 | Formic acid | 10.7 | 15.8 | 107 | Water | 10.2 | 13.0 |
| 53 | Freon 11 | 14.4 | 9.0 | 108 | Xylene, ortho | 13.5 | 12.1 |
| 54 | Freon 12 | 16.8 | 5.6 | 109 | Xylene, meta | 13.9 | 10.6 |
| 55 | Freon 21 | 15.7 | 7.5 | 110 | Xylene, para | 13.9 | 10.9 |

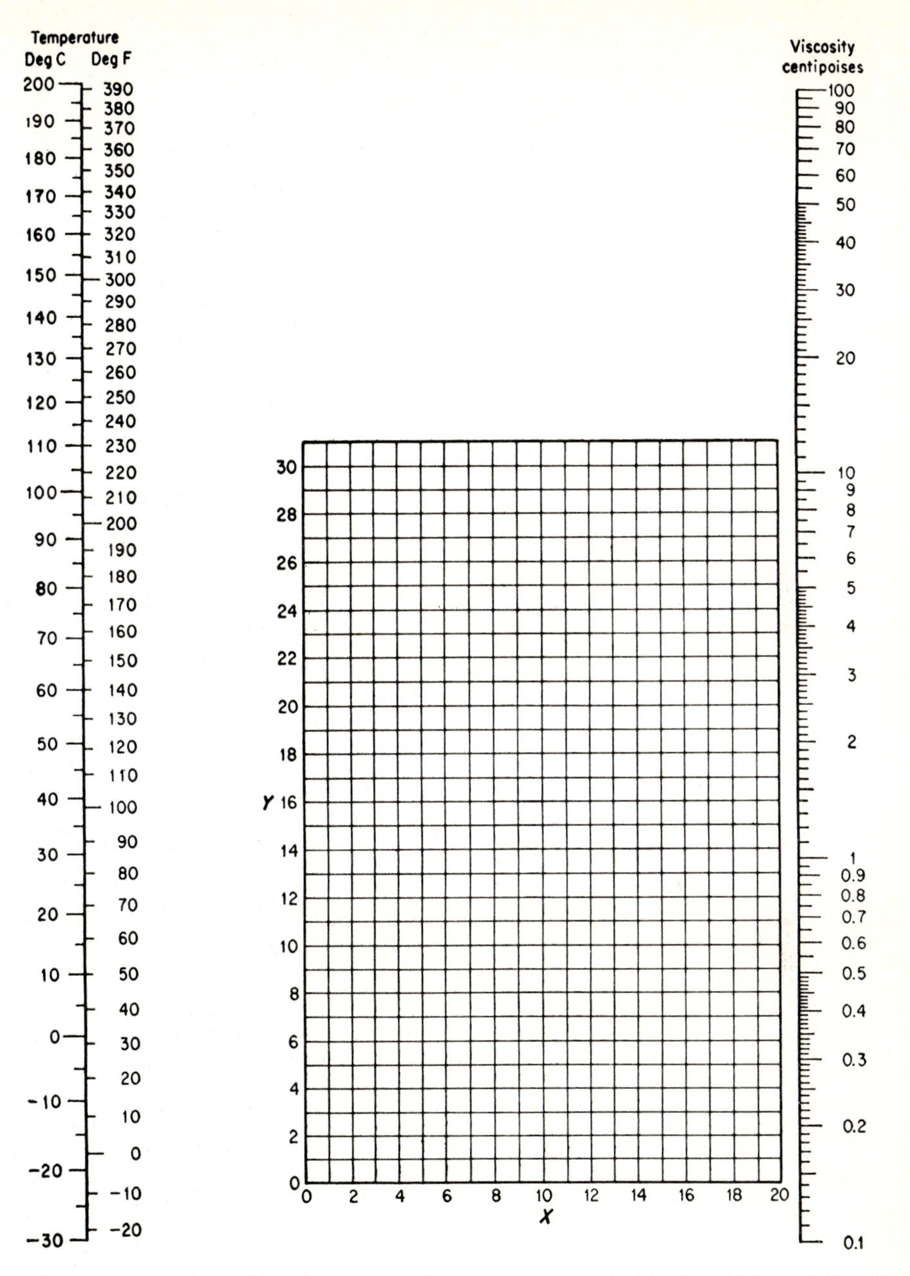


**Figure C-2** Viscosities of liquids at 1 atm. (For coordinates see Table C-3.) [°F = 1.8 **K** − 459.7; cP × 0.001 = kg/(m)(s).]

## Table C-4 Density, viscosity, and thermal conductivity of water*

| Temperature, °F | Density of liquid water, $lb/ft^3$ | Viscosity of water, cP | Thermal conductivity of water, $Btu/(h)(ft^2)(°F/ft)$ |
|---|---|---|---|
| 32 | 62.42 | 1.794 | 0.320 |
| 40 | 62.43 | 1.546 | 0.326 |
| 50 | 62.42 | 1.310 | 0.333 |
| 60 | 62.37 | 1.129 | 0.340 |
| 70 | 62.30 | 0.982 | 0.346 |
| 80 | 62.22 | 0.862 | 0.352 |
| 90 | 62.11 | 0.764 | 0.358 |
| 100 | 62.00 | 0.684 | 0.363 |
| 110 | 61.86 | 0.616 | 0.367 |
| 120 | 61.71 | 0.559 | 0.371 |
| 130 | 61.55 | 0.511 | 0.375 |
| 140 | 61.38 | 0.470 | 0.378 |
| 150 | 61.20 | 0.433 | 0.381 |
| 160 | 61.00 | 0.401 | 0.384 |
| 170 | 60.80 | 0.372 | 0.386 |
| 180 | 60.58 | 0.347 | 0.388 |
| 190 | 60.36 | 0.325 | 0.390 |
| 200 | 60.12 | 0.305 | 0.392 |
| 210 | 59.88 | 0.287 | 0.393 |
| 212 | 59.83 | 0.284 | 0.393 |

* $°F = 1.8\ K - 459.7$
$lb/ft^3 \times 16.018 = kg/m^3$
$cP \times 0.001 = kg/(m)(s)$
$Btu/(h)(ft^2)(°F/ft) \times 1.7307 = J/(s)(m^2)(K/m)$

## Table C-5 Thermal conductivity of metals*

| Metal | *k* as $Btu/(h)(ft^2)(°F/ft)$ At 32°F | At 212°F | At 572°F |
|---|---|---|---|
| Aluminum | 117 | 119 | 133 |
| Brass (70–30) | 56 | 60 | 66 |
| Cast iron | 32 | 30 | 26 |
| Copper | 224 | 218 | 212 |
| Lead | 20 | 19 | 18 |
| Nickel | 36 | 34 | 32 |
| Silver | 242 | 238 | |
| Steel (mild) | | 26 | 25 |
| Tin | 36 | 34 | |
| Wrought iron | | 32 | 28 |
| Zinc | 65 | 64 | 59 |

* $Btu/(h)(ft^2)(°F/ft) \times 1.7307 = J/(s)(m^2)(K/m)$

**Table C-6 Thermal conductivity of nonmetallic solids***

| Material | Temperature, °F | $k$, Btu/(h)(ft$^2$)(°F/ft) |
|---|---|---|
| Asbestos-cement boards | 68 | 0.43 |
| Bricks: | | |
| Building | 68 | 0.40 |
| Fire clay | 392 | 0.58 |
| | 1832 | 0.95 |
| Sil-O-Cel | 400 | 0.042 |
| Calcium carbonate | | |
| (natural) | 86 | 1.3 |
| Calcium sulfate | | |
| (building plaster) | 77 | 0.25 |
| Celluloid | 86 | 0.12 |
| Concrete (stone) | | 0.54 |
| Cork board | 86 | 0.025 |
| Felt (wool) | 86 | 0.03 |
| Glass (window) | | 0.3–0.61 |
| Rubber (hard) | 32 | 0.087 |
| Wood (across grain): | | |
| Oak | 59 | 0.12 |
| Maple | 122 | 0.11 |
| Pine | 59 | 0.087 |

* Btu/(h)(ft$^2$)(°F/ft) × 1.7307 = J/(s)(m$^2$)(K/m)

## Table C-7 Thermal conductivity of liquids*

| Liquid | Temperature, °F | $k$, Btu/(h)(ft$^2$)(°F/ft) |
|---|---|---|
| Acetic acid, 100% | 68 | 0.099 |
| 50% | 68 | 0.20 |
| Acetone | 86 | 0.102 |
| | 167 | 0.095 |
| Benzene | 86 | 0.092 |
| | 140 | 0.087 |
| Ethyl alcohol, 100% | 68 | 0.105 |
| | 122 | 0.087 |
| 40% | 68 | 0.224 |
| Ethylene glycol | 32 | 0.153 |
| Glycerol, 100% | 68 | 0.164 |
| | 212 | 0.164 |
| 40% | 68 | 0.259 |
| *n*-Heptane | 86 | 0.081 |
| Kerosene | 68 | 0.086 |
| Methyl alcohol, 100% | 68 | 0.124 |
| | 122 | 0.114 |
| 40% | 68 | 0.234 |
| *n*-Octane | 86 | 0.083 |
| Sodium chloride brine, 25% | 86 | 0.330 |
| Sulfuric acid, 90% | 86 | 0.210 |
| 30% | 86 | 0.300 |
| Toluene | 86 | 0.086 |
| Water | 32 | 0.320 |
| | 200 | 0.392 |

* Btu/(h)(ft$^2$)(°F/ft) × 1.7307 = J/(s)(m$^2$)(K/m)

**Table C-8 Thermal conductivity of gases***

| Gas | Temperature, °F | $k$, Btu/(h)(ft$^2$)(°F/ft) |
|---|---|---|
| Air | 32 | 0.0140 |
| | 212 | 0.0183 |
| | 392 | 0.0226 |
| Ammonia | 32 | 0.0128 |
| | 122 | 0.0157 |
| Carbon dioxide | 32 | 0.0085 |
| | 212 | 0.0133 |
| Chlorine | 32 | 0.0043 |
| Hydrogen | 32 | 0.100 |
| | 212 | 0.129 |
| Methane | 32 | 0.0175 |
| | 122 | 0.0215 |
| Nitrogen | 32 | 0.0140 |
| | 212 | 0.0180 |
| Oxygen | 32 | 0.0142 |
| | 212 | 0.0185 |
| Sulfur dioxide | 32 | 0.0050 |
| | 212 | 0.0069 |
| Water vapor | 200 | 0.0159 |
| | 600 | 0.0256 |

* Btu/(h)(ft$^2$)(°F/ft) × 1.7307 = J/(s)(m$^2$)(K/m)

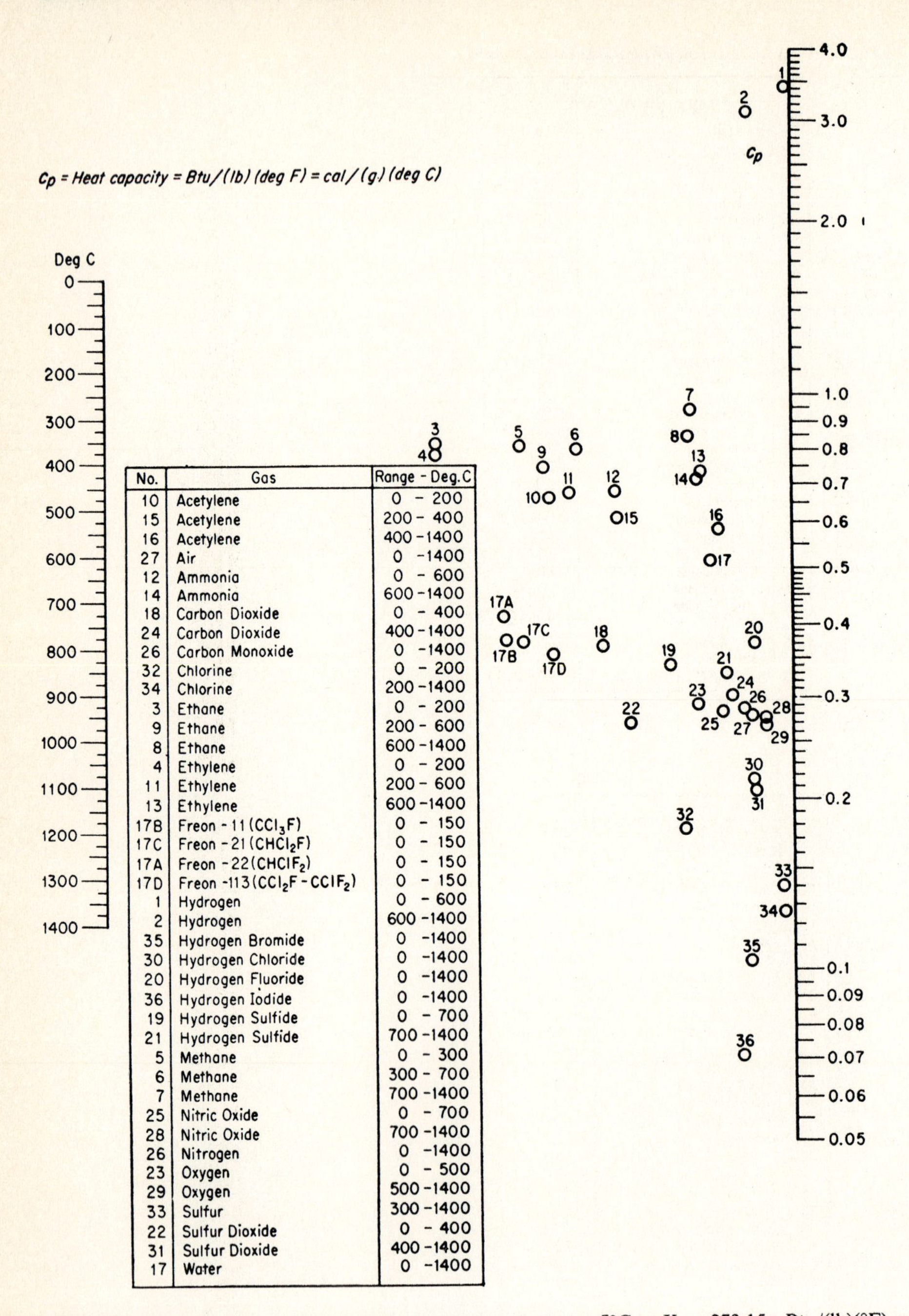

| No. | Gas | Range - Deg. C |
|---|---|---|
| 10 | Acetylene | 0 - 200 |
| 15 | Acetylene | 200 - 400 |
| 16 | Acetylene | 400 -1400 |
| 27 | Air | 0 -1400 |
| 12 | Ammonia | 0 - 600 |
| 14 | Ammonia | 600 -1400 |
| 18 | Carbon Dioxide | 0 - 400 |
| 24 | Carbon Dioxide | 400 -1400 |
| 26 | Carbon Monoxide | 0 -1400 |
| 32 | Chlorine | 0 - 200 |
| 34 | Chlorine | 200 -1400 |
| 3 | Ethane | 0 - 200 |
| 9 | Ethane | 200 - 600 |
| 8 | Ethane | 600 -1400 |
| 4 | Ethylene | 0 - 200 |
| 11 | Ethylene | 200 - 600 |
| 13 | Ethylene | 600 -1400 |
| 17B | Freon - 11 ($CCl_3F$) | 0 - 150 |
| 17C | Freon - 21 ($CHCl_2F$) | 0 - 150 |
| 17A | Freon - 22 ($CHClF_2$) | 0 - 150 |
| 17D | Freon - 113 ($CCl_2F - CClF_2$) | 0 - 150 |
| 1 | Hydrogen | 0 - 600 |
| 2 | Hydrogen | 600 -1400 |
| 35 | Hydrogen Bromide | 0 -1400 |
| 30 | Hydrogen Chloride | 0 -1400 |
| 20 | Hydrogen Fluoride | 0 -1400 |
| 36 | Hydrogen Iodide | 0 -1400 |
| 19 | Hydrogen Sulfide | 0 - 700 |
| 21 | Hydrogen Sulfide | 700 -1400 |
| 5 | Methane | 0 - 300 |
| 6 | Methane | 300 - 700 |
| 7 | Methane | 700 -1400 |
| 25 | Nitric Oxide | 0 - 700 |
| 28 | Nitric Oxide | 700 -1400 |
| 26 | Nitrogen | 0 -1400 |
| 23 | Oxygen | 0 - 500 |
| 29 | Oxygen | 500 -1400 |
| 33 | Sulfur | 300 -1400 |
| 22 | Sulfur Dioxide | 0 - 400 |
| 31 | Sulfur Dioxide | 400 -1400 |
| 17 | Water | 0 -1400 |



**Figure C-3** Heat capacities ($c_p$) of gases at 1 atm pressure: [°C = K − 273.15; Btu/(lb)(°F) × 4186.8 = J/(kg)(K).]

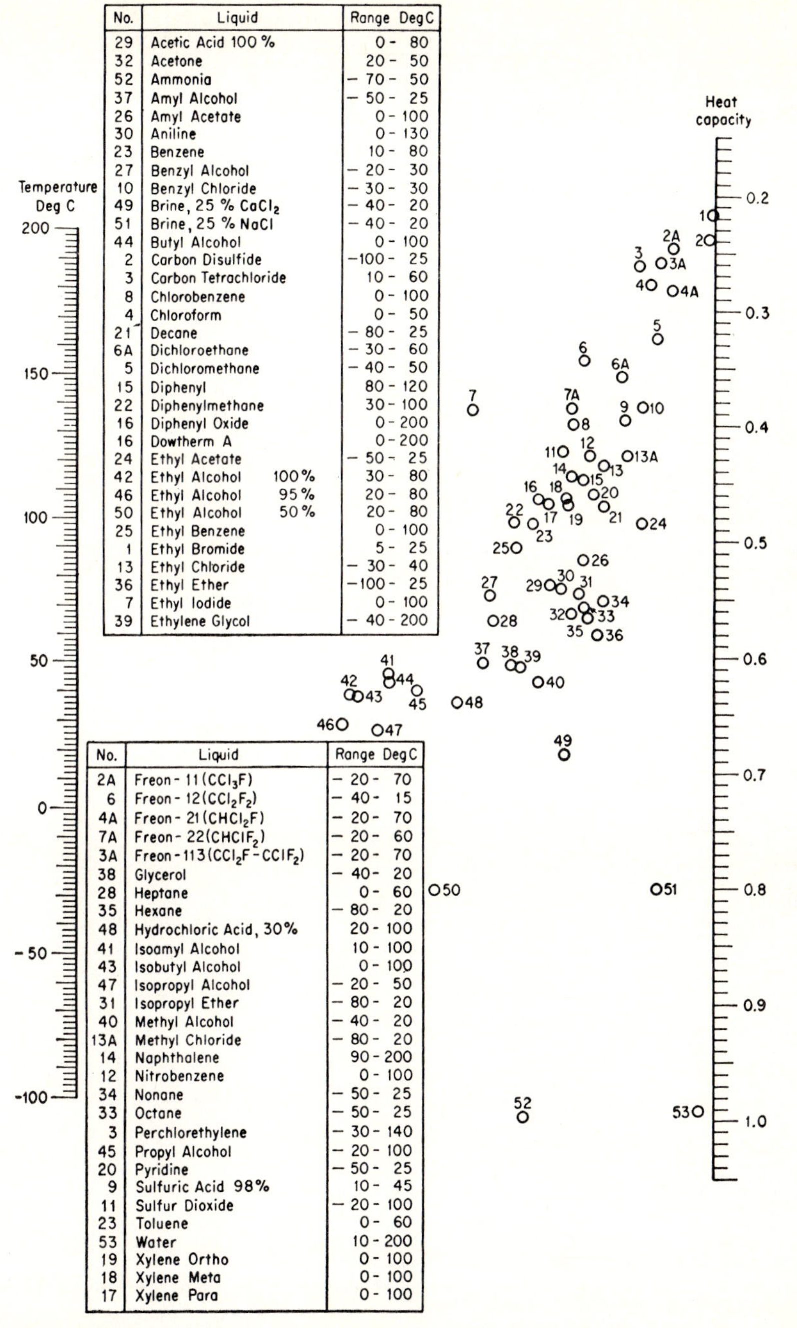


| No. | Liquid | Range Deg C |
|---|---|---|
| 29 | Acetic Acid 100% | 0 - 80 |
| 32 | Acetone | 20 - 50 |
| 52 | Ammonia | − 70 - 50 |
| 37 | Amyl Alcohol | − 50 - 25 |
| 26 | Amyl Acetate | 0 - 100 |
| 30 | Aniline | 0 - 130 |
| 23 | Benzene | 10 - 80 |
| 27 | Benzyl Alcohol | − 20 - 30 |
| 10 | Benzyl Chloride | − 30 - 30 |
| 49 | Brine, 25% $CaCl_2$ | − 40 - 20 |
| 51 | Brine, 25% NaCl | − 40 - 20 |
| 44 | Butyl Alcohol | 0 - 100 |
| 2 | Carbon Disulfide | −100 - 25 |
| 3 | Carbon Tetrachloride | 10 - 60 |
| 8 | Chlorobenzene | 0 - 100 |
| 4 | Chloroform | 0 - 50 |
| 21 | Decane | − 80 - 25 |
| 6A | Dichloroethane | − 30 - 60 |
| 5 | Dichloromethane | − 40 - 50 |
| 15 | Diphenyl | 80 - 120 |
| 22 | Diphenylmethane | 30 - 100 |
| 16 | Diphenyl Oxide | 0 - 200 |
| 16 | Dowtherm A | 0 - 200 |
| 24 | Ethyl Acetate | − 50 - 25 |
| 42 | Ethyl Alcohol 100% | 30 - 80 |
| 46 | Ethyl Alcohol 95% | 20 - 80 |
| 50 | Ethyl Alcohol 50% | 20 - 80 |
| 25 | Ethyl Benzene | 0 - 100 |
| 1 | Ethyl Bromide | 5 - 25 |
| 13 | Ethyl Chloride | − 30 - 40 |
| 36 | Ethyl Ether | −100 - 25 |
| 7 | Ethyl Iodide | 0 - 100 |
| 39 | Ethylene Glycol | − 40 - 200 |

| No. | Liquid | Range Deg C |
|---|---|---|
| 2A | Freon-11 ($CCl_3F$) | − 20 - 70 |
| 6 | Freon-12 ($CCl_2F_2$) | − 40 - 15 |
| 4A | Freon-21 ($CHCl_2F$) | − 20 - 70 |
| 7A | Freon-22 ($CHClF_2$) | − 20 - 60 |
| 3A | Freon-113 ($CCl_2F-CClF_2$) | − 20 - 70 |
| 38 | Glycerol | − 40 - 20 |
| 28 | Heptane | 0 - 60 |
| 35 | Hexane | − 80 - 20 |
| 48 | Hydrochloric Acid, 30% | 20 - 100 |
| 41 | Isoamyl Alcohol | 10 - 100 |
| 43 | Isobutyl Alcohol | 0 - 100 |
| 47 | Isopropyl Alcohol | − 20 - 50 |
| 31 | Isopropyl Ether | − 80 - 20 |
| 40 | Methyl Alcohol | − 40 - 20 |
| 13A | Methyl Chloride | − 80 - 20 |
| 14 | Naphthalene | 90 - 200 |
| 12 | Nitrobenzene | 0 - 100 |
| 34 | Nonane | − 50 - 25 |
| 33 | Octane | − 50 - 25 |
| 3 | Perchlorethylene | − 30 - 140 |
| 45 | Propyl Alcohol | − 20 - 100 |
| 20 | Pyridine | − 50 - 25 |
| 9 | Sulfuric Acid 98% | 10 - 45 |
| 11 | Sulfur Dioxide | − 20 - 100 |
| 23 | Toluene | 0 - 60 |
| 53 | Water | 10 - 200 |
| 19 | Xylene Ortho | 0 - 100 |
| 18 | Xylene Meta | 0 - 100 |
| 17 | Xylene Para | 0 - 100 |

**Figure C-4** Heat capacities of liquids. [°C = K − 273.15; Btu/(lb)(°F) × 4186.8 = J/(kg)(K).]

## Table C-9 Specific gravities of liquids

The values presented in the following table are based on the density of water at 4°C and a total pressure of 1 atm.

$$\text{Specific gravity} = \frac{\text{density of material at indicated temperature}}{\text{density of liquid water at 4°C}}$$

Density of liquid water at 4°C = 1.0000 g/cm$^3$ = 62.43 lb/ft$^3$ = 1000.0 kg/m$^3$

| Pure liquid | Formula | Temperature, °C | Specific gravity |
|---|---|---|---|
| Acetaldehyde | $CH_3CHO$ | 18 | 0.783 |
| Acetic acid | $CH_3CO_2H$ | 0 | 1.067 |
| | | 30 | 1.038 |
| Acetone | $CH_3COCH_3$ | 20 | 0.792 |
| Benzene | $C_6H_6$ | 20 | 0.879 |
| *n*-Butyl alcohol | $C_2H_5CH_2CH_2OH$ | 20 | 0.810 |
| Carbon tetrachloride | $CCl_4$ | 20 | 1.595 |
| Ethyl alcohol | $CH_3CH_2OH$ | 10 | 0.798 |
| | | 30 | 0.791 |
| Ethyl ether | $(CH_3CH_2)_2O$ | 25 | 0.708 |
| Ethylene glycol | $CH_2OH.CH_2OH$ | 19 | 1.113 |
| Glycerol | $CH_2OH.CHOH.CH_2OH$ | 15 | 1.264 |
| | | 30 | 1.255 |
| Isobutyl alcohol | $(CH_3)_2CHCH_2OH$ | 18 | 0.805 |
| Isopropyl alcohol | $(CH_3)_2CHOH$ | 0 | 0.802 |
| | | 30 | 0.777 |
| Methyl alcohol | $CH_3OH$ | 0 | 0.810 |
| | | 20 | 0.792 |
| Nitric acid | $HNO_3$ | 10 | 1.531 |
| | | 30 | 1.495 |
| Phenol | $C_6H_5OH$ | 25 | 1.071 |
| *n*-Propyl alcohol | $CH_3CH_2CH_2OH$ | 20 | 0.804 |
| Sulfuric acid | $H_2SO_4$ | 10 | 1.841 |
| | | 30 | 1.821 |
| Water | $H_2O$ | 4 | 1.000 |
| | | 100 | 0.958 |

## Table C-10 Specific gravities of solids

The specific gravities as indicated in this table apply at ordinary atmospheric temperatures. The values are based on the density of water at 4°C.

$$\text{Specific gravity} = \frac{\text{density of material}}{\text{density of liquid water at 4°C}}$$

Density of liquid water at 4°C = 1.0000 g/cm$^3$ = 62.43 lb/ft$^3$ = 1000.0 kg/m$^3$

| Substance | Specific gravity |
|---|---|
| Aluminum, hard-drawn | 2.55–2.80 |
| Brass, cast-rolled | 8.4–8.7 |
| Copper, cast-rolled | 8.8–8.95 |
| Glass, common | 2.4–2.8 |
| Gold, cast-hammered | 19.25–19.35 |
| Iron, gray cast | 7.03–7.13 |
| wrought | 7.6–7.9 |
| Lead | 11.34 |
| Nickel | 8.9 |
| Platinum, cast-hammered | 21.5 |
| Silver, cast-hammered | 10.4–10.6 |
| Steel, cold-drawn | 7.83 |
| Tin, cast-hammered | 7.2–7.5 |
| White oak timber, air-dried | 0.77 |
| White pine timber, air-dried | 0.43 |
| Zinc, cast-rolled | 6.9–7.2 |

## Table C-11 Properties of saturated steam*

Values in table based on zero enthalpy of liquid water at 32°F

| Temperature, °F | Absolute pressure, lb/in$^2$ | Volume of vapor, ft/lb$^3$ | Enthalpy | | Latent heat, of evaporation, Btu/lb |
|---|---|---|---|---|---|
| | | | Liquid, Btu/lb | Vapor, Btu/lb | |
| 32 | 0.0885 | 3306 | 0.00 | 1075.8 | 1075.8 |
| 35 | 0.0999 | 2947 | 3.02 | 1077.1 | 1074.1 |
| 40 | 0.1217 | 2444 | 8.05 | 1079.3 | 1071.3 |
| 45 | 0.1475 | 2036.4 | 13.06 | 1081.5 | 1068.4 |
| 50 | 0.1781 | 1703.2 | 18.07 | 1083.7 | 1065.6 |
| 55 | 0.2141 | 1430.7 | 23.07 | 1085.8 | 1062.7 |
| 60 | 0.2563 | 1206.7 | 28.06 | 1088.0 | 1059.9 |
| 65 | 0.3056 | 1021.4 | 33.05 | 1090.2 | 1057.1 |
| 70 | 0.3631 | 867.9 | 38.04 | 1092.3 | 1054.3 |
| 75 | 0.4298 | 740.0 | 43.03 | 1094.5 | 1051.5 |
| 80 | 0.5069 | 633.1 | 48.02 | 1096.6 | 1048.6 |
| 85 | 0.5959 | 543.5 | 53.00 | 1098.8 | 1045.8 |
| 90 | 0.6982 | 468.0 | 57.99 | 1100.9 | 1042.9 |
| 95 | 0.8153 | 404.3 | 62.98 | 1103.1 | 1040.1 |
| 100 | 0.9492 | 350.4 | 67.97 | 1105.2 | 1037.2 |
| 105 | 1.1016 | 304.5 | 72.95 | 1107.3 | 1034.3 |
| 110 | 1.2748 | 265.4 | 77.94 | 1109.5 | 1031.6 |
| 115 | 1.4709 | 231.9 | 82.93 | 1111.6 | 1028.7 |
| 120 | 1.6924 | 203.27 | 87.92 | 1113.7 | 1025.8 |
| 125 | 1.9420 | 178.61 | 92.91 | 1115.8 | 1022.9 |
| 130 | 2.2225 | 157.34 | 97.90 | 1117.9 | 1020.0 |
| 135 | 2.5370 | 138.95 | 102.90 | 1119.9 | 1017.0 |
| 140 | 2.8886 | 123.01 | 107.89 | 1122.0 | 1014.1 |
| 145 | 3.281 | 109.15 | 112.89 | 1124.1 | 1011.2 |
| 150 | 3.718 | 97.07 | 117.89 | 1126.1 | 1008.2 |
| 155 | 4.203 | 86.52 | 122.89 | 1128.1 | 1005.2 |
| 160 | 4.741 | 77.29 | 127.89 | 1130.2 | 1002.3 |
| 165 | 5.335 | 69.19 | 132.89 | 1132.2 | 999.3 |
| 170 | 5.992 | 62.06 | 137.90 | 1134.2 | 996.3 |
| 175 | 6.715 | 55.78 | 142.91 | 1136.2 | 993.3 |
| 180 | 7.510 | 50.23 | 147.92 | 1138.1 | 990.2 |
| 185 | 8.383 | 45.31 | 152.93 | 1140.1 | 987.2 |
| 190 | 9.339 | 40.96 | 157.95 | 1142.0 | 984.1 |
| 195 | 10.385 | 37.09 | 162.97 | 1144.0 | 981.0 |
| 200 | 11.526 | 33.64 | 167.99 | 1145.9 | 977.9 |
| 210 | 14.123 | 27.82 | 178.05 | 1149.7 | 971.6 |
| 212 | 14.696 | 26.80 | 180.07 | 1150.4 | 970.3 |
| 220 | 17.186 | 23.15 | 188.13 | 1153.4 | 965.2 |
| 230 | 20.780 | 19.382 | 198.23 | 1157.0 | 958.8 |
| 240 | 24.969 | 16.323 | 208.34 | 1160.5 | 952.2 |
| 250 | 29.825 | 13.821 | 218.48 | 1164.0 | 945.5 |

**Table C-11 Properties of saturated steam** (*Continued*)

| Temperature, °F | Absolute pressure, lb/in$^2$ | Volume of vapor, ft$^3$/lb | Enthalpy | | Latent heat of evaporation, Btu/lb |
|---|---|---|---|---|---|
| | | | Liquid, Btu/lb | Vapor, Btu/lb | |
| 260 | 35.429 | 11.763 | 228.64 | 1167.3 | 938.7 |
| 270 | 41.858 | 10.061 | 238.84 | 1170.6 | 931.8 |
| 280 | 49.203 | 8.645 | 249.06 | 1173.8 | 924.7 |
| 290 | 57.556 | 7.461 | 259.31 | 1176.8 | 917.5 |
| 300 | 67.013 | 6.466 | 269.59 | 1179.7 | 910.1 |
| 310 | 77.68 | 5.626 | 279.92 | 1182.5 | 902.6 |
| 320 | 89.66 | 4.914 | 290.28 | 1185.2 | 894.9 |
| 330 | 103.06 | 4.307 | 300.68 | 1187.7 | 887.0 |
| 340 | 118.01 | 3.788 | 311.13 | 1190.1 | 879.0 |
| 350 | 134.63 | 3.342 | 321.63 | 1192.3 | 870.7 |
| 360 | 153.04 | 2.957 | 332.18 | 1194.4 | 862.2 |
| 370 | 173.37 | 2.625 | 342.79 | 1196.3 | 853.5 |
| 380 | 195.77 | 2.335 | 353.45 | 1198.1 | 844.6 |
| 390 | 220.37 | 2.0836 | 364.17 | 1199.6 | 835.4 |
| 400 | 247.31 | 1.8633 | 374.97 | 1201.0 | 826.0 |

* °F = 1.8 K − 459.7
Pressure as lb/in$^2$ × 6894.76 = Pa
ft$^3$/lb × 0.06243 = m$^3$/kg
Btu/lb × 2326.00 = J/kg

## Table C-12 Steel pipe dimensions*

| Nominal pipe size, in | Outside diameter, in | Schedule no. | Inside diameter, in | Wall thickness, in |
|---|---|---|---|---|
| $\frac{1}{8}$ | 0.405 | 40† | 0.269 | 0.068 |
| | | 80‡ | 0.215 | 0.095 |
| $\frac{1}{4}$ | 0.540 | 40 | 0.364 | 0.088 |
| | | 80 | 0.302 | 0.119 |
| $\frac{3}{8}$ | 0.675 | 40 | 0.493 | 0.091 |
| | | 80 | 0.423 | 0.126 |
| $\frac{1}{2}$ | 0.840 | 40 | 0.622 | 0.109 |
| | | 80 | 0.546 | 0.147 |
| $\frac{3}{4}$ | 1.050 | 40 | 0.824 | 0.113 |
| | | 80 | 0.742 | 0.154 |
| 1 | 1.315 | 40 | 1.049 | 0.133 |
| | | 80 | 0.957 | 0.179 |
| $1\frac{1}{4}$ | 1.660 | 40 | 1.380 | 0.140 |
| | | 80 | 1.278 | 0.191 |
| $1\frac{1}{2}$ | 1.900 | 40 | 1.610 | 0.145 |
| | | 80 | 1.500 | 0.200 |
| 2 | 2.375 | 40 | 2.067 | 0.154 |
| | | 80 | 1.939 | 0.218 |
| $2\frac{1}{2}$ | 2.875 | 40 | 2.469 | 0.203 |
| | | 80 | 2.323 | 0.276 |
| 3 | 3.500 | 40 | 3.068 | 0.216 |
| | | 80 | 2.900 | 0.300 |
| $3\frac{1}{2}$ | 4.000 | 40 | 3.548 | 0.226 |
| | | 80 | 3.364 | 0.318 |
| 4 | 4.500 | 40 | 4.026 | 0.237 |
| | | 80 | 3.826 | 0.337 |
| 5 | 5.563 | 40 | 5.047 | 0.258 |
| | | 80 | 4.813 | 0.375 |
| 6 | 6.625 | 40 | 6.065 | 0.280 |
| | | 80 | 5.761 | 0.432 |
| 8 | 8.625 | 40 | 7.981 | 0.322 |
| | | 80 | 7.625 | 0.500 |
| 10 | 10.75 | 40 | 10.020 | 0.365 |
| | | 60 | 9.750 | 0.500 |
| 12 | 12.75 | 30 | 12.090 | 0.330 |
| | | 60 | 11.626 | 0.562 |

* Based on ASA Standards B36.10–1939. (in × 0.0254 = m).

† Schedule 40 designates former "standard" pipe.

‡ Schedule 80 designates former "extra-strong" pipe.

## Table C-13 International atomic weights

| Substance | Symbol | Atomic weight | At. no. | Substance | Symbol | Atomic weight | At. no. |
|---|---|---|---|---|---|---|---|
| Actinium | Ac | 227.03 | 89 | Mercury | Hg | 200.59 | 80 |
| Aluminum | Al | 26.98 | 13 | Molybdenum | Mo | 95.94 | 42 |
| Americium | Am | 243 | 95 | Neodymium | Nd | 144.24 | 60 |
| Antimony | Sb | 121.75 | 51 | Neon | Ne | 20.179 | 10 |
| Argon | A | 39.948 | 18 | Neptunium | Np | 237 | 93 |
| Arsenic | As | 74.92 | 33 | Nickel | Ni | 58.70 | 28 |
| Astatine | At | 210 | 85 | Niobium | | | |
| Barium | Ba | 137.33 | 56 | (Columbian) | Nb | 92.91 | 41 |
| Berkelium | Bk | 247 | 97 | Nitrogen | N | 14.007 | 7 |
| Beryllium | Be | 9.012 | 4 | Nobelium | No | 259 | 102 |
| Bismuth | Bi | 208.98 | 83 | Osmium | Os | 190.2 | 76 |
| Boron | B | 10.81 | 5 | Oxygen | O | 16 | 8 |
| Bromine | Br | 79.904 | 35 | Palladium | Pd | 106.4 | 46 |
| Cadmium | Cd | 112.41 | 48 | Phosphorus | P | 30.975 | 15 |
| Calcium | Ca | 40.08 | 20 | Platinum | Pt | 195.09 | 78 |
| Californium | Cf | 251 | 98 | Plutonium | Pu | 244 | 94 |
| Carbon | C | 12.011 | 6 | Polonium | Po | 209 | 84 |
| Cerium | Ce | 140.12 | 58 | Potassium | K | 39.098 | 19 |
| Cesium | Cs | 132.91 | 55 | Praseodymium | Pr | 140.91 | 59 |
| Chlorine | Cl | 35.457 | 17 | Promethium | Pm | 145 | 61 |
| Chromium | Cr | 52.00 | 24 | Protactinium | Pa | 231 | 91 |
| Cobalt | Co | 58.93 | 27 | Radium | Ra | 226.03 | 88 |
| Copper | Cu | 63.55 | 29 | Radon | Rn | 222 | 86 |
| Curium | Cm | 247 | 96 | Rhenium | Re | 186.2 | 75 |
| Dysprosium | Dy | 162.50 | 66 | Rhodium | Rh | 102.91 | 45 |
| Einsteinium | Es | 254 | 99 | Rubidium | Rb | 85.47 | 37 |
| Erbium | Er | 167.26 | 68 | Ruthenium | Ru | 101.1 | 44 |
| Europium | Eu | 152.0 | 63 | Samarium | Sm | 150.4 | 62 |
| Fermium | Fm | 257 | 100 | Scandium | Sc | 44.96 | 21 |
| Fluorine | F | 19.00 | 9 | Selenium | Se | 78.96 | 34 |
| Francium | Fr | 223 | 87 | Silicon | Si | 28.09 | 14 |
| Gadolinium | Gd | 157.25 | 64 | Silver | Ag | 107.868 | 47 |
| Gallium | Ga | 69.72 | 31 | Sodium | Na | 22.990 | 11 |
| Germanium | Ge | 72.59 | 32 | Strontium | Sr | 87.62 | 38 |
| Gold | Au | 197.0 | 79 | Sulfur | S | 32.06 | 16 |
| Hafnium | Hf | 178.49 | 72 | Tantalum | Ta | 180.95 | 73 |
| Helium | He | 4.003 | 2 | Technetium | Tc | 97 | 43 |
| Holmium | Ho | 164.93 | 67 | Tellurium | Te | 127.60 | 52 |
| Hydrogen | H | 1.008 | 1 | Terbium | Tb | 158.93 | 65 |
| Indium | In | 114.82 | 49 | Thallium | Tl | 204.37 | 81 |
| Iodine | I | 126.90 | 53 | Thorium | Th | 232.04 | 90 |
| Iridium | Ir | 192.2 | 77 | Thulium | Tm | 168.93 | 69 |
| Iron | Fe | 55.85 | 26 | Tin | Sn | 118.69 | 50 |
| Krypton | Kr | 83.80 | 36 | Titanium | Ti | 47.90 | 22 |
| Lanthanum | La | 138.91 | 57 | Tungsten (Wolfram) | W | 183.85 | 74 |
| Lawrencium | Lr | 260 | 103 | Uranium | U | 238.03 | 92 |
| Lead | Pb | 207.2 | 82 | Vanadium | V | 50.94 | 23 |
| Lithium | Li | 6.941 | 3 | Xenon | Xe | 131.30 | 54 |
| Lutetium | Lu | 174.97 | 71 | Ytterbium | Yb | 173.04 | 70 |
| Magnesium | Mg | 24.31 | 12 | Yttrium | Y | 88.91 | 39 |
| Manganese | Mn | 54.94 | 25 | Zinc | Zn | 65.38 | 30 |
| Mendelevium | Md | 258 | 101 | Zirconium | Zr | 91.22 | 40 |

APPENDIX

# D

# ANSWERS TO PROBLEMS

## Chapter 2

**2-1** 88 kg

**2-2** (*a*) Unit operation
(*b*) Chemical technology
(*c*) Unit operation
(*d*) Chemical technology

**2-3** 29,060

**2-4** 2250 $g/cm^3$

**2-5** 0.735 lb mol

**2-6** $825

**2-7** 22,440 (dimensionless)

**2-8** Cal/(s)(°C)(cm)

**2-9** (*a*) No
(*b*) $32,600 profit

**2-10** $1.71 \times 10^{25}$ molecules/oz mol
44 oz in an ounce mole of $CO_2$

**2-11** 1.20 (dimensionless)

**2-12** 3.40 (dimensionless)

**2-13** 9 lbm lost

**2-14** Need computer printout
Ht = 1.00 m, vol = 2.09 $m^3$, 50% full
Ht = 2.00 m, vol = 4.19 $m^3$, 100% full

## Chapter 3

**3-1** 18.37 lb/100 lb

**3-2** 0.072 lb $H_2O$/lb dry air

**3-3** 52.7 lb

**3-4** (*a*) 69.9 lb
(*b*) 94.5 lb
(*c*) 92.6 g
(*d*) 0.934 lb

**3-5** 687 kg oxygen needed

**3-6** 2500 kg seawater needed/h
**3-7** 0.543 lb $H_2O$/lb dry sand
**3-8** \$0.48/100 lb of initial wet sand
**3-9** 0.0273 kg/h
**3-10** \$7500/d
**3-11** 800 kg/h
**3-12** 1.44 cents/kg of initial wet sand
**3-13** (*a*) 712 $ft^3$
(*b*) 785 $ft^3$
(*c*) 792 $ft^3$
**3-14** 0.035 lb $H_2O$/lb dry air
**3-15** 0.0897 lb/$ft^3$
**3-16** 0.0560 lb/$ft^3$
**3-17** 140.3°C
**3-18** 3364 cal/g mol
**3-19** 1570 Btu/lb
**3-20** 279 lb $CO_2$/100 lb coal
**3-21** 36.0 lb
**3-22** \$1.94
**3-23** 0.0812 lb mol
**3-24** 20.8%
**3-25** 18.9% $CO_2$, 3.4% $O_2$, and 77.7% $N_2$
**3-26** 2043°C
**3-27** 351,000 cal
**3-28** 16.5% $CO_2$
**3-29** 1908°C
**3-30** 9.2% $H_2O$
**3-31** Need computer printout
Temperature = 200 K, $C_p$ = 0.652 kJ/(kg)(K) for $CO_2$
Temperature = 700 K, $C_p$ = 1.144 kJ/(kg)(K) for $CO_2$

## Chapter 5

**5-1** (*a*) Turbulent
(*b*) Streamline
**5-2** 11.9 ft 3.63 m
**5-3** (*a*) $4 \times 10^{-3}$ poise
(*b*) $26.88 \times 10^{-5}$ lb/(ft)(s)
(*c*) 0.968 lb/(ft)(h)
(*d*) 0.499 cS
(*e*) $4.99 \times 10^{-3}$ stokes
**5-4** 56.8°F
**5-5** 188 ft · lbf/lbm 561 J/kg
**5-6** 2.1 lb/ft.$^2$ 101 Pa
**5-7** 1.315 ft · lbf/lbm
**5-8** 0.091 ft · lbf/lbm 0.272 J/kg
**5-9** 1.1 hp
**5-10** 10,300 lb/h 4670 kg/h
**5-11** 16,700,000 gal/day
**5-12** 8.57 ft
**5-13** 4.26 hp 3170 W
**5-14** 57.3 kPa
**5-15** 1661 lb/ft$^2$ 79.5 kPa

**5-16** 0.0058 (dimensionless)
**5-17** Need computer printout
$Q = 0.007243\ m^3/s$

## Chapter 6

**6-1** 36.8 Btu/(h)($ft^2$)
**6-2** 240 Btu/h
**6-3** (*a*) 89 Btu/(h)($ft^2$)
(*b*) 763°F
**6-4** 19.7 Btu/(h)(ft) 68,000 J/(h)(m)
**6-5** 14,140 Btu/(h)(ft)
**6-6** 268 Btu/(h)($ft^2$)(°F) 1520 J/(s)($m^2$)(K)
**6-7** 1780 Btu/(h)($ft^2$)(°F)
**6-8** 17,350 Btu/(h)(ft)
**6-9** 54.3 ft 16.6 m
**6-10** 259 Btu/(h)($ft^2$)(°F)
**6-11** (*a*) 189 Btu/(h)($ft^2$)(°F) 1076 J/(s)($m^2$)(K)
(*b*) 24 ft 7.3 m
**6-12** 186.5 Btu/h
**6-13** 6450 Btu/h $6.805 \times 10^6$ J/h
**6-14** 33,400 Btu
**6-15** $8400
**6-16** 105,000 J/(h)(m)
**6-17** 513 lb/h 233 kg/h

## Chapter 7

**7-1** (*a*) 163.9°F
(*b*) 211.8°F
(*c*) 203.8°F 368.6 K
(*d*) 998.1 Btu/lb $2.322 \times 10^6$ J/kg
**7-2** (*a*) 1140.3 Btu/lb
(*b*) 117.89 Btu/lb
(*c*) 1146.5 Btu/lb
**7-3** 262 Btu/(h)($ft^2$)(°F)
**7-5** 205°F 369 K
**7-6** 206.7°F B.P., 4.9°F B.P.R
**7-7** 16.8 lb/$in^2$, absolute 115.7 kPa
**7-8** 1063 Btu/lb
**7-9** 410 Btu/(h)($ft^2$)(°F) 6860 lb/h
**7-10** 8.8 lb/$in^2$, absolute 60.7 kPa
**7-11** Temp in vapor space of first effect = 217°F. Pressure in vapor space of first effect = 13.3 lb/$in^2$, absolute. Temp in vapor space of second effect = 152°F. Pressure in vapor space of second effect = 2.22 lb/$in^2$, absolute. Area in each effect = 424 $ft^2$. Evaporation from first effect = 4860 lb/h. Evaporation from second effect = 5140 lb/h. Economy = 1.92
**7-12** (*a*) Constant in 1, 2, and 3
(*b*) Decreases in 1, increases in 2, constant in 3
(*c*) Constant
**7-13** Need computer printout. For feed concentration of 6 wt % and product concn. of 11 wt %, 45.45 kg $H_2O$ is evaporated per 100 kg of feed

## Chapter 8

**8-1** 0.552 mol fraction benzene
**8-2** 1.79
**8-3** 0.665 mol fraction methanol
**8-4** 0.725 mol fraction *a*
**8-5** 48%
**8-6** 3.0
**8-7** 0.717 mol fraction
**8-8** 77.6%
**8-9** (*a*) 97.5°C; 0.046 mol fraction dimethylaniline
(*b*) 32.4 kg/100 kg $H_2O$ in distillate
**8-10** (*a*) 11
(*b*) Sixth plate from bottom
(*c*) 8540 lb/h
(*d*) 4910 lb/h
(*e*) 5.44 ft
(*f*) Stripping section is larger
**8-11** 583 kg mole/hr

## Chapter 9

**9-1** 0.0188 atm
**9-2** 0.418 mol HCl/mol air
**9-3** 1.05
**9-4** 2.4
**9-5** 80.6%
**9-6** 91.5%
**9-7** 8.97%
**9-8** 7 kg

## Chapter 10

**10-1** 0.0178 kg $H_2O$/kg dry air
**10-2** (*a*) 29.4 mmHg
(*b*) 83.5°F
(*c*) 30.9%
(*d*) 33.6%
**10-3** 93.4 Btu
**10-4** (*a*) 0.0155 lb $H_2O$/lb dry air
(*b*) 0.247 Btu/(°F)(lb dry air)
(*c*) 31.3 $ft^3$/kg dry air
(*d*) 75°F
**10-5** 0.010 lb $H_2O$/lb dry air
**10-6** (*a*) 0.0075 lb $H_2O$/lb dry air
(*b*) 24%
**10-7** 0.011 lb $H_2O$/lb dry air
**10-8** 0.4 kg $H_2O$/100 kg dry air
**10-9** 19,500 Btu/h

## Chapter 11

**11-1** 0.0267 kg $H_2O$/kg dry sand
**11-2** 7.1 lb $H_2O$/h
**11-3** 0.44 lb $H_2O$/lb dry solid
**11-4** 3.55 lb/($ft^2$)(h)
**11-5** 13.9 h
**11-6** 23,000 $ft^3$/h
**11-7** $5.97/100 kg
**11-8** 60 trays

## Chapter 12

**12-1** 3600 lb   1630 kg
**12-2** 40 $ft^2$   3.7 $m^2$
**12-3** 397 $ft^3$
**12-4** 0.25
**12-5** 2.4 h

## Chapter 13

**13-1** 32,700 lb/d
**13-2** 6060 kg
**13-3** 93%
**13-4** 11,150 kg
**13-5** 279 Btu/$ft^3$
**13-6** $0.066/lb
**13-7** 5.1%
**13-8** $1.98 \times 10^8$ Btu
**13-9** 316 Btu/$ft^3$ at 1 atm and 60°F
**13-10** 80.4%
**13-11** Need computer printout. For argon in fresh feed, there are 8.00 metric tons/day

## Chapter 14

**14-1** $1200
**14-2** 25%
**14-3** The third design
**14-4** (*a*) $66,300
(*b*) $63,300
(*c*) $62,200
**14-5** Author: W. A. N. Severance
**14-6** 4.3 in
**14-7** 320 $ft^3$/min
**14-8** 2.3 ft
**14-9** 40 kg/h
**14-10** 0.194 lb/$ft^3$
**14-11** The third design
**14-12** The second design

# INDEX